Radar for Fully Autonomous Driving

For a listing of recent titles in the
Artech House Radar Library,
turn to the back of this book.

Radar for Fully Autonomous Driving

Matt Markel
Editor

ARTECH
HOUSE
BOSTON | LONDON
artechhouse.com

Library of Congress Cataloging-in-Publication Data
A catalog record for this book is available from the U.S. Library of Congress

British Library Cataloguing in Publication Data
A catalog record for this book is available from the British Library.

ISBN-13: 978-1-63081-896-8

Cover design by Charlene Stevens

685 Canton Street
Norwood, MA 02062

10 9 8 7 6 5 4 3 2 1

To everyone working to make cars safely and effectively drive themselves

Contents

Foreword

One of the fastest growing areas of radar development is the area of autonomous systems in support of autonomous driving systems. A quick Google Scholar search yields more than 26,000 citations since only 2017! This is in direct comparison to the extremely broad category *airborne radar* with 22,000 citations during the same period. The large number of global commercial, academic, and government entities actively engaged in this area also speaks to its growing market appeal.

While the commercial and military appeal of fully autonomous driving systems is huge from a safety, economic, and logistics perspective, it is one of the most severely constrained radar design problems ever encountered. In addition to extreme constraints on form factor (size, weight, and power) and especially cost, the reliability and safety constraints are frankly unprecedented. Only using the most advanced design methods (e.g., digital engineering) and advanced technologies (e.g., MIMO radar and advanced embedded AI), is there any chance of robustly meeting all constraints?

In *Radar for Fully Autonomous Driving*, the authors bring together for the first time a comprehensive and detailed overview of the entire design process, from system and operational safety requirements, to detailed technical solutions to meet those challenges. Moreover, the book is written by engineers and scientists (and businesspeople!) actively working on the bleeding edge. From advanced low C-SWAP enabling RF MMICs, the latest low C-SWAP embedded computing, to novel radar signal processing architecture and machine intelligence, *Radar for Fully Autonomous Driving* is your one-stop destination to the latest developments.

—Joe Guerci
April 2022

Acknowledgments

While numerous friends, colleagues, and family members have helped and supported me in creating this work, I feel that a few must be recognized for their exceptional contributions. First, I would like to deeply thank all of the authors for giving their time and talents to this work, and their families for giving them the latitude to do so. Second, I would like to recognize Tim Campbell and Gary Clayton, who not only coauthored two chapters with me but also provided valuable and insightful advice on the project, keeping me focused and inspired. Third, I would like to thank my family in Saint Louis (Linda, Judy, Dave, Jon, and Arianna) for their unending love and support. Finally, and most significantly, my wife, Traci. During the writing of this work, we moved from California to Texas, bought and sold our houses, packed and unpacked, and both made major career changes. Without her strength, patience, encouragement, and partnership, this project would not have succeeded, and I am deeply in her gratitude.

—Dr. Matt Markel
April 2022

Introduction

In August of 2017 I started a new career with Waymo, a leader in the self-driving technology space, to join and eventually lead the team developing their radars. Before that, my entire career had been in defense, primarily designing and testing advanced radar and electronic warfare systems. I worked on radio frequency (RF) systems that could be held in one hand and on radars with antennas from the size of a dollar bill to more than 14 feet tall. I enjoyed my work, and I was doing well. So why would I change fields—especially to something as different as self-driving or *fully autonomous* driving as we say today? Aren't those two fields as far apart conceptually and culturally as you can get?

It turns out that within both the defense and self-driving arenas, there is an unyielding passion binding together and unifying those in that area. In defense, we are defending our land, people, and service members; we are protecting our way of life. With fully autonomous vehicles, we are striving to save and improve lives. In the United States, approximately 37,000 people die every year from accidents on our roadways,[1] and we agree that is far too many. Nearly all of those accidents are due to human errors resulting from distraction, fatigue, and intoxication. Fully autonomous driving technology promises to save lives by removing humans from this equation, creating a safe and efficient way for people to get to where they want to go. Those working in this space feel this passion and calling, and it isn't any less intense than those working in defense. I found that I wanted to join the autonomy field to try to accelerate their efforts.[2]

Even though there isn't a universally specified architecture or set of components, when we think of fully autonomous vehicles, there is a consensus that the system is composed of several canonical constituents: a *sensing and perception system* that perceives signals from the external environment and turns those signals into information about the scene, a *localization system* that determines where the vehicle is, a *navigation system* that determines where and how to journey to the destination, a *decision system* that determines the subsequent actions of the vehicle, and a series of control systems that actuate the various vehicle steering, braking, and acceleration capabilities. Historically the sensing and perception systems focused primarily on lidar and camera inputs, but recently imaging radar has become increasingly utilized by the system to achieve, or surpass, parity with the safety of a human driver.

1. NHTSA report. To put that figure in context, we lose around 41,000 humans per year to breast cancer.
2. There was another reason as well: I have always believed in pushing yourself, and sometimes this requires you to leave your comfort zone (e.g., defense) to embark on a new career. In 2021, I continued this "challenge yourself" mindset and after developing a world-class radar team at Waymo, joined Ghost Locomotion to create the optimal radars for full autonomy in personally owned cars.

This is primarily caused by the expansion of the autonomous vehicles' operational design domains (ODDs) to include scenarios with poorer weather, and situations more stressing to the perception system, such as high-speed roads and complicated urban environments.

That is the motivation behind this book.

Radar for fully autonomous vehicles is a new field in the still-burgeoning fully autonomous vehicle space. Unlike more traditional radar fields (e.g., aerospace, defense, weather), the very basics such as setting requirements, the amount of verification, and the choice of technologies that achieve the desired performance (without losing the affordability necessary for widescale adoption) are all being explored and developed now as we go.

More and more groups are developing radar technology and products for this field, and system integrators must sort through increasingly complex radar offerings. To exacerbate this, the community has little comprehensive literature for introduction, reference, and context. We created this work to remedy that, utilizing the expertise of the best and brightest experts working in the automotive radar space today.

This wasn't easy. The number of possible automotive radar topics is large, quickly becoming overwhelming from both the development and the organizational standpoint. There are issues that are still considerably in flux, such as safety requirements for radars on fully autonomous vehicles. And there are areas where useful results have been gathered but not fully evaluated over all possible scenarios, like the effects of rain and road spray on radar attenuation. Finally, we needed to decide whether there should be a strong integration of the chapters or let them represent the different viewpoints and styles of the authors.

In the end, we decided to include a broad representation of what we considered to be the most relevant topics. We also chose to let each chapter and subject area suggest what best suits the reader's needs and, in doing so, avoid a strong integration across chapters in favor of a more anthological approach. The reader can view each chapter as a "master class" with the author, seeing the world through their eyes and understanding what we felt best served the readers. With this approach, each chapter is a stand-alone work; where beneficial, they reference one another but can be read in any order desired. We also decided not to include a basic primer on radar theory (you won't find a discussion about why radars have range and Doppler ambiguities or the relation between gain and effective area of an antenna here), but there are countless worthwhile tutorials for those topics available both in textbooks and online. The reader isn't expected to be a radar expert, but it will be beneficial to have had some exposure to radar concepts or be willing to spend some time with tutorials along the way while reading this work.

We organized the material into three major parts. Part 1, Radar Technologies for Autonomous Vehicles, focuses on the fundamental concepts present in the automotive radar landscape today. Chapter 1 sets the stage for radar's use in fully autonomous vehicles through a history of automotive radar and autonomy followed by discussions on common waveforms (i.e., linear frequency modulated waveforms) and processing approaches. A significant portion of the increase in the capability and performance of automotive radars over the past several years has been achieved through advanced algorithms, which nearly always beget the need for increased computational power. This chapter concludes with a treatment of the ongoing debate between central and distributed processing approaches. Chapter 2 paints a more thorough picture of the automotive radars by cleverly comparing them to those used in the well-known defense and aerospace fields, with unique insight into the differences in requirements, installation, and products between the areas. Chapter 3 explores phase and code modulation, a promising waveform suite that is an alternative to the popular frequency modulation and discusses the tradeoffs between the two classes. Chapter 4 provides a treatment of multiple input/multiple output angle of arrival processing, a prevalent technology that increases the effective aperture size without increasing the number of receiver channels, usually at the expense of timeline and increased processing complexity. Chapter 5 explores the benefits and mechanization of using synthetic aperture radar for automotive applications. Finally, Chapter 6 presents a deep dive into the silicon transceiver technologies, challenges, and advancements that enable much of the automotive radar space to achieve high performance without sacrificing affordability.

In Part 2, Challenges and Solutions for the Automotive Environment, we explore many in-depth aspects of the driving scenario that complicate radar performance. This section contains a significant amount of real-world data and detailed mathematical analysis that provide the reader with a firm foundation in these concepts.

Chapter 7 analyzes two significant challenges in the automotive scene: the need to process a very large dynamic range and the complications that come from radar reflections (multipath) and then presents several families of solutions to those challenges. Chapter 8 takes a detailed look at radio frequency interference (RFI), which occurs when the emissions from one radar deleteriously affect another radar. It derives several key relationships in this space, provides numerous examples, and discusses multiple types of solutions available to the radar designer. Chapter 9 presents a validation of some classical analyses and several new results regarding radar performance in weather. We believe it is essential that those designing, specifying, or acquiring radars for fully autonomous vehicles understand these challenges to ensure safe performance across the intended range of operational scenarios.

Part 3, Integration and System Considerations, explores radar-related concepts that transcend typical radar design and form a bridge to the rest of the fully autonomous vehicle. Safety is of paramount importance to fully autonomous vehicles. Chapter 10 walks the reader through the many safety standards in existence today, some of the pitfalls of implementing them, and several best practices from the authors' experience that avoid these pitfalls—all through the eyes of the radar development. Finally, Chapter 11 describes the test environment throughout the radar's buildup, integration, and installation on a vehicle.

At every turn in the development of this work, we have endeavored to focus on the essential information to the reader—the knowledge that enables them to better develop, specify, integrate, or purchase a radar for fully autonomous vehicles. I sincerely desire that you find this work to be informative and valuable. We believe that the developments of the radar community provide the final technology needed for ubiquitous deployment of full driving autonomy and the many boons that it will bring to our societies.

PART I

Radar Technologies for Autonomous Vehicles

CHAPTER 1

Modern Radar Sensors in Advanced Automotive Architectures

Mark Steigemann and Cicero S. Vaucher

Innovations made by semiconductor manufacturers are one of the driving forces for future highly automated cars. Although radar sensing technology is already present in several car generations, the ability to sense better and longer distances made huge steps forward in the last 5 to 10 years. There will be more cars with radars, more radar sensors per car, and more radar integrated circuits (ICs) per sensor. Likewise, as for radar sensors, the number of cameras and higher automation in the future means the amount of light detection and ranging (lidar) sensors will also increase. The in-vehicle communication architecture of a modern mid- to high-end car already features approximately a hundred or more control units. With the growing amount of sensor nodes, the requirements will further increase. High performance compute engines are now available and will offer higher performance and new ways of distributing data. This chapter looks beyond state-of-the-art radar sensing dominated by radar sensor edge processing toward distributed (cooperative approaches) and fully centralized advanced driver-assistance systems (ADAS) architectures. Radar sensors are playing a pivotal role in the complete suite of available sensors. This is due to low cost, weather, and light condition independency, and last but not least their ability to directly measure object velocity.

Using the example of a radar sensor, the chapter discusses evolving performance needs as well as the semiconductor technology capabilities needed. Signal processing at the sensor edge offers various advantages. High performance compute engines can process a vast amount of data at a central place in the car. With the rise of artificial intelligence (AI) and machine learning algorithms, the question at hand is how to best feed those engines to get more performance and higher resolution. Distributing more data from radar sensor to different processing entities in the car requires a holistic approach for security and functional safety. The chapter concludes with an overview of cost for various levels of centralization of radar sensor processing, in dependency of the performance gain. An outlook on how radar sensors may evolve in the future given the increasing performance requirements concludes the chapter.

1.1 Inspiration for More Advanced Systems

1.1.1 Traffic Density and Fatal Accident Rate

Worldwide, each year about 1.3 million humans die in traffic accidents.[1] Approximately 90 to 94% of those accidents are caused by human error. Human errors can vary: wrong perception of the situation, distraction from driving tasks, tiredness, or weather conditions, to name just a few.

As the traffic density steadily increases, driver assistance using advanced sensing systems based on camera, lidar, and radar technology is becoming mandatory. Recent research has shown that ADAS functions help to reduce fatal accidents by up to one-third.

1.1.2 Human Factor

Humans can observe the most complex driving situations with just visual and auditory sensors. Combined with training, rich experience, and the ability to anticipate, these complex situations can be mastered. However, humans can be distracted easily by many additional events happening at the same time. The human's system can be divided into three phases: time to sense, time to think, and time to act. The human system can improve this time further by using anticipation. Based on rich experience of similar situations or training, the system is prepared that something might happen. The diagram in Figure 1.1 shows these phases based on a braking situation of a car following another car. The timing of the phases is based on many trials of human's reaction times. Clearly, the anticipation capability can dramatically improve the reaction time. This capability is one of the key advantages of humans over autonomous driving system and will be surely one of the main challenges of future autonomous driving (AD) system using artificial intelligence.

1.1.3 Autonomous Driving Levels

Researchers forecast that by 2025 approximately 8 million autonomous or semi-autonomous vehicles will be on the road. The Society of Automotive Engineers (SAE) defines six levels of driving automation ranging from 0 (fully manual) to 5 (fully autonomous). These levels have been adopted by the U.S. Department of Transportation [1].

Level 0 (*no driving automation*) is where historically most vehicles belong. The driver will fully perform the dynamic driving task. There might be some systems to support the driver. The most prominent system might be the emergency braking system, which supports drivers to achieve better breaking results. These systems are not driving the vehicle, so they will fall in Level 0.

Systems in Level 1, the lowest level of automation, usually feature one assistance system that can either perform autonomous braking or steering or accelerating. Even if the vehicle has two of these features implemented, if they are not connected and interacting with each other, the vehicle is classified as Level 1. In Level 2, which is often

1. Association for Safe International Road Travel (ASIRT). https://www.asirt.org/safe-travel/road-safety-facts.

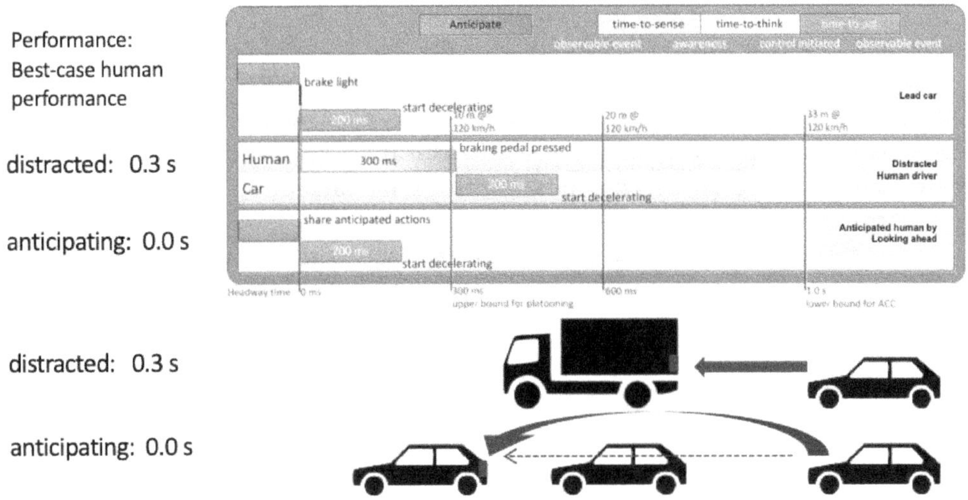

Figure 1.1 Braking reaction times under different circumstances.

named *partial driving automation*, the vehicle can control both steering and accelerating and decelerating. The automation falls short of self-driving because a human sits in the driver's seat and can take control of the car at any time. Tesla Autopilot and Cadillac (General Motors) Super Cruise systems both qualify as Level 2. Levels 0–2 have one important common characteristic: The driver must always supervise the driving task. Human intervention may be needed to ensure safety at any time.

To reach the next levels of autonomous driving (3–5), a high hurdle must be surpassed. While in Levels 0–2, the driver is always in control. In Level 3, the driver only needs to take back control if the system cannot control the driving anymore. The driver is not actually driving the vehicle; the system must actively give back control to the driver, and it must assure that the driver has enough time to sense and understand the situation to perform the required action. Implementing Level 3 partial automation at an affordable cost for broad mass market cars has turned out to be unobtainable, at least to date. While Level 3 automation appeared for years to be within reach as the next step in automation for almost all car makers, it has been realized that a Level 3 car must be as autonomous as a Level 4 or 5 car. Consider the case of the Audi A8L. In 2017, Audi (Volkswagen) announced that the next generation of the A8—their flagship sedan—would be the world's first production Level 3 vehicle. Audi delivered on their promise. The 2019 Audi A8L arrived in commercial dealerships, featuring Traffic Jam Pilot, which combines a lidar scanner with advanced sensor fusion and processing power (plus built-in redundancies in case a component fail). However, the regulatory process in the United States shifted from federal guidance to state-by-state mandates for autonomous vehicles. So, for the time being, the A8L is still classified as a Level 2 vehicle in the United States and will ship without key hardware and software required to achieve Level 3 functionality. In Europe, however, Audi wanted to roll out the full Level 3 A8L with Traffic Jam Pilot. However, in Europe, the Level 3 system would not require a driver to monitor the road and the liability in case of an accident when engaged would transfer to the automaker. Audi gave up their plans to offer a Level 3 autonomous car.

A Level 5 car can drive under all conditions—a driver is not needed—whereas a Level 4 car can perform all driving tasks in a predefined area or set of conditions. One example of a Level 5 vehicle is the driverless robot taxi that can operate in a certain area. These fleet cars use a large suite of radar, lidar, and cameras to reach sufficient quality of data to perform safe and secure operation. Fleet-operated cars usually use a car sharing model with a 24/7 time of operation due to high cost of the autonomous (AD) systems.

Traditional car makers today are concentrating on bringing Level 2 driving features to perfection with respect to safety, performance, and ease of use, before jumping to Levels 3–5. Automotive radar is playing a pivotal role for the next generation of advanced driving functions with its unique capability to measure directly in four dimensions: range, position X and Y, and the relative velocity of an object. Figure 1.2 maps the various driving features on the autonomy levels based on a complete sensor suite.

1.2 The Evolving Automotive Radar Landscape

Automotive radar sensors are expanding from traditional road applications, such as automatic cruise control, to urban scenarios including pedestrian detection and self-parking [2]. A major difference from a road to an urban scenario is the number of unique targets that must be detected, identified, and tracked.

For reliable operation and wide deployment in urban and road scenarios, the following requirements apply:

- *Unambiguous multiple-target detection.* This is discussed in detail in the next section on fast chirp sequence waveform.
- *Precise range, speed and angular positioning of each target.* This translates into:
  - Increased measurement bandwidth, for increased range resolution;
  - Precise waveform generation for accurate Doppler estimation; in this context *precise* means a sequence of identical (frequency-modulated) pulses equally spaced in time;

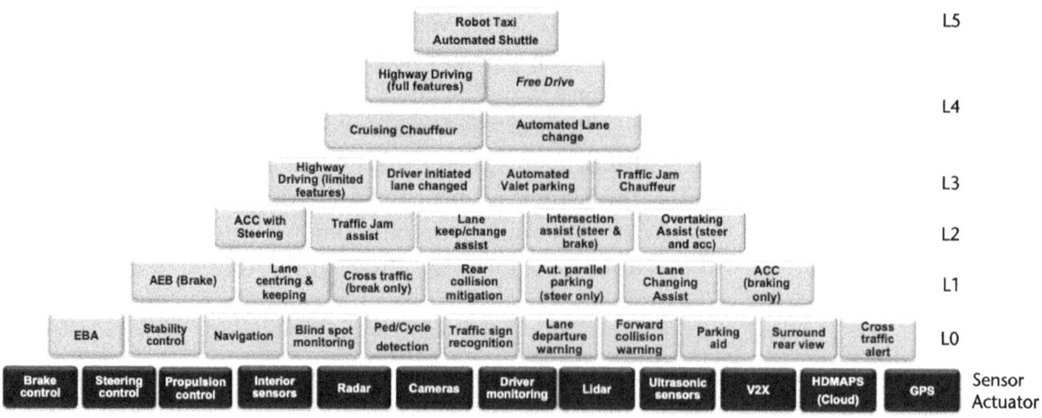

Figure 1.2 Driving functions pyramid.

  - Larger antenna aperture for high angular resolution and accuracy, possibly combined with super-resolution algorithms and multiple input/multiple output (MIMO) techniques.
- *Programmable short-, mid-, and long-range functionality.* In practice, this requires a programmable waveform generation circuitry providing adequate performance level for all operational modes.
- *Low sensor cost, for market penetration into lower and mid-class segments.* High integration level in a mainstream complementary metal oxide semiconductor (CMOS) technology provides lowest IC production costs. In addition, CMOS technology potentially leads to low(er) power IC implementations, further enabling cost saving due to smaller sensor heat-sink requirements.

Figure 1.3 summarizes requirements for automotive sensors.

1.3 Fast Chirp Sequence Radar Sensing

Urban scenarios are characterized by a large number of relevant objects to be unambiguously detected and tracked. Earlier automotive radar waveforms operated with a reduced number of linear frequency-modulated (FM) sweeps (chirps), typically an up-sweep and a down-sweep with a sweep interval on the order of milliseconds.

Before proceeding, there are a few advantageous properties of an FM carrier to consider [3]. First, as the transmit signal power is active during the complete observation interval, the link budget and therefore the maximum achievable range are positively impacted. Second, the frequency modulation can, in principle, be generated in a relatively straightforward way, by applying a varying voltage to the tuning voltage input of a (free-running) voltage-controlled oscillator (VCO), although contemporary systems incorporate a phase-locked loop to control the VCO momentaneous frequency. Third, by using frequency modulation for sensing, amplitude linearity aspects on the transmitter sections are irrelevant and saturated amplifiers are used, benefitting the overall sensor power efficiency.

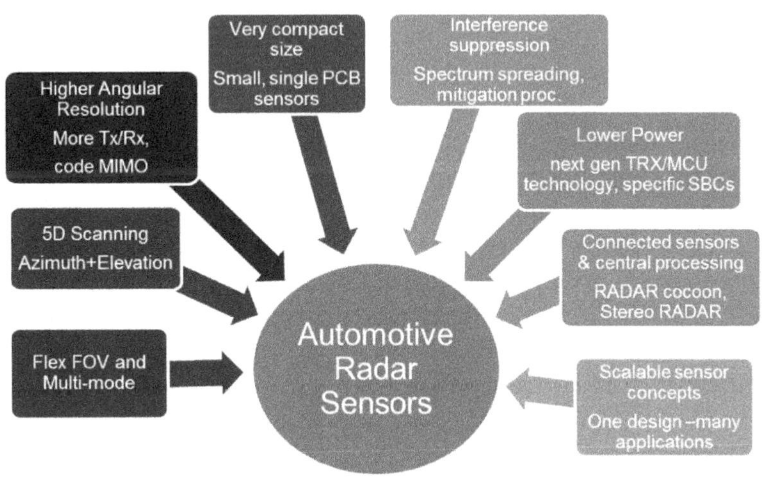

Figure 1.3 Schematic representation of automotive radar requirements.

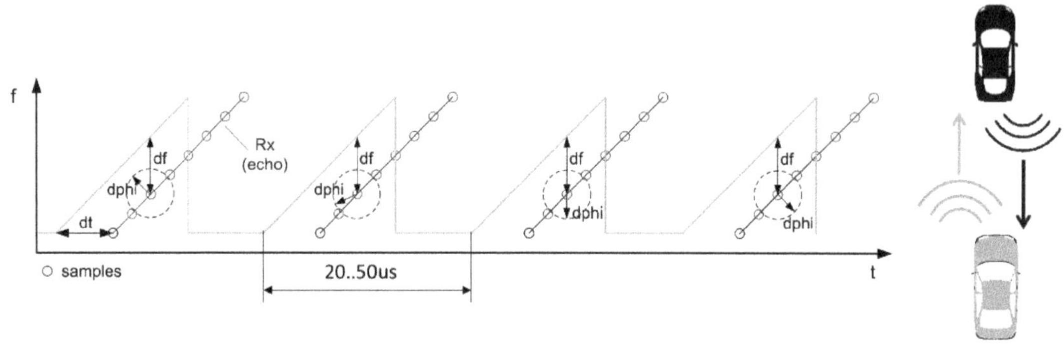

Figure 1.4 Fast chirp sequence waveform.

Returning to the millisecond-long (slow-chirp) up-sweep and down-sweep waveform, a practical advantage is that the intermediate frequency (IF) range was relatively small (up to 100kHz), leading to reduced requirements on the analog-to-digital converter (ADC) sampling frequency and on the signal processing due to the relatively small data volume to be analyzed. However, the slow-chirp waveform suffered from a so-called *range-velocity ambiguity*, as the target range information on the IF spectrum could be significantly altered by Doppler frequency shifts [4]. By usage of the up- and down-sweeps it was possible to solve the range and velocity dependency for a limited number of relevant targets. However, the problem becomes quickly intractable as the number of targets increases, as is typically the case in urban scenarios.

The fast chirp sequence waveform principle [5, 6] provides a solution for the range-velocity ambiguity, and it has intrinsic multitarget detection capability.[2] The principle of the measurement is presented in Figure 1.4. It consists of a sequence of linearly-modulated frequency sweeps, with a precise time interval between each sweep. The period of each sweep is typically on the order of 20 us to 50 us, and there are hundreds of frequency sweeps on a measurement cycle; we will refer to each frequency sweep as a *chirp*, with each chirp being a pulse of a linearly-modulated carrier signal. Figure 1.4 shows the reflected signal from the target arrives delayed in relation to the instantaneous transmit signal by a time dt. The receiver processes the incoming reflected signal by mixing it with the transmit signal. The resulting IF signal is a tone with frequency (indicated as df) and phase (indicated as dphi). The frequency indicates the range to the target, whereas observation of the behavior of the phase along the pulse sequence provides the relative velocity of the target: the higher the rate of change of the phase, the larger the relative velocity.

The fast-chirp sequence signal processing approach is depicted in Figure 1.5. It represents a hypothetical system with six RX channels (A/D), each associated with a respective RX antenna. Incoming data from the ADCs are processed in *range* Fast, Fourier Transforms (FFT), with the FFT results stored in layers on the *data cube*. At the end of the chirp sequence, the data cube is complete with range information

2. This is achieved with separation of the range assessment on a first FFT operation, with the velocity being obtained with a second FFT operation along each range gate, after completion of the sequence.

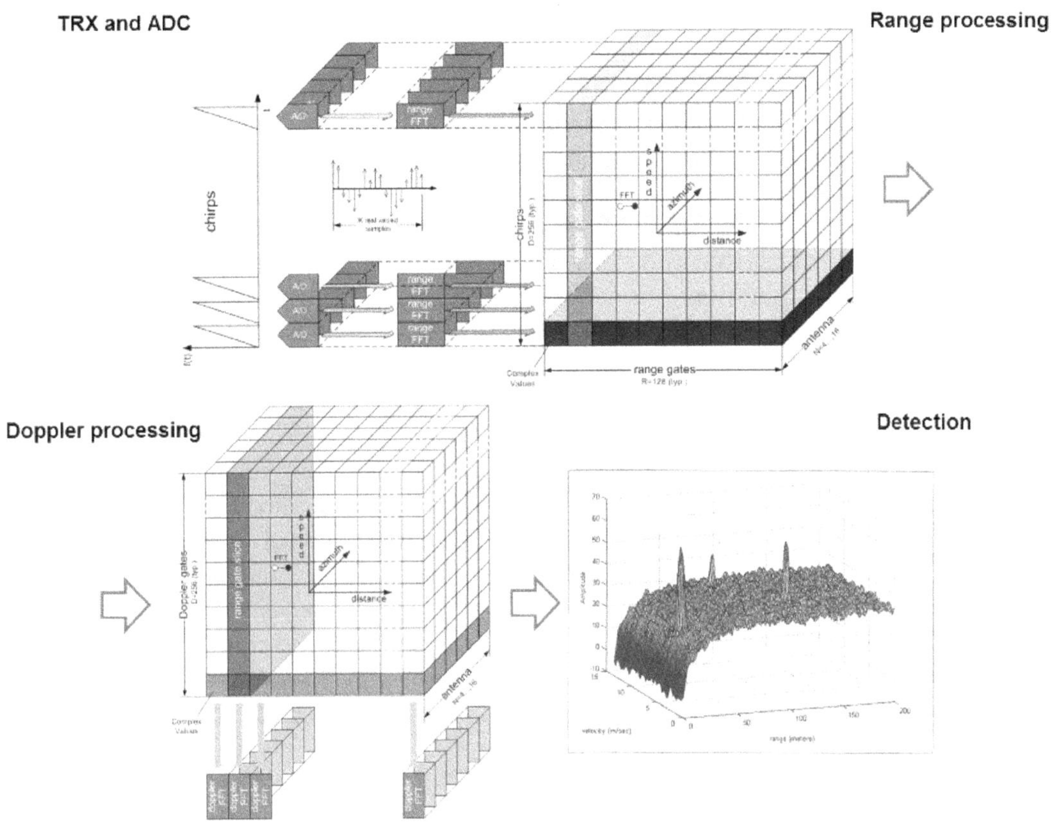

Figure 1.5 Fast chirp sequence signal processing.

(amplitude and phase of targets). The range data is then further processed along each range gate (range gate slice), with the changes in phase along the sequence finally mapping into the velocity (Doppler) information. For each antenna one gets then a two-dimensional range-velocity plot as depicted on the *detection* graph. Three targets are clearly visible in the example shown. The angular position of the targets in relation to the sensor (*azimuth information*) can then be found by a third FFT processing along the antenna dimension.

1.4 RFCMOS Car Radar Transceiver

In the previous section, we reviewed the principles of the fast-chirp sequence waveform and its benefits for automotive sensing in urban scenarios. A key element for correct operation is precise timing relationships along the complete chirp sequence. This aspect applies to the frequency modulation linearity, to the repetition period between chirps, and also to stability of the ADC sampling moments along the whole measurement interval. From this perspective, there is a natural convergence of the timing control element, of the chirp generation phase locked loop (PLL), and of the ADCs to be integrated on a single die. High performance ADCs, featuring high conversion bandwidth, linearity, and sample rates are naturally implemented in advanced CMOS nodes. Therefore, the ADCs were a key element driving the migration from SiGe radar transceivers (which normally did not include the ADC) toward integration into pure CMOS technologies. In addition, azimuth and elevation angular measurements demand multiple transmit and receive elements operating simultaneously. Also here, the high integration density provided by advanced CMOS nodes (40 nm and lower) comes at an advantage, leading to relatively small die areas and package sizes. Finally, CMOS technologies potentially allow integration of an MCU close to the front-end circuitry, which can be advantageous for autonomous execution of calibration routines and functional safety checks.

Figure 1.6 depicts the block diagram of NXP's TEF8102 product, a fully-integrated RFCMOS radar front end for operation on the 76 to 81 GHz frequency band that is representative of the state of the art today. It contains three transmit (TX) and four receive (RX) channels that can be simultaneously operated. The chip is optimized for fast chirp modulation, with its timing engine supports several radar operational modes with relatively simple user-defined configurations. The product was developed in compliance to ISO26262 standard and supports automotive safety integrity level (ASIL) B applications. The package measures 7.5 mm × 7.5 mm, enabling small sensor designs, with the only external functional element being a 40 MHz crystal resonator. Last but not least, the chip supports usage in cascaded configuration, addressing applications where more than three TXs and four RXs must be used simultaneously. This operation mode will be discussed in more detail on Section 1.6.

1.5 Elements of a Radar Module

An explorative radar module is pictured in Figure 1.6 on the right. The TEF8102 transceiver is the dark square at bottom-right location. The three TX ports are located opposite the four RX ports. This arrangement facilitates high TX to RX isolation at sensor level, including transceiver package as well as antenna isolation. This is an important aspect, as TX-to-RX coupling can add TX noise at the RX ports and effectively degrade receiver noise performance. The 40 MHz crystal

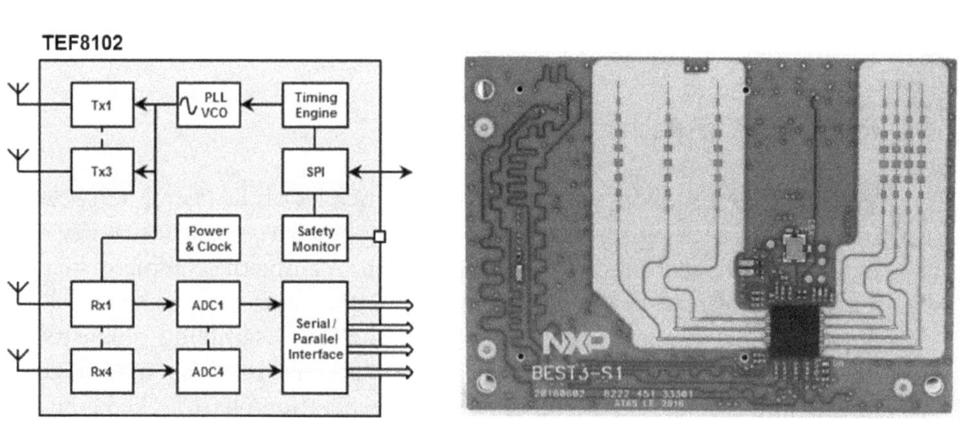

Figure 1.6 Dolphin 76 to 81 GHz car-radar transceiver in 40 nm CMOS technology. Demonstration board including TX-MIMO functionality is shown on the right.

resonator is visible above the transceiver, in addition to a few supply decoupling capacitors. The antenna design follows a Franklin array implementation [7]. The elongated design consists of resonant patches connected in series. The construction compresses the radiation pattern in the elevation direction, as depicted in Figure 1.7, effectively creating a *fan-beam* with high antenna gain along the azimuth direction, peaking at 14.8 dBi for broadside. This pattern would suit a short- to mid-range application where a relatively wide field of view in azimuth is required.

A first-order detection performance of a radar sensor can be assessed with a level diagram. A representative level diagram for the sensor in Figure 1.6 is presented in Figure 1.8. It is based on a sequence of 512 chirps, each chirp with an acquisition bandwidth of 300 MHz and a period of 30 us. The effective noise bandwidth of the measurement is rounded to 100 Hz, without loss of generality; in practice, it will depend on the complete measurement length and on the windowing used for the range and Doppler FFTs. The x-axis expresses the IF frequency at the input of the ADC. Vertical lines are added to facilitate the mapping of the IF frequency into associated range, from 1m up to 128m, which is this example corresponds to the maximum detectable range with an ADC sampling at 20 MHz.

The three solid lines indicate the system response for a range of practical targets at a given distance from the sensor. They represent a small plastic pole (−30 dBsqm), a pedestrian (−5 dBsqm), and a large truck (30 dBsqm). The analog IF filter contains a high-pass filter with a corner frequency of around 200 kHz, in this example. The filter is important to suppress strong signals at low frequencies that can overload the ADC, for example the truck signal at distances smaller than 5m. For ranges mapping into the pass band of the IF filter, all targets show a −40 dB/decade response as predicted by the classic radar equation.

The bottom part of the graph indicates system noise contributions. The RX thermal noise reaches around −100 dB from 100 kHz to 5 MHz. The dashed lines indicate the effect of TX phase noise as reflected from a bumper at 5 cm from the sensor, and from trucks at 10m, 50m, and 150m [8–9]. The noise performance is clearly determined by RX thermal noise along the IF spectrum, except in the presence of the truck at approximately 10m, which would increase the effective noise figure by approximately 3 dB close to 300 kHz and 1.5 MHz frequencies. The

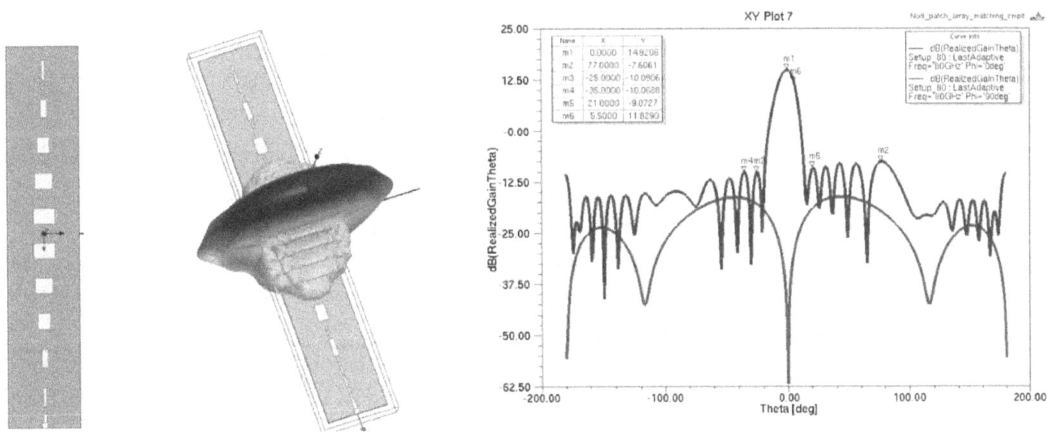

Figure 1.7 High-gain Fan Beam antenna for increased range and road clutter attenuation.

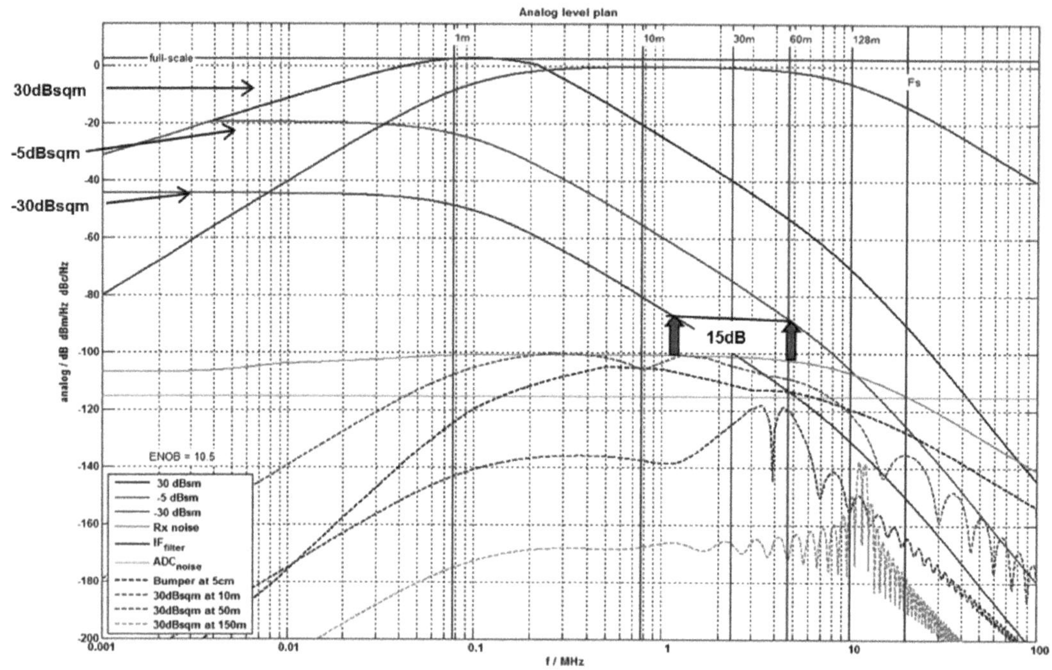

Figure 1.8 Analog level plan for 300 MHz in 30 us, $N = 512$, ADC sample rate 20 MHz, NBW = 100 Hz.

noise level decreases quickly as the truck moves away from the sensor, in the same proportion as the target return signal itself.

The maximum detection range depends on a target radar cross-section and on the minimum SNR required by the signal processing detection algorithms. Considering a very conservative minimum SNR of 15 dB, the plastic pole can be detected up to around 12m and the pedestrian up to 60m; SNRs of 0 dB are found at approximately 30m and 130m for the pole and pedestrian, respectively.

Qualitatively speaking, increasing the detection range can be accomplished with the following measures:

- Increase output power to the TX antennas; this may be limited by maximum power (density) regulations in some geographic regions.
- Decrease noise figure on the receiver stages. Improving ADC SNR in case it is impacting the effective noise figure on the receiver chain.
- Decrease the equivalent noise bandwidth of the measurement sequence. This means longer measurements (e.g., increased number of chirps) and a larger amount of data for Doppler FFT processing.
- Perform digital beam forming on the RX signals before detection. In a typical system with four RX paths, this operation can boost the SNR by up to 6 dB when compared to a single channel.
- Increase the TX and RX antenna gains in dBi. This creates a fundamental trade-off between azimuth and elevation field of views and maximum attainable range.
- Operate the sensor with detection algorithms reliably operating at lower SNRs.

1.6 Angular Resolution Increase: MIMO Example and Cascaded Application

In the previous section, we discussed the range detection performance of the generic sensor depicted in Figure 1.6. Also of importance is *Doppler resolution*, or the ability to separate closely located objects moving at different relative velocities, and angular resolution or the ability to separate objects along the radar field of view, in azimuthal as well as on elevation positioning with respect to the sensor. We will consider techniques to increase angular resolution in this section.

The angular resolution, as a function of the antenna aperture and of the mmWave carrier wavelength lambda can be roughly expressed as

$$\text{ang,res} \approx 70 \cdot \text{lambda/aperture [degrees]},$$

for broadside direction. In the sensor of Figure 1.6, the RX antennas are located at half a wavelength from each other, to avoid grating lobes in the angular domain. Therefore, the total RX antenna aperture is approximately two lambda, with an associated angular resolution of only approximately 35 degrees when a single TX antenna is used. To increase the effective RX aperture, the TX antennas are located at two lambda from each other [10]. This creates a MIMO configuration with 12 virtual RX channels, with a factor three increase in effective antenna aperture and therefore an associated angular resolution of approximately 12 degrees at broadside direction.

The MIMO principle, angular resolution formula including the effect of the angle of arrival phi, and the measured resolution are presented in Figure 1.9. There is a good agreement between measurements and predicted theoretical resolution.

While an improvement by a factor three in angular resolution is significant, 12 degrees is still one order of magnitude off from latest resolution targets on the order of 1 degree or less. To reach this value, the aperture must be increased by a proportional amount, ideally retaining a lambda/two spacing between antennas. It means, in practice, that either the number of receiver channels within a package

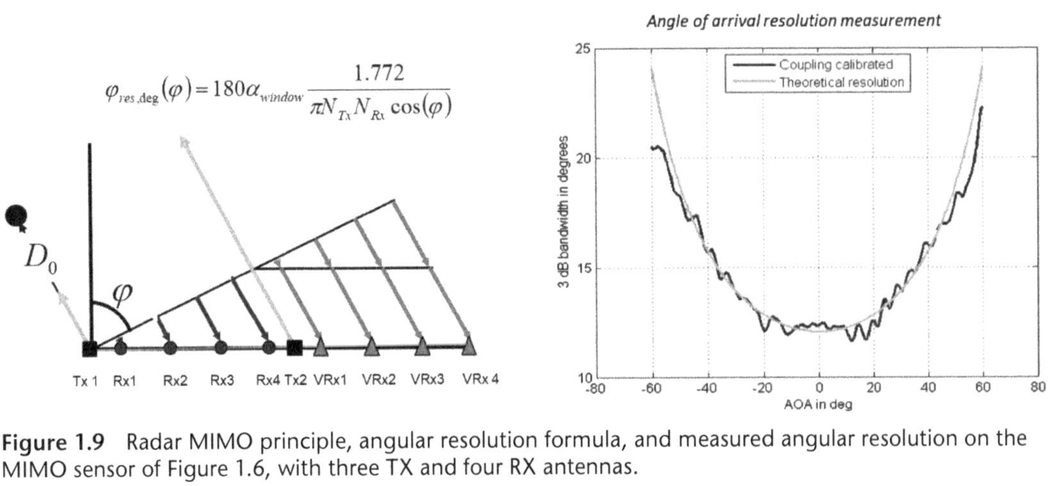

Figure 1.9 Radar MIMO principle, angular resolution formula, and measured angular resolution on the MIMO sensor of Figure 1.6, with three TX and four RX antennas.

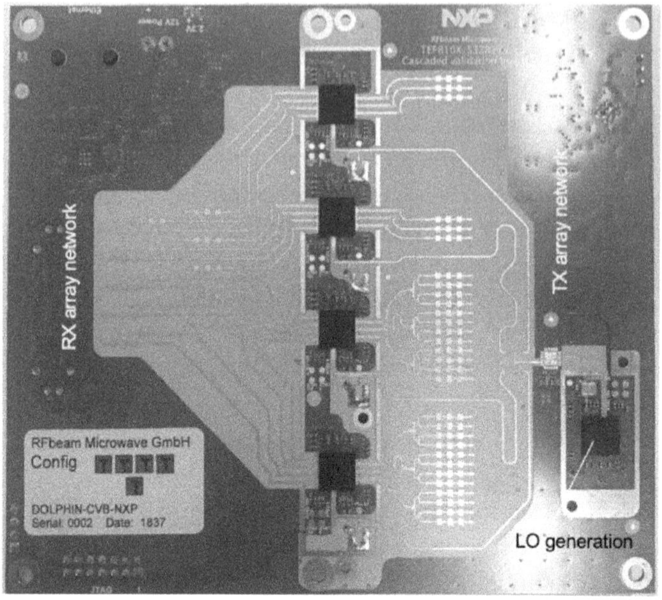

Figure 1.10 Cascaded system built with TEF8102s, four operating as RX/TX elements, and one as an LO generator and timing synchronizer.

must increase, or that different chips (e.g., TEF8102) must be configured to operate simultaneously and coherently. This is often referred to as a *cascaded configuration.*

For a cascaded configuration to provide the expected theoretical behavior, it is crucial that the local oscillator (LO) signal at all distributed receiver channels be stable with temperature after an initial calibration of possible systematic offsets. It is also of fundamental importance that the sampling moments of all distributed ADCs be stable across temperature and operational conditions. Phase deviations on portions of the LO distribution network or on the sampling moment of a given ADC will be superposed into associated IF signals, and therefore may lead to decreased system angular resolution and possible errors on direction of arrival estimations.

Figure 1.10 depicts an experimental cascaded system built with five TEF8102 ICs. It was used to prove the stability of the LO distribution concept as well as the timing consistency of the ADC's operating across the four ICs performing TX and RX operations. The chirp generation and leading timing engine operations are performed by a fifth TEF8102, visible on the bottom right of Figure 1.10.

1.7 Vehicle Network and Compute Considerations

1.7.1 Vehicle Network Architecture Evolution

Today, a fully connected passenger vehicle can be thought of as a networked computer on wheels that can have within its architecture more than a hundred control units and many sensors, interconnected by wires. The vehicle is also rapidly increasing the software content on board and communicating with other vehicles, the cloud and user devices. The software (SW) complexity of a modern car, in line of codes, is in an order of 20 to 30 times more complex than SW needed to operate the Hubble telescope, the NASA Mars Rover, and the Space Shuttle combined.[3] The cost of radar sensors is important, because up to twenty radar sensors may be employed for the higher autonomy levels, assuming radar is taking up the role of ultrasonic sensors over time.

Switched Ethernet (SwEth) prepares the communication system for Scalability, enabling the addition of sensors over the model platform lifetime. Furthermore, it allows traffic engineering (parameters latency, bandwidth, throughput), providing switching (time synchronization, prioritization, traffic shaping, admission control) and data security. In view of communication protocols, Ethernet is the first choice for high-speed communication in the vehicle as it is a matured technology based on IP protocols, the ecosystem has existed for a long time, and the industry is familiar with its usage; the automotive domain is merely an extension of its application space. Ethernet connections in a car address the need to rapid communication and data sharing between sensors and central compute units.

The car architecture is under a constant evolution and there are further factors that are leading to a constant change of architecture. The networked IP architecture in vehicles is becoming more and more an attractive target for hackers. Cloud-connected cars must protect its life-saving functions from external attacks. In the past years, there have been quite a few prominent hacks of cars, showing their vulnerability to those attacks. Vehicles today have internal gateways securing safety critical subnetworks, and communication between network units such as a radar sensor and a central electronic control unit (ECU) must be protected by authentication and encryption [11].

A further driver for new architectures is what vehicle manufactures call *functions on demand*. In this new concept, a vehicle could be already equipped with a large, fixed range of radar sensors, simplifying the variety in vehicle platforms as a fixed set of radar sensors would be installed on every vehicle. In a usage model the customer can order a certain function, for example Highway Pilot when ordering the vehicle at the dealer. During production only the respective SW packets need to be installed to implement this function. Another scenario goes a step further. Customer can book on demand additional functions for a limited time period additional function. Using the example of the Highway Pilot, such feature might be booked at the beginning of a longer holiday trip.

ADAS processing is based on a variety of sensors. Traditionally the architecture of ADAS sensors has been edge based. For example, with radar this means the radar signal processing is fully carried out on the sensor itself, from waveform generation, transmission to reception, signal processing object detection, and tracking. Object level data can be sent out using standard, low data rate communication networks, such as CAN-FD, where the object level data or tracked object data fit well within the data rate of up to 2 Mbit/sec.

In contrast, one can think of a different architecture, where all signal processing is done on a central unit. The radar sensor itself would output only raw sampling data toward a central compute unit. In this case, the communication bandwidth

3. https://informationisbeautiful.net/visualizations/million-lines-of-code/.

requirements would increase into the Gbit/sec range for each sensor in the car, with the requirement of high-speed communication interfaces such as 1,000 BaseT1 Ethernet connections. Figure 1.11 illustrates the two cases and the main motivations.

A further variant, the zonal-based architecture (depicted in Figure 1.12) exists between these two extremes. Zonal-based architecture for ADAS applications are derived from the zonal-based body networks in cars, which define certain zones in the car, where functions and related processing are combined and physically located. In the case of a radar application, one can think of a zonal node for the front and the rear of the car, each being a separate zonal node. The front zone may be responsible for the control, synchronization, and the data aggregation of all front-facing sensors, such as corner sensors in front rear and front right, and forward-looking radar sensors for automated emergency braking (AEB), automated cruise control (ACC), or higher levels of autonomy.

1.7.2 Distributed Versus Centralized Processing

The conceptual idea of centralizing the radar processing is not new; vision-based processing using camera data made a similar development over the past 5 to 10 years. Early camera modules hosted a large part of the camera processing called the image sequence processor (ISP), but now this is done centrally. In contrast to cameras, radar sensors are active sensors that form a signal, transmit it into the environment, and receive the reflected electromagnetic waves. This requires constant control of the analog front end to ensure signal uniformity through calibration of transmitters over temperature and time. Unlike many other automotive applications, radar processing steps are not standardized; outside of the frequency regulations that define the use of a certain frequency band as well as the interaction with adjacent bands (emissions), no standard signal processing scheme exists. The use of IF bandwidth, chirp sequences as well as post-processing steps varies vastly in the automotive radar market. Figure 1.13 describes a generic approach of the

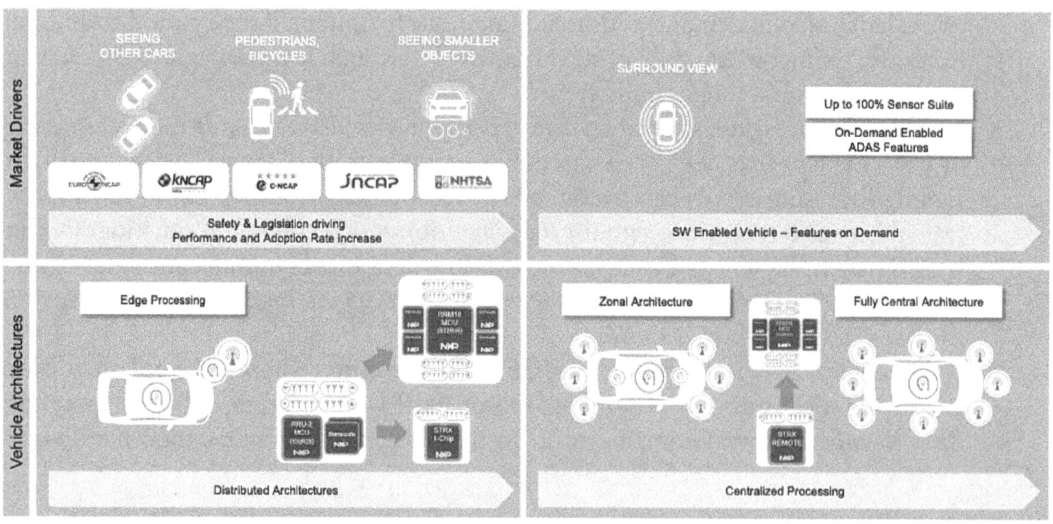

Figure 1.11 Trend toward centralized processing.

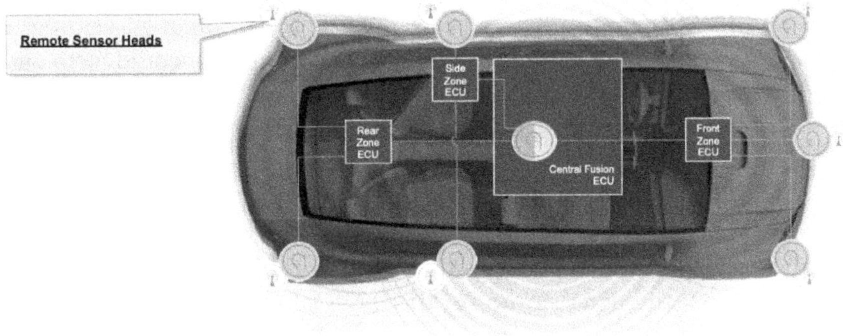

Figure 1.12 Zonal ADAS architecture.

main processing steps that are needed from availability of ADC samples in digital domain until object clustering, tracking, and output via a standardized car in-vehicle network. The remaining part of the chapter explores options and consequences for centralizing various of those processing steps.

"Where to cut the processing?" is one of the central questions to be answered. Let's assume a radar system architecture that will perform only a subset of these processing steps locally in the radar edge sensor. The remainder of this chapter looks into several aspects that will determine an efficient partitioning.

One of the major points is the increase of radar sensor performance. Several decision steps in radar processing leading to a reduction of data, which is on one hand an advantage, as it leads to reduced data rate and thus, simpler, and cheaper in-vehicle networking (IVN) connections, less memory storage at various locations like Ethernet switches, gateways, and central ECU. On the other hand, reduction

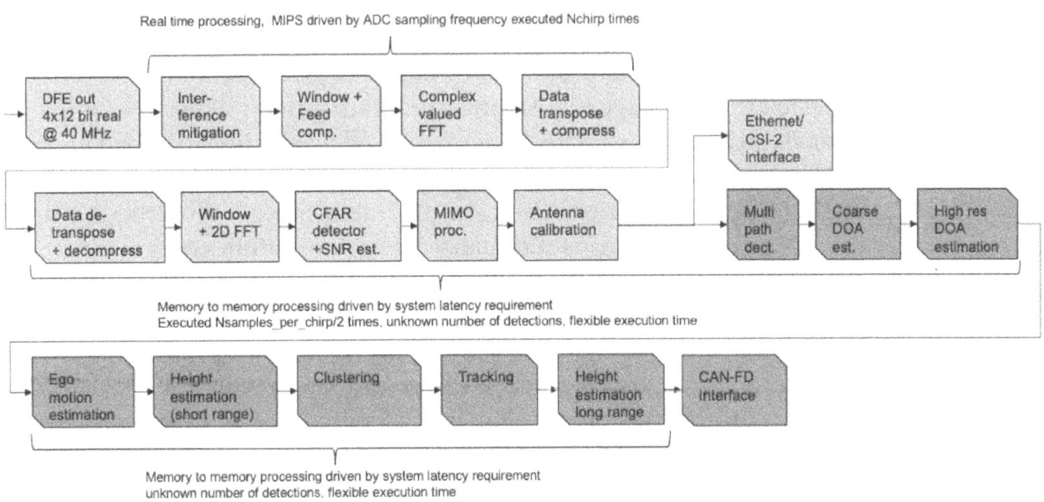

Figure 1.13 Radar signal processing.

of data is an irreversible step. Once removed, data cannot be reconstructed. With larger compute capacities, faster in-vehicle bus architectures being available, it is time to rethink if there are better processing options available, leading to better radar data with much higher performance. An obvious step would be to output raw time domain sampled data right after the digitization. Since radar data are usually sparse data, only few samples hold real information. One could assume there is a good amount of compression of this data possible. However, in time domain it is impossible to determine samples with relevant information from those that contain irrelevant data. Also, in time domain data many target returns may be small or below the noise floor, as this is before the radar applies various signal processing steps that increase the signal-to-noise ratio (SNR). Eventually, using highly complex compressions schemes (e.g., matching pursuit (MP) algorithms) might lead to success achieving reasonable compression rates with low amount of compression artifacts. MP is a sparse approximation algorithm that finds the best matching projections of a multidimensional field of data using a trained dictionary D. One of the main problems of MP is the computational complexity of the encoder [12]. In contrast, after conversion into frequency domain, by means of FFT transformation, it is possible to apply more efficient and straightforward compression techniques to the radar FFT cubes.

Possible approaches to compression could use efficient number coding, like block floating point or logarithmic coding. A different group of compression techniques target the intrachannel dynamic range and uses adaptive scaling techniques. Usually, the spectra of antenna channels are highly correlated. This characteristic can be used also for efficient compression techniques. The so-called *channel-to-channel* (C2C) compression scheme benefits from the antenna correlation effect, which allows to treat channels with the same compression settings.

C2C compression operates on spectral values, which are recoded as block floating point data. The common exponent defines the joint signal amplitude. The individual mantissa values are requantized for lower resolution. The specific mantissa treatment will result in additional quantization noise, which will lead to amplitude errors, which is within a half mantissa step times the exponent. As the error scales with the signal level, there will be a low impact on the effective SNR. The relevance of this error is not high if it is above the detection threshold. As outlined in Figure 1.14, there will be a phase error because of mantissa quantization; however, this effect is expected to be averaged by Doppler processing that follows.

A further class of compression techniques is based on identification and possible removal of clutter in the radar cube. Clutter refers to stationary echoes in radar signal that might be signal reflections from ground. Identification of clutter is critical because in some specific scenarios stationary objects need to be recognized.

The choice of a suitable compression technique will depend largely on the application field of the radar. Further important factors are the number of chirps and IF bandwidth. Good adaptive compression algorithms can deliver compression factors of 1:3 to 1:6 without notable unwanted artifacts and low compute latency. Ideally, a compression shall be reversible for the Doppler processing with low processing requirements, resulting in little to no extra delays. For internal memory organization fixed compression ratios are preferred. Content dependent compression ratios are usually challenging as they lead to nonpredictable memory needs and word

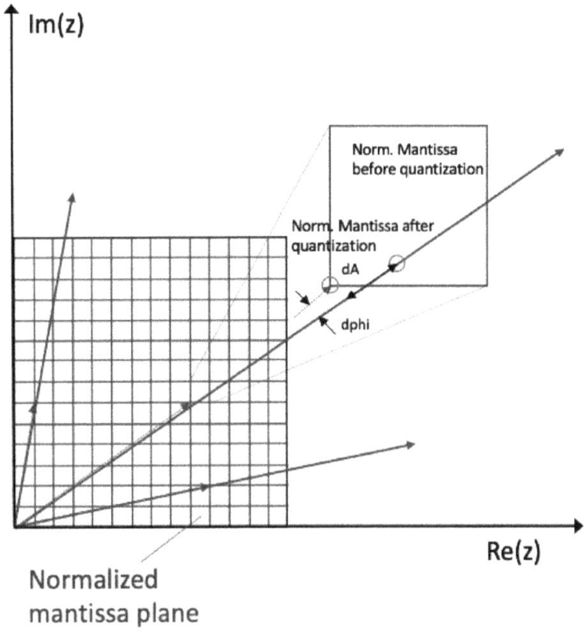

Figure 1.14 Magnitude/Phase Error Channel-to-Channel Coding Compression.

width. A fixed compression scheme is usually efficient for most of the signals, and the compression-decompression task is nearly lossless for average strong signals. Figure 1.15 shows a 3:1 fixed mantissa compression for a 256 FFT signal with a SNR of 38.57 dB. There is no coding effect visible in the resulting signal.

Returning to the signal processing flow introduced Figure 1.13, there are more potential partitions of the signal processing. Figure 1.16 is based on a typical radar use case with four RX channels, a sample rate 20 MS/sec, and a 12-bit representation of each data word.

With a compression factor of two the range compressed, often referred to as FFT1 data, can be reduced to about 640 Mbit/sec. The net compression rate comparing raw ADC data at 960 Mbit/sec and compressed FFT1 data is 1.5 since FFT data use 16 bit data word representation versus 12 bit for ADC output.

A further important aspect of distributed systems is the requirement of local buffering. Output data are present as data bursts, based on the coherent processing interval (CPI) collection time (often called the *chirp time*), and associated processing

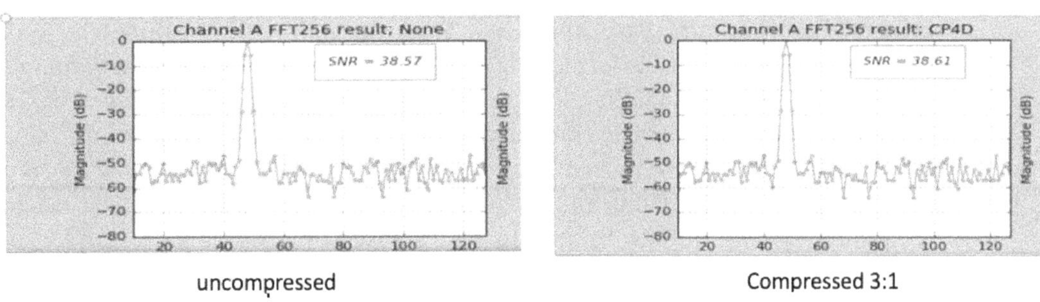

Figure 1.15 SNR impact FFT compression.

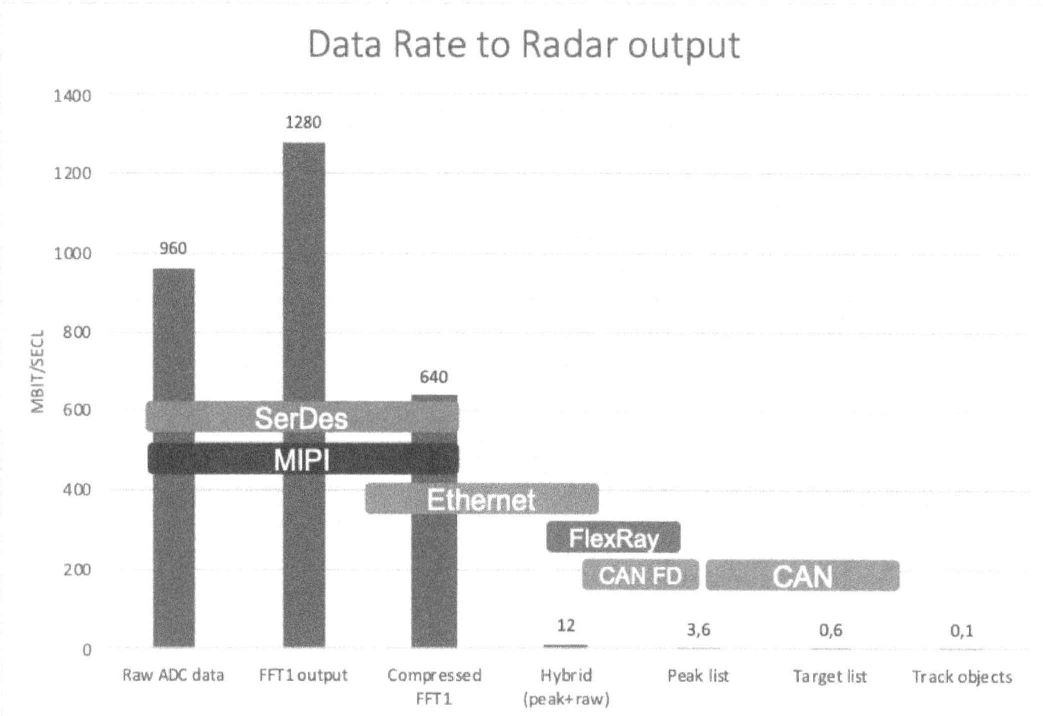

***Hybrid 1** peak list + raw data (multiple detections per point source, each detection comes with vector of raw data)*

Figure 1.16 Typical data rates for each stage in radar signal processing.

time. With a typical radar system cycle (collection and processing) of 40 ms and a chirp time of 20 ms, the full set of data is available within the first 50% of the radar system cycle.[4] Thus, the instantaneous data rate occurring during data sampling can be higher than available network bandwidth. Additional local buffering of data is needed to avoid overflow of the pipeline. A first-in first-out (FiFo) buffer can be used, but the buffering of data in the sensor will result in additional latency. The minimal latency directly depends on the ratio between input and output rate of the buffer system. The higher the output rate, less buffering is needed, and the latency is smaller. The transport latency caused by protocol handling and transport of data (e.g., Ethernet protocol) can be seen as neglectable compared to latency of buffering.

The impact of the buffering latency can be significant. As an example, let's assume an input rate of 960 Mbps and an output rate of 800 Mbps (the assumed maximum Ethernet data rate capability). With the following formula, the max buffer size (MBS) and the buffering delay can be calculated, based on the available network transmission capability. The read-write delay (rwd) is assumed to be 30 us (1 chirp), and the duty cycle time (dct) is the active measurement time during a scan period. The term $rate_{\text{in}} \cdot rwd$ adds one additional sample buffer, in case continuous chirping is done. Continuous chirping refers to an *endless* chirp period. In that case, the minimum buffer size must be increased by one sample.

4. In reality, this time is even shorter because the chirp time contains also setup times of chirp engine.

$$MBS = \left(rate_{\text{in}} - rate_{\text{out}}\right) \cdot dct + rate_{\text{in}} \cdot rwd \tag{1.1}$$

and

$$txl = \frac{MBS}{rate_{\text{out}}} \tag{1.2}$$

With a network transmission rate of 800 Mbps, a buffer delay of 4.2 ms can be achieved, which is likely an acceptably small addition to the total radar cycle time of 40 ms. If the network transmission rate drops to, for example 480 Mbps, the buffer delay will increase to 14 ms which is a significant increase of latency and would drastically reduce the capabilities of the radar system. Figure 1.17 depicts the total radar processing latency based on the time needed to receive the first data-to-object detection.

There are several key relationships involved in this trade.

- The time needed for range compression (FFT) is determined by the chirp time (duty cycle) and the performance of real-time FFT processing.
- The time needed for Doppler FFT and power density is determined by the real-time processing performance and additional processing steps (like multiple-input-output (MIMO)) and the FFT size, channel count, number of range bins, and constant false alarm rate (CFAR) method.
- The time needed for direction of arrival calculation often depends on the number of detections.
- The time needed to establish track is based on number of consecutive detections (typically ≥ 5) required, which is often the longest time in this process.

The total latency in the signal processing chain will be in the order of six scan periods with approximately 160 ms in our example with a 40 ms radar cycle period. Faster scan periods can reduce latencies. Considering that the buffer latency txl occurs every scan period and a tracking algorithm typically needs a history of around five consecutive detections, the total radar processing latency will be 5 · txl longer in a central configuration for the same compute capability, or the signal processing needs to be accelerated to compensate the additional buffer delay (Figure 1.18).

Returning to the previous example of buffer latency being 4.2 ms based on an 800 Mbit/sec network transmission capability and an input rate of 960 Mbit/sec, the buffer delay is adding up to 5 · 4.2 ms = 21 ms. The central compute speed would need to increase to compensate the delay to maintain equivalent performance to

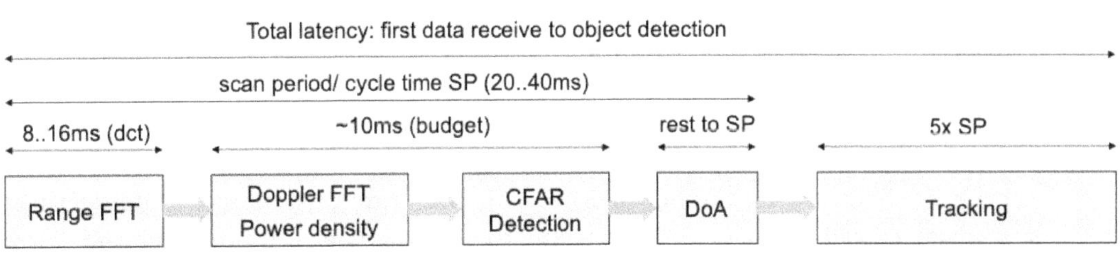

Figure 1.17 Radar processing latency.

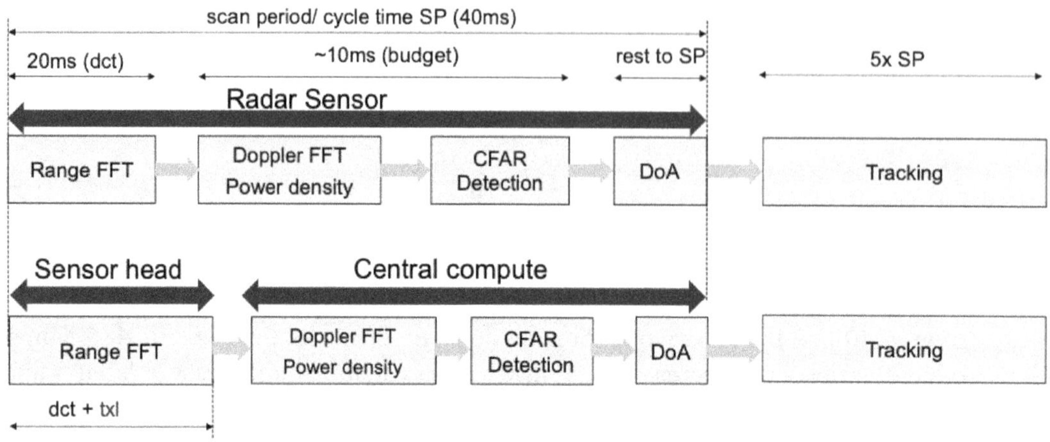

Figure 1.18 Impact of buffering latency.

the edge-processing case. In the worst case example of only 640 Mbit/sec network transaction capability, the delay to compensate by central compute would be 70 ms. Compensating this delay would need to nearly quadruple the compute performance, leading to significant cost and power consumption increase.

Another option to avoid such delays is to incorporate one of the earlier-discussed compression techniques. Assuming a compression factor of three, 960 Mbit/sec of the previous example can be compressed to 320 Mbit/sec which could be transferred even without buffering.

1.7.3 Conclusion

Many motivating reasons exist to move toward more centralized architectures for advanced driving functions, yet not all of them are easy to achieve. Reduced complexity in a radar sensor by moving processing steps from edge to central domain can enable a more standardized radar sensor. With the growing number of radars in new car architectures this can help to reduce complexity and drive scalability of AD architectures. We have discussed some possible variants for signal processing partitioning between the edge and central processing that can facilitate this.

Higher data rates, because of less edge processing, can lead to more expensive network architectures need to transfer Gbit/sec from each sensor to a central compute unit. Those networks' physical layers and gateways can significantly drive cost and power consumption. Breaking up the signal processing chain can lead to significant processing delays that may impact AD functions like autonomous emergency breaking due to additional delays. Advanced compression techniques that mitigate artifacts in range spectrum can be used to reduce data rates and allow reduction of delays, and these are present on today's advances radar application chipsets (e.g., the NXP S32R45). Table 1.1 summarizes the benefits and disadvantages of centralized and distributed radar architectures.

Autonomous driving will continue to advance with more functionality, higher service level, and better quality of service. To facilitate this, network architectures for advanced radar sensors will continue to develop as well with better and faster network components and more powerful central compute units.

Table 1.1 Benefits of and Disadvantages of Radar System Architectures

Attribute	*Centralized Radar System*	*Distributed Radar System*
Car network architecture	Less signal processing in sensor	Low speed network requirements
Data transfer	Advanced data compression required	Small size object lists only
Software (SW) maintenance	Centralized SW maintenance for OEMs	Distributed SW in edge sensor
Power	Higher power consumption due to high speed network PHY	More power needs to be distributed to edge for processing
Advanced processing	Path toward advanced signal processing using raw data in central processing	Central processing limited to object list delivered by edge sensor

1.8 Summary

In this chapter, we have presented the need and taxonomy for autonomous vehicles and established that radar is a key enabler for this autonomy. We have discussed CMOS technology as an enabler for the deployment of automotive radar to all market segments and have introduced a 76–81 GHz transceiver in 40 nm CMOS that can be used in single configuration for cost-sensitive sensors as well as in cascaded configuration for higher angular resolution applications. We have explored the challenges of radar signal processing, data movement, and various compute architectures. During the past years, major advancements in automotive radar capabilities have been made, many of them enabled through algorithmic research and investments in mixed-signal devices. As autonomy continues to advance and flourish, work in radar architectures, algorithms, and device technologies will continue as well, fueling further advances and applications.

1.9 Acknowledgments

The authors are indebted to many colleagues at NXP Semiconductors for their contributions, with special mention due to Kees Gehrels, Anton de Graauw, Maik Brett, Curt Hillier, Gustavo Adolfo Guarin, Timo van Roermond, Steffen Mueller, Mingda Huang, Feike Jansen, and Kostas Doris.

References

[1] "SAE J3016™: Taxonomy and Definitions for Terms Related to On-Road Motor Vehicle Automated Driving Systems," www.SAE.org.

[2] Meinel, H. H., "Evolving Automotive Radar—From the Very Beginnings into the Future," *The 8th European Conference on Antennas and Propagation (EuCAP 2014)*, The Hague, Netherlands, April 6–11, 2014, pp. 3107–3114.

[3] Brooker, G. M., "Understanding Millimeter Wave FMCW Radars," *Proceedings of the 1st International Conference on Sensing Technology*, Palmerston North, New Zealand, November 21–23, 2005, pp. 152–157.

[4] Richards, M. A., *Fundamentals of Radar Signal Processing*, Second Edition, New York: McGraw-Hill Education, 2014.

[5] Rohling, H., and C. Moeller, "Radar Waveform for Automotive Radar Systems and Applications," *Proceedings of 2008 IEEE Radar Conference*, Rome, Italy, May 26–30, 2008, pp. 1–4.

[6] Rohling H., and M. Kronauge, "New Radar Waveform Based on a Chirp Sequence," *Proceedings of the 2014 International Radar Conference,* Lille, France, October 13–17, 2014, pp. 1–4.

[7] Menzel, W., and A. Moebius, "Antenna Concepts for Millimeter-Wave Automotive Radar Sensors," *Proceedings of the IEEE*, Vol. 100, No. 7, July, 2012, pp. 2372–2379.

[8] Beasley, P. D. L., "The Influence of Transmitter Phase Noise on FMCW Radar Performance," *Proceedings of the 3rd European Radar Conference 2006*, Manchester, U.K., pp. 331–334.

[9] Schnabel, R., et al., "Reflection, Refraction, and Self-Jamming," *IEEE Microwave Magazine*, May 2012, pp. 107–117.

[10] Feger, R., et.al., "A 77-GHz FMCW MIMO Radar Based on an SiGe Single-Chip Transceiver," *IEEE Transactions on Microwave Theory and Techniques*, Vol. 57, No. 5, May 2009, pp. 1020–1035.

[11] Mueller, S., van Roermund, T., and M. Steigemann, "Communication Beyond Road Vehicles—Road to Automated Driving" in *Advanced Microsystems for Automotive Applications*, Springer, 2016, pp. 55–66.

[12] Mallat, S. G., and Z. Zang, "Matching Pursuits with Time-Frequency Dictionaries," *IEEE Transactions on Signal Processing*, Vol. 41, December 1993, pp. 3397–3415.

CHAPTER 2

Design Considerations for Automotive Radar

Shunqiao Sun, Donnie Smith, and Matt Markel

Millimeter-wave technology has enabled significant advances in capability for automotive radar. The typical frequency band of millimeter-wave automotive radar is 76–81 GHz, and these high frequencies allow for antennas small enough to fit behind the bumpers of the vehicle. Also, the wide available bandwidth enables high target range resolution [1] (typically through frequency modulation) and velocity estimation at much lower cost than light detection and ranging (lidar) technology.

In this chapter, we provide a brief history of radar and its applications, an overview of how automotive radar requirements are derived, and some key architectural aspects that must be considered in radar design, such as installation on the vehicle and use of the radar timeline. We also provide a review of frequency modulated waveforms and a common processing method for these waveforms called *stretch* and the typical data products provided by an automotive radar.

While many reading this work may have some background in automotive radar already, we believe that a significant portion of readers may not, but instead have experience in radar applications from the aerospace industry. To help bridge this gap, we have—where we felt it was illustrative—highlighted key similarities or differences between automotive radars and those from the airborne community.

2.1 Radar Requirements

Radar was originated and almost entirely developed for aerospace applications. From the Chain Home system monitoring the coastline of Great Britain during World War II to the AN/APG-81 fire control radar in the nose of every F-35 fighter jet, radar has been developed by the aerospace industry for a multitude of applications, including surveillance, early warning, fire control, tracking, and altimetry, to name just a few. Applications outside aerospace certainly exist, such as meteorology and radioastronomy, but it is aerospace that has made developing radar applications into the reliable and established discipline that it is today.

Automotive radar, on the other hand, is a nascent field with rapidly changing and often poorly understood requirements, due to increasingly complex demands

Figure 2.1 The fifth generation Waymo driver operating a Jaguar I-Pace. Note the radar modules on the roof, above the front wheels, and below the taillights. (Source: Waymo.)

and applications. The automotive industry in general is in the process of a transformational change in how it manages technological complexity with the addition of engine and drivetrain monitoring sensors and software, hybrid drivetrains, various levels and applications of automated driving, integrated infotainment systems, and other electronics. As a result, automotive development is increasingly adopting the techniques created by the aerospace industry for developing requirements and producing technological solutions.

Which isn't to say that there haven't been extraordinary successes in automotive radar during the past several decades. A number of experimental and demonstration systems were developed starting from the 1970s in a number of bands, including 16, 35, 47, and 60 GHz. Production applications began in the 1990s with Mercedes's DISTRONIC 77 GHz adaptive cruise control (ACC) system (Figure 2.2) and Eaton's VORAD CWS 24 GHz blind spot detection system (Figure 2.3). The 2000s and 2010s saw an explosion in both driver warning and automatic cruise control from nearly every major manufacturer and universal adoption of the narrowband 24 GHz and wider 77 GHz bands. Systems came primarily from the traditional automotive suppliers like Autoliv, Bosch, Continental, Delphi (now Aptiv), and Denso, colloquially known as the ABCDs. The race to full self-driving capability starting in the early 2010s coincided with a growing selection of integrated 77 GHz MMIC (millimeter-wave integrated circuit) solutions from developers including Infineon, NXP, Texas Instruments, and ADI that quickly began decreasing the cost per channel and increasing the capability of affordable radar systems. These devices typically have one to four transmit and receive channels per chip, and most can be cascaded to build systems with higher channel counts.

The advancing capabilities of automotive radar systems didn't happen in a vacuum; using multiple sensor modalities is a key feature of any modern automated driving system. The most common sensors integrated with radar are lidar and camera. In general, these sensors have much better range resolution, in the case of lidar, and angular resolution in the case of both. The advantages that radar brings to the table are longer range, the ability to measure doppler, or range rate, and better performance in adverse weather conditions. The combination of these sensors provides a rich and complementary set of measurements, but brings the additional challenge of combining that data, a discipline known as *sensor fusion*. In order to be used effectively, the data from multiple sensors must be combined so that objects in the scene can be described with data from each sensor. Typically, this requires segmenting sensor data into separate objects, followed by associating data describing the same object across sensors. Difficulties include the lack of range measurement, and thus spatial position, in camera frames, and the drastic differences in resolution between radar and lidar making association difficult between closely spaced objects. These problems are active areas of research and development.

The goals of automated driving solutions are as numerous as the sensors that power them. Early cruise control systems merely held a constant speed and required no remote sensing, relying on the driver to avoid collision by braking and steering. ACC systems responded to surrounding traffic by automatically braking and accelerating. Lane centering systems applied small steering corrections to keep vehicles centered in their lane. More advanced systems combined the functionality of ACC and lane centering with navigation to allow vehicles to drive themselves under the supervision of a human driver, who still had the responsibility to take over in the event that the system failed to correctly respond to its environment. The ultimate aim of the self-driving industry is to produce a system that requires no intervention from a human driver whatsoever. These systems are all collectively known as advanced driver-assistance systems (ADAS), and the spectrum of capabilities has been classified into five levels (see Table 2.1) by the Society of Automotive Engineers (SAE).

Amazing developments are yet to come in the capabilities of automotive radar. The primary areas of development are in increasing channel count, compute performance, and cost. Increasing the channel count, both on the transmit and receive side, enables better spatial resolution and the use of radar as an "imaging" sensor.

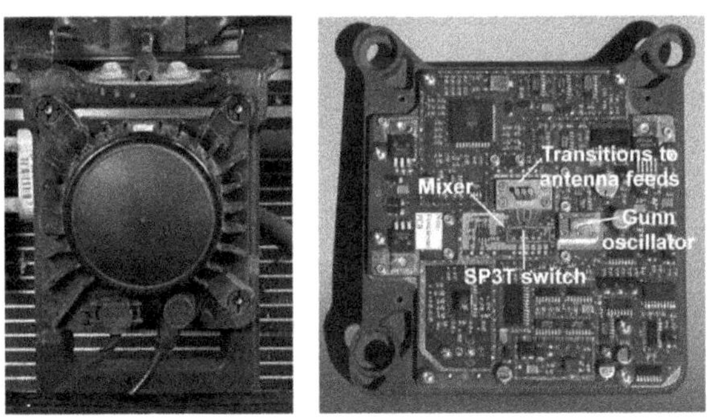

Figure 2.2 Front and back side of first commercial automotive ARS 100 radar by Mercedes Benz. (© 2021 IEEE. Reprinted, with permission, from [2].)

Figure 2.3 A 24 GHz VORAD radar mounted at the front side of a Greyhound bus. (© 2021 IEEE. Reprinted, with permission, from [2].)

Table 2.1 SAE Autonomy Levels

Level 0	*Level 1*	*Level 2*	*Level 3*	*Level 4*	*Level 5*
No Automation	*Driver Assistance*	*Partial Automation*	*Conditional Automation*	*High Automation*	*Full Automation*
Zero autonomy; the driver performs all automation tasks	Vehicle is controlled by the driver, but some driving assist features may be included in the vehicle design	Vehicle has combined automated functions, like acceleration and steering, but the driver must remain engaged with the driving task and monitor the environment at all times	Driver is a necessity, but it not required to monitor the environment. The driver must be ready to take control of the vehicle at all times with notice	The vehicle is capable of performing all driving functions under certain conditions. The driver may have the option to control the vehicle	The vehicle is capable of performing all driving functions under all conditions. The driver may have the option to control the vehicle

With growing channel counts comes additional computational complexity, much of which is FFT and vector operations that can be executed more efficiently by specialized hardware (e.g., digital signal processors (DSPs) and field programmable gate arrays (FPGAs)), than by traditional central processing units (CPUs). As advanced systems come closer to production, cost is always a major factor. Economies of scale have enabled vast reductions in cost during the past decade and will likely continue to do so. Channel counts can be reduced with techniques like MIMO and sparse arrays, but typically require more advanced signal processing and computation to implement.

2.2 The Spectrum for Automotive Radar

Unlike aerospace and defense radars that have multiple bands to choose from (VHF, UHF, S, X, etc.) (shown in Table 2.2), automotive radars are limited to 76–81 GHz.[1] Depending on the requirement for range resolution, automotive radar designers are free to choose multiple channels within the 76–81 GHz band. Further, this band allows automotive radars to have small form factor, which allows the incorporation of automotive radars behind vehicle bumpers and windshield. Other frequency bands, such as 120–260 GHz, are currently under discussion for possible automotive radar applications [4–6], bringing multiple technology challenges but also a considerable increase in the electrical size (size in wavelengths) for the same form factor.

As the number of vehicles equipped with automotive radar increases, so does the probability of encountering mutual interference between these radar units

1. Historically, automotive radars could operate at both 24 GHz and 77 GHz. Due to the overlap with industrial, scientific, and medical (ISM) band, the 24 GHz ultrawide band (UWB) will be phased out in 2022 in both the United States and Europe [3].

Table 2.2 Radio Frequency Bands and Application

Band	*Frequency*	*Wavelength*	*Applications*
HF	3–30 MHz	100–10m	Over-the-horizon radar, RFID
VHF	30–300 MHz	10–1m	FM radio, television broadcast
UHF	0.3–1 GHz	100–30 cm	Television broadcast, ground penetrating radar
L	1–2 GHz	30–15 cm	Satellite navigation
S	2–4 GHz	15–7.5 cm	Airport surveillance and weather radar
	4–8 GHz	7.5–3.75 cm	Airport surveillance and weather radar
X	8–12 GHz	3.75–2.5 cm	Space communication, civil and military radars
K_u	12–18 GHz	2.5–1.67 cm	Police traffic-speed radar
K	18–27 GHz	1.67–1.11 cm	21.65–26.65 GHz ultrawide band (UWB) radar 24.05–24.25 GHz narrow band ISM
K_a	27–40 GHz	1.11–0.75 cm	Satellite communication
V	40–75 GHz	7.5–4 mm	Radar for human-machine interface
W	75–110 GHz	4–2.7 mm	76–77 GHz long range automotive radar 77–81 GHz medium/short range automotive radar
mm	100–300 GHz	3–1 mm	120–260 GHz possible automotive radar application

because of the increased spectral congestion [7]. To achieve a high range resolution, conventional automotive radar typically use frequency modulated continuous wave (FMCW) waveforms that occupy a large continuous bandwidth, significantly increasing the probability of interference. The automotive radar mutual interference is a challenging issue that needs to be addressed; without interference mitigation, automotive radar sensors suffer from performance degradation [8]. This topic is discussed in detail later in this book in Chapter 8, Radar Interference.

2.3 Range (Distance) Required for Automotive Radar

When compared to ground-based or airborne surveillance radars, automotive radars have smaller size (inches by inches), shorter ranges of interest (within a few hundred meters), lower power (tens of watts), and are much more driven by cost. In addition, they operate in a highly dynamic and cluttered environment with significant multipath and are often integrated in suboptimal locations such as behind the vehicle bumper, grille, or windshield. Figure 2.4 shows a typical radar perception configuration for a vehicle capable of some levels of autonomous driving.

Four short-range radar (SRR) sensors with an azimuth field of view (FOV) [−75°, 75°] and a detection range of 45m are deployed at the four corners of the vehicle for blind spot detection and cross traffic alert. Two mid-range radar (MRR) sensors with an azimuth FOV of [−40°, 40°] and a detection range of 100m are deployed at both front and rear sides for lane change assist (LCA) and automatic emergency braking (AEB). A long-range radar (LRR) sensor with an azimuth FOV of [−15°,

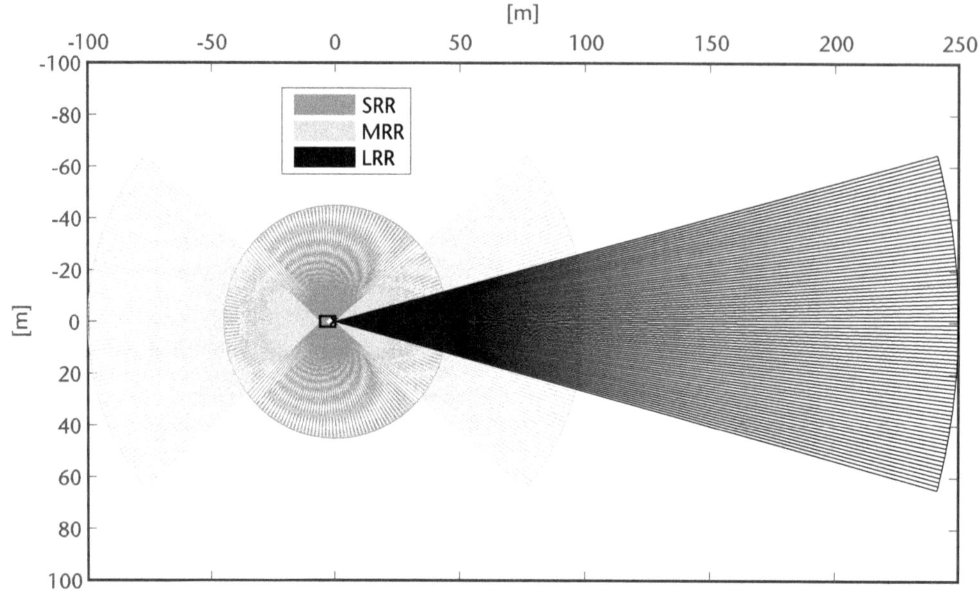

Figure 2.4 One potential radar perception configuration for autonomous driving contains four SRRs, deployed at the four corners of the car providing 360° coverage, two MRRs for both forward and rear looking, and one LRR for forward looking. For perspective, the width of a highway lane in the United States is 3.7m. (© 2022 IEEE. Reprinted, with permission, from [7].)

15°] and a detection range of 250m is deployed at the front for ACC. To support the perception task in trucks that need a longer distance for braking, the LRR is desired to detect targets in range of 300–500m. All radar sensors in a vehicle are connected to an electronic control unit (ECU) for further processing, such as radar tracking and sensor fusion [9]. The automotive radar classification is summarized in Table 2.3.

The maximum unambiguous range of automotive radar is

$$R_{\max} = \frac{cT_{\mathrm{PRI}}}{2} \tag{2.1}$$

Table 2.3 Automotive Radar Classification*

Radar Type	*Long-Range Radar*	*Mid-Range Radar*	*Short-Range Radar*
Range (m)	10–250	1–100	0.15–40
Azimuth FOV (degree)	±15°	±40°	±80°
Elevation FOV (degree)	±5°	±5°	±1
Applications	Adaptive cruise control	Lane-change assist, cross traffic alert, blind-spot detection, rear-collision warning	Park assist, obstacle detection, precrash

*The detailed numbers vary with application and radar designer, but this represents the broad classes in use today. Radars used for fully autonomous vehicles typically contain aspects of all of these classes of radars.

where c is speed of light, and T_{PRI} is the pulse/chirp repetition interval (PRI). In automotive radar, a typical chirp duration is $T_{PRI} = 0.2$ ms, corresponding to $R_{\max} =$ 30 km. In other words, automotive radars with hundreds of meters detection range typically don't suffer from range ambiguity.

On the other hand, in typical automotive radar scenarios the velocity may be ambiguous (see (2.5) and an example of velocity ambiguity in Section 2.6) due to the high speeds observed and high carrier frequency. Similar to airborne radars, automotive radars resolve the velocity ambiguity through the use of multiple coherent processing intervals (CPIs, also called a *radar frame*) with slightly different PRIs.

In the highly dynamic automotive scenario, however, the positions of both radar-mounted vehicle and objects may often change rapidly [7]. As a result, it is challenging to coherently process the target's echo measurements from multiple CPIs. A typical duration of a frame in automotive radar is around 50 ms, corresponding to detection update rate of 20 Hz [10]. In such short duration, the early generations of automotive radars for ADAS reported a maximum of 64–200 detections. With high resolution imaging radar, the number of cells that can be selected for angle estimation from the 2-D range and Doppler spectrum is around 10,000 in a single frame for a typical road scenario [11]. To achieve point clouds for autonomous driving, angle estimation needs to be performed thousands of times in a single frame, which is a great challenge for imaging radar with hundreds of virtual array elements. Computationally efficient, high resolution angle finding algorithms are highly desirable for real-time implementation in automotive radar. In the highly dynamic automotive radar scenario, only few or single array snapshots are available for angle estimation [12]. Here, the array snapshot is defined as the array response of all virtual array elements corresponding to the same range-Doppler bin [6]. On the other hand, it is challenging to apply the subspace based high resolution angle estimation algorithms, such as MUSIC [13] and ESPRIT [14] that requires multiple snapshots in automotive radar. Usually, forward-backward spatial smoothing [14, 15] technique can be applied to obtain multiple snapshots in automotive radar at a cost of sacrificing the antenna array aperture.

2.4 Automotive Radar Installation

Many airborne radars must deal with extreme temperatures and supersonic flight, driving the radome to aerodynamically tolerant shapes that complicate radar calibration. Although the temperatures and wind stresses automotive radars experience are not as severe as those encountered by an airborne radar, there are still peculiar challenges of the automotive environment that complicate design. For both airborne and automotive radars, the "as-installed" performance can differ significantly from that of the radar in a lab or anechoic chamber. We will briefly touch on some key aspects regarding automotive radar installation.

Because of the relatively low velocities for automotive radars, we can use less expensive and higher performance materials for the radome. Often automotive radars use a polycarbonate (plastic) material for the radome, which has a dielectric constant of 2.8, slightly better than fiberglass or cyanate ester and around one-half that of pyroceram, a common material for missile radomes. In addition, the plastics are lightweight (around 1.2 g/cm^3), and because automotive velocities are considerably lower, the can be formed to a space-saving planar shape for the radar, adding little to the weight of the vehicle. That being said, forward-facing planar shapes can be subject to deformation when driving at freeway speeds, complicating the calibration approach for spatial mainlobe and sidelobes performance.

A key difference between automotive radars for fully autonomous driving and those for driver-assist functions is in the placement of the radar. For fully autonomous applications, the radars are often installed in prominent locations, giving them the clearest, unobstructed view of the environment (refer back to Figure 2.1). Driver-assist radars, on the other hand, are often hidden behind the vehicle's fascia or front grille, to preserve the brand image of the vehicle. This installation, however, can penalize the driver-assist radar's performance through loss of sensitivity (the painted fascia is essentially a second radome through which the radar's energy must pass, subjecting it to additional loss) and complications in angle estimation from the additional material and reflections. In addition, the low grille placement can increase the likelihood that the radar is obstructed by another vehicle in front of it.

Finally, automotive radars must deal with the impacts of the ground-level outside environment. While the radomes for airborne and fully autonomous vehicles are routinely cleaned, that isn't always the case for all automotive radars. Since these radars are difficult to access, cleaning them is often ignored, allowing dust and dirt to build up on the radomes over time to degrade sensitivity.

2.5 Automotive Radar Considerations for Scanning the FOV

An important choice in radar design for fully autonomous vehicles is in the balance between the FOV (the angular area the radar has responsibility over) and the instantaneous field of view (IFOV, the angular area the radar can observe at any particular point in time). Historically, radars used in the defense and aerospace fields partitioned radars that dealt with airborne vehicles (e.g., airplanes, helicopters, missiles) into two broad categories: surveillance or acquisition radars, and target-tracking or fire-control radars. The reason for the development of these different types of radars was that it was difficult and costly to create a single radar that could surveil a wide area with the update rate needed while still providing the tracking accuracy needed to take whatever actions were appropriate. The operational concept was that the acquisition radar would pass tracks to the target-tracking radar, which would be able to acquire the target and put it into its own track by only having to search a small uncertainty region whose size was based on the track covariances of the acquisition radar [17].

Depending on the architecture chosen for a fully autonomous vehicle, the automotive radar may take on each of these roles. The most common architectures today use radar in a surveillance mode, where the radar independently finds, fixes, and tracks targets for presentation to the overall perception system. Here the radar serves both the acquisition and target-tracking functions described and must choose how to determine the FOV and IFOV to meet the system needs. As with everything in radar, there are several tradeoffs.

One method of surveilling the FOV is to use a staring radar, where the IFOV and FOV are matched. Here the radar doesn't scan in angle, it sees the entire area of its responsibility all the time. Because of this, the staring radar usually provides the highest update rate to the perception stack. In addition, because it always sees the exact same area all the time, it is ideal for gathering data for playback in a simulation offline. However, sensitivity can be a challenge for the staring radar, since the wide patterns of both the transmit and receive antennas will come at the expense of gain and therefore signal-to-noise ratio (SNR). In addition, depending on the FOV it may be challenging to achieve the wide patterns with low sidelobes over the desired frequency range of operation. Finally, the staring radar may suffer more from the dynamic range in the scene (for more details, see Chapter 7 on the automotive environment), as its wide antenna allows more large objects in the scene to be seen and processed.

An alternate method is with a scanning radar, where the IFOV is smaller than the FOV. To cover the entire FOV the antenna could be electrically or mechanically steered, although electrically steered is more common because it avoids moving parts. Here the effects of the scene dynamic range are lessened, but for some situations it could still be challenged when a large object (e.g., truck, sign) is close to an object of interest, like a car or pedestrian. Unlike the staring radar, the scanning radar has to manage its timeline to provide the appropriate update rate across the FOV. From a sensitivity standpoint, the time spent in each beam position can be shorter than for the staring radar because of the higher antenna gains; however, there is a lower limit on the time at each beam position that is set by the required Doppler resolution.

The choice of how to scan the radar is another area of tradeoffs. The simplest approach is with a fixed scan sequence, which steps the IFOV through the FOV in a predictable and repeatable manner. This is also a good choice for replay of radar data offline, as the beam position is not dynamic to what is present or observed in the scene. A more involved approach is with a dynamic scan sequence, where the choice of where to position beams is determined in real time by the radar. Dynamic scanning can provide better performance by allocating more timeline to higher value areas of the scene; for example, the radar may identify a stopped vehicle on the shoulder and choose to allocate more timeline to observing this vehicle in anticipation of a pedestrian rapidly exiting and potentially entering the street. In practice, the balance between looking for new objects versus revisiting high value areas isn't always straightforward, but if designed properly can be beneficial. With dynamic scanning the ability to use radar data for offline simulation and evaluation can be compromised, because it may have been looking in one direction when the data were collected and in simulation later it may want to be looking in another area, which now has no data to provide.

If we extend the decision process for how to scan from the local radar control to the higher-level autonomous vehicle system software stack, we can move the radar function from being an acquisition or surveillance radar to a target-tracking radar. Here the radar no longer needs to internally guarantee the scanning of the entire FOV, but instead only the areas where the system software is directing it to evaluate. This method provides the most flexibility to the system software, and it usually requires the most complexity. As with dynamic scanning, use of the radar data for offline simulation can be challenging.

2.6 Frequency Modulation Waveforms and the Radar Data Cube

A frequency modulated continuous wave (FMCW) waveform, also referred to as *chirp*, is a complex sinusoid whose frequency increases linearly with time $t \in [0,T]$, that is,

$$f_{\mathrm{T}}(t) = f_c + \frac{B}{T}t \tag{2.2}$$

where B is the signal bandwidth and f_c is the carrier frequency. FMCW radar transmit chirps in a periodic fashion, with a period referred to as pulse repetition interval (PRI). FMCW is currently the most prolific waveform used in automotive radars, offering large time-bandwidth products and a straightforward processing approach (*stretch*) that enjoys the (cost, power, complexity) benefits of a lower ADC rate at the expense of a limited instantaneous range coverage. The frequency of a FMCW signal over multiple periods, with PRI equal to T, is shown in Figure 2.5. The target echo at the radar receiver contains a delayed and attenuated copy of the transmitted chirp. For a target at range R, moving with radial speed of v, the delay

$$T = \frac{2(R + vt)}{c} \tag{2.3}$$

where time t spans multiple periods and c is the speed of light. The received signal is mixed with the transmitted chirp, which results in a complex sinusoid, known as the *beat signal*. The beat signal frequency equals $f_b = f_R + f_D$, where $f_R = 2RB/T_c$ is the range frequency and $f_D = (2v/c)f_c$ is the Doppler frequency. The process of obtaining the beat signal is implemented in the radio-frequency (RF) domain by a mixer followed by a band pass filter (BPF) with maximum cutoff frequency $f_b^{\max}$; the latter filter is used to remove signals with frequencies outside the band of interest, which also places a limit to the maximum detectable range. For automotive radars, this part of the range matched filter is typically performed in the analog domain and is followed by the digitization.

Estimation of the beat signal frequency is implemented in the digital domain, after the sampling of the beat signal. In automotive scenarios, the maximum detectable range, $R_{\max}$, is hundreds of meters. It holds that $2R_{\max}/c \ll$ the pulsewidth T and thus the range frequency $f_R \ll$ the signal bandwidth B. Since it typically holds that the Doppler frequency $f_D \ll f_R$, the beat frequency is much smaller than B and therefore, a low-speed analog-to-digital converter (ADC) can be used to sample the beat signal. The time over one period is usually referred to as *fast time*, while time in steps equal to the period is referred to as *slow time*. Thus, if we sample the beat signal and put the samples of each chirp in the columns of a matrix, then the row indices of that matrix correspond to fast time and the column indices to slow time (see Figure 2.5). In automotive scenarios, $f_D \ll f_R$, therefore f_D can be taken

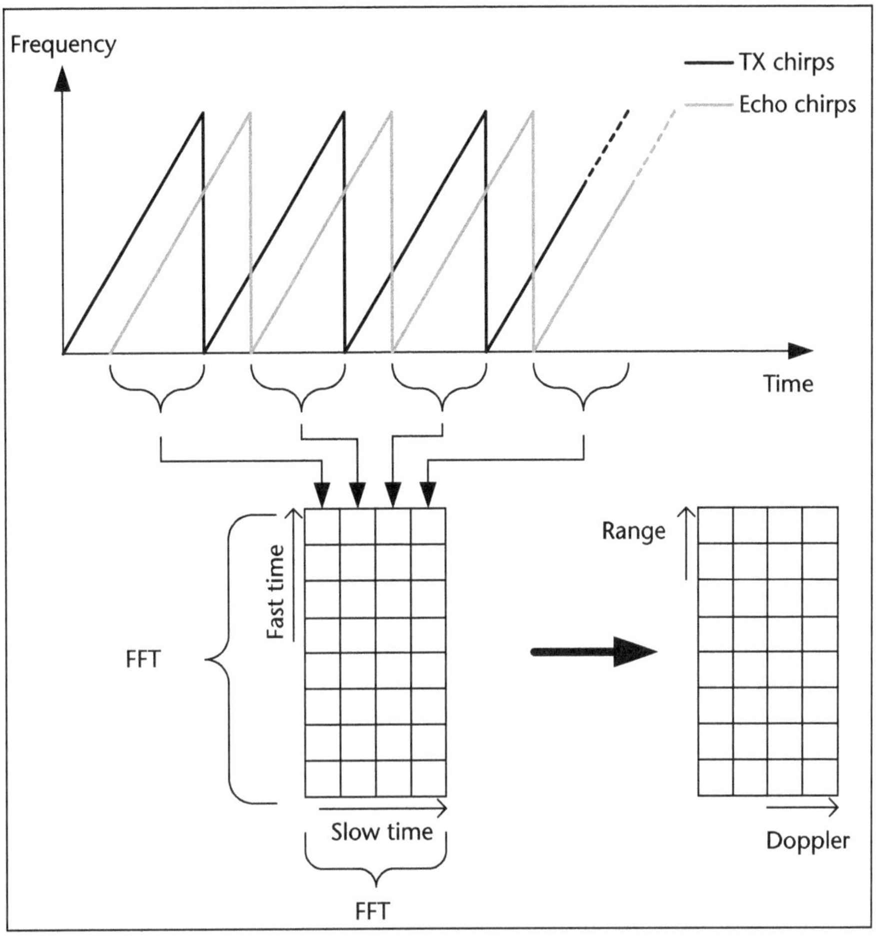

Figure 2.5 FMCW radar chirps. The range and Doppler estimation are performed using a 2-D FFT. (© 2022 IEEE. Reprinted, with permission, from [7].)

as constant within each chirp. Therefore, by applying FFTs on the sampled beat signal along fast time, one can identify f_R, based on which the target's range can be obtained as

$$R = \frac{cf_R T}{(2B)} \tag{2.4}$$

To obtain the target's Doppler frequency, a second FFT operation is subsequently carried out along the slow time (the range frequency f_R is the same across slow time). The pulse repetition frequency (PRF) is $f_{\mathrm{PRF}} = 1/T_{\mathrm{PRI}}$. To avoid Doppler ambiguity, it is desired that $f_{\mathrm{PRF}} \geq 2f_D$. Thus, the maximum unambiguous detectable radial speed of FMCW radar is

$$v_{\max} = \frac{c}{4f_c T_{\mathrm{PRI}}} \tag{2.5}$$

The application of these two computationally inexpensive FFTs is equivalent to a 2-D FFT of the beat signal in fast and slow time, and the result is called *range-Doppler spectrum* or *range-Doppler map*. Range and Doppler detection can be performed using conventional thresholding-based methods applied to 2-D range-Doppler spectrum, such as the constant false alarm rate (CFAR) detector [18], or the recently proposed deep neural network-based detector [19]. Via the 2-D FFT, the targets can be separated in the range and Doppler domain. Since the number of targets within the same range-Doppler bin is usually small when moving, angle finding can be carried out using sparse sensing techniques, such as compressive sensing.

The 2-D FFT operation used for beat frequency estimation can be computed with low-cost digital signal processors (DSPs) or field programmable gate arrays (FP-GAs). The range resolution depends on the beat frequency resolution. As the latter is estimated based on a signal of time duration approximately equal to T, the beat frequency resolution is $1/T$, and thus the range resolution is $c/(2B)$. One can see how the range resolution benefits from high bandwidth. Thus, the low hardware cost, coupled with the high range resolution that can be achieved, make FMCW radar very popular in automotive radar applications.

As an example, a LRR FMCW radar operating at carrier frequency $f_c = 77$ GHz, has a typical pulse duration of $T = 50$ μs and bandwidth of $B = 150$ MHz. Assume the maximum detectable range is 250m. Then, the maximum range frequency is $f_R^{\max} = 5$ MHz. For a moving target with a maximum speed of $v = 120$ miles per hour (mph), the maximum Doppler frequency is $f_D^{\max} = 27.4$ KHz. The beat signal can be sampled with a low-cost ADC with sampling rate as low as $f_s = 12$ MHz. For PRI equal to T, and based on the previously mentioned, the PRF is $f_{\text{PRF}} = 1/T_{\text{PRI}} = 20$ KHz. It can be seen that in this example $f_{\text{PRF}} < 2f_D^{\max}$, and therefore, the Doppler frequency will be aliased, resulting in Doppler ambiguity. In practice, multiple CPIs using chirp sequences with different PRIs are transmitted to resolve the Doppler ambiguity.

As discussed, FMCW radar has the desirable ability to estimate target range and Doppler at low cost. However, FMCW radar also faces several challenges. Those include frequency sweep nonlinearity [19–21], arising due to phase noise of the local oscillator (LO), and imperfect phase-locked loop (PLL) circuit and voltage control oscillator (VCO). Sweep nonlinearity has a significant impact on radar range estimation [20].

A challenge common to all radars is to maintain high dynamic range, which is the ratio between the maximum and minimum possible successfully received power. As path loss is inversely proportional to the fourth-order of the range, targets at long distances typically yield weaker reflections as compared to targets at closer distances. Therefore, high dynamic range is required to accommodate targets at a wide range of distances (see Chapter 7, Radar Challenges from the Automotive Scene, for a detailed treatment of this topic). In FMCW radar, as the waveforms are transmitted continuously, the receiver always receives the signals of the transmitters via a direct path, referred to as *leakage*. Leakage may saturate the low-noise amplifier (LNA) in the front end of the radar receiver [21]. It can also cause high peaks at the first few bins of the range FFT, thus reducing the radar dynamic range.

In many pulse radars, sensitivity time control (STC) has been widely used to control the gain of targets at near and far distances by exploiting the relationship between time delay and target range. As discussed, the range frequency of the beat signal is proportional to the target range. Therefore, in FMCW radar, an analog gain control logic can be implemented in the frequency domain to reduce the receiver gain of close range targets, while continually increasing the gain as the target's range increases [22]. An automatic gain control (AGC) logic can be implemented in a BPF so that the receiver gain increases within the detectable range [25]. The leakage between transmit and receive antennas can be suppressed via sufficient isolation between transmit and receive antennas through antenna design in the analog domain [24, 26]. Further, the transmit antenna leakage effect in the mixer output can be suppressed via a BPF with lowest cutoff frequency corresponding to a range of tens of centimeters to 1 meter. The BPF also helps suppress the multibounce signals among automotive radar, radome, and vehicle bumper. Finally, in the complete receiving chain, an N bit ADC can provide an additional $6N$ dB dynamic range. It should be noted that in the signal processing chain of FMCW radar, a range FFT length of N_R and a Doppler FFT length of N_D can provide a signal-to-noise ratio (SNR) improvement of $10\log_{10}(N_R N_D)$ dB [27], which significantly benefits both detection and angle of arrival estimation.

2.7 Outputs from Automotive Radar

In broad strokes, the purpose of automotive radar is similar to that of aerospace surveillance radars. Both radiate energy in the RF spectrum into the world, and by clever comparison of the received echos with the transmitted waveform, are able to form kinematic measurements of objects in the FOV. The primary measurements, as discussed in the previous section on FMCW processing, are range, range rate (derived from the Doppler shift), azimuth, and, depending on application, elevation. These quantities are all coordinates of the radar data cube, as shown in Figure 2.6, and are thus first order descriptors in any type of output processing. Additional quantities that may be derived from the data cube include radar cross section (RCS) and extent, in either the spatial coordinates or in Doppler.

The range measurement gives the distance from the radar system to the target. A key concept in radar is resolution, or the ability to discriminate between two signals. There are multiple ways to characterize range resolution, but one common method is to use the peak to first null distance (the so-called Rayleigh bandwidth) in the range domain:

$$\frac{c}{2 \cdot B} \tag{2.6}$$

although the actual resolution may be degraded by beam spreading due to windowing and other signal processing artifacts. The maximum instrumented range is given by

$$\frac{c \cdot samples}{4 \cdot B} = \frac{c \cdot f_s}{4 \cdot \gamma} \tag{2.7}$$

The range rate measurement gives the component of the target velocity in the range direction. This measurement is derived from measuring the phase rate across

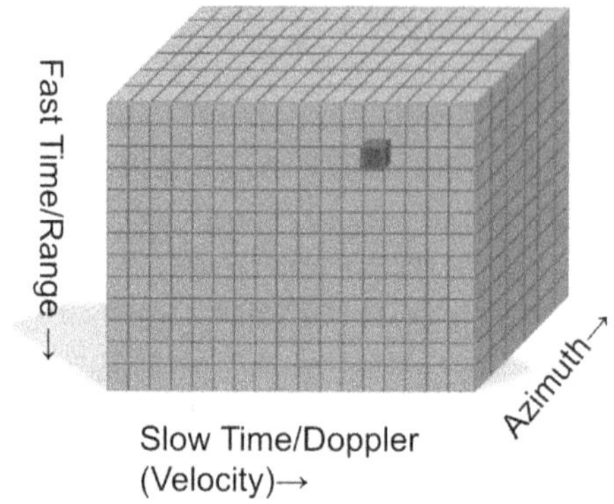

Figure 2.6 The Radar Data Cube.

multiple chirps, or pulses, and is commonly thought of as a series of discrete samples of the phase taken at a sampling frequency equal to the PRF. This directly measures the Doppler frequency shift, which is proportional to the range rate, with a constant of proportionality equal to $-\lambda/2$. Similar to range resolution, the Rayleigh bandwidth measure of the range rate resolution is

$$\frac{\lambda}{2 \cdot T_{dwell}} \tag{2.8}$$

where T_{dwell} is the length of the entire dwell. A further complication with range rate measurements is aliasing, due to the finite sampling frequency (PRF). The unambiguous range rate swath is given by $PRF \cdot \lambda/2$, and targets moving with range rate $\alpha + N \cdot swath$ for all integers N will alias to α. N is typically not observable from a single measurement but can be inferred by associating multiple measurements over time. The spatial angle measurement is also the result of a discrete Fourier transform (DFT), although in this case the frequencies measured are proportional to the sine of the desired quantity, rather than the quantity itself. Because of this, bin sizes are not constant when measured in angular units. Assuming a uniform linear array (ULA) with N samples, bin sizes are given by

$$\frac{2}{N \cdot \cos(\theta)} \tag{2.9}$$

in radians. Note that the resolution gets worse as the beam is scanned off of boresight. However, in practice, for a radar with a 90° FOV, this effect increases bin sizes by only 41% at the azimuthal limits. An additional complexity of the azimuth measurement is that it is actually a measure of the angle the reflected wave makes with the direction along the receive array, which, combined with the range measurement, corresponds to a circle of potential sources, as shown in Figure 2.7. The *azimuth* is only the true azimuth for the one point on this circle with elevation 0. Radar systems with two-dimensional arrays that can measure both azimuth and elevation are more expensive due to the higher channel count but are nevertheless becoming more popular as they enable precise determination of both angles.

In general, resolution is a much higher priority in automotive radar than aerospace, owing to the significantly denser environment. One can envision scenarios

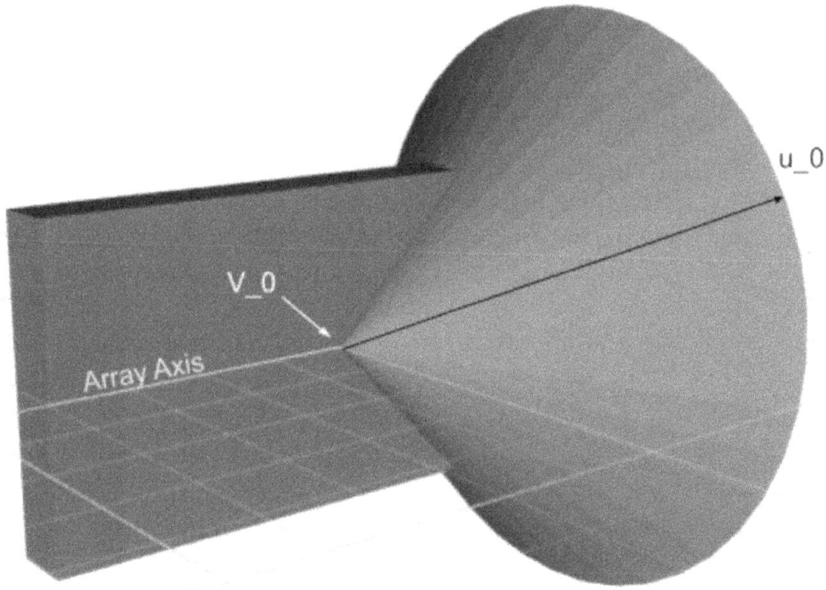

Figure 2.7 Measurements from horizontal linear array (denoted the Array Axis in this figure) assume zero elevation and are reported as the azimuth angle of arrival, but may arise from any point at the base of cone u_0. The true azimuth and elevation are coupled into the array *azimuth* measurement. (© 2020 IEEE. Reprinted, with permission, from [27].)

in aerospace in which resolution is needed to separately detect many objects in a small area, but the automotive problem often requires high resolution in all dimensions over the entire FOV.

Traditionally, CFAR algorithms have been used to identify peaks in the radar data cube that have significantly higher energy than the surrounding noise floor. This method yields a single point in range/range rate/angle space, along with a measure of power and signal to noise ratio. Increasingly, emphasis has been placed on using more contextual information by examining other pixels in the neighborhood of peaks. For objects that are larger than the resolutions in any of the three dimensions, these pixels can be used to estimate the size of objects, or even to classify them. A common use case is to identify micro-Doppler signatures, such as Doppler spreading due to wheels, or swinging limbs on pedestrians. Microdoppler signatures for various human motions are shown in Figure 2.8.

Moving beyond peaks and the local region around them, as the cost of resolution falls, it becomes possible to build *imaging* radars with sufficient resolution and sidelobe control to produce images of the entire FOV. Unfortunately, the volume of the data cube generally becomes substantial: a typical bin count of 256 range, 256 Doppler, and 128 azimuth at 50 Hz with 32 bits per voxel comes to 13.4 Gbits/sec, and including elevation information further increases this value. One approach to alleviating this burden is to project the data cube onto a smaller subspace. The standard approach is to project orthogonally along one of the three axes. Projecting through the azimuth dimension yields a range/doppler image, through the doppler dimension yields a range/azimuth image, and through the range dimension yields a doppler/azimuth image. Each of these images has different strengths:

- Range/Doppler images have been used since early in the history of radar. Controlling the impulse response to achieve high resolution is easiest in these two dimensions, so this type of image plays to the strengths of radar. The downside is that pixels are not localized in space, and Doppler imagery can be very nonintuitive, making these images difficult to interpret for the uninitiated.
- Range/azimuth images are easier to interpret because the pixels are localized in space (in two dimensions, at least). However, they require very good azimuth resolution and impulse response control. As channels become cheaper, these

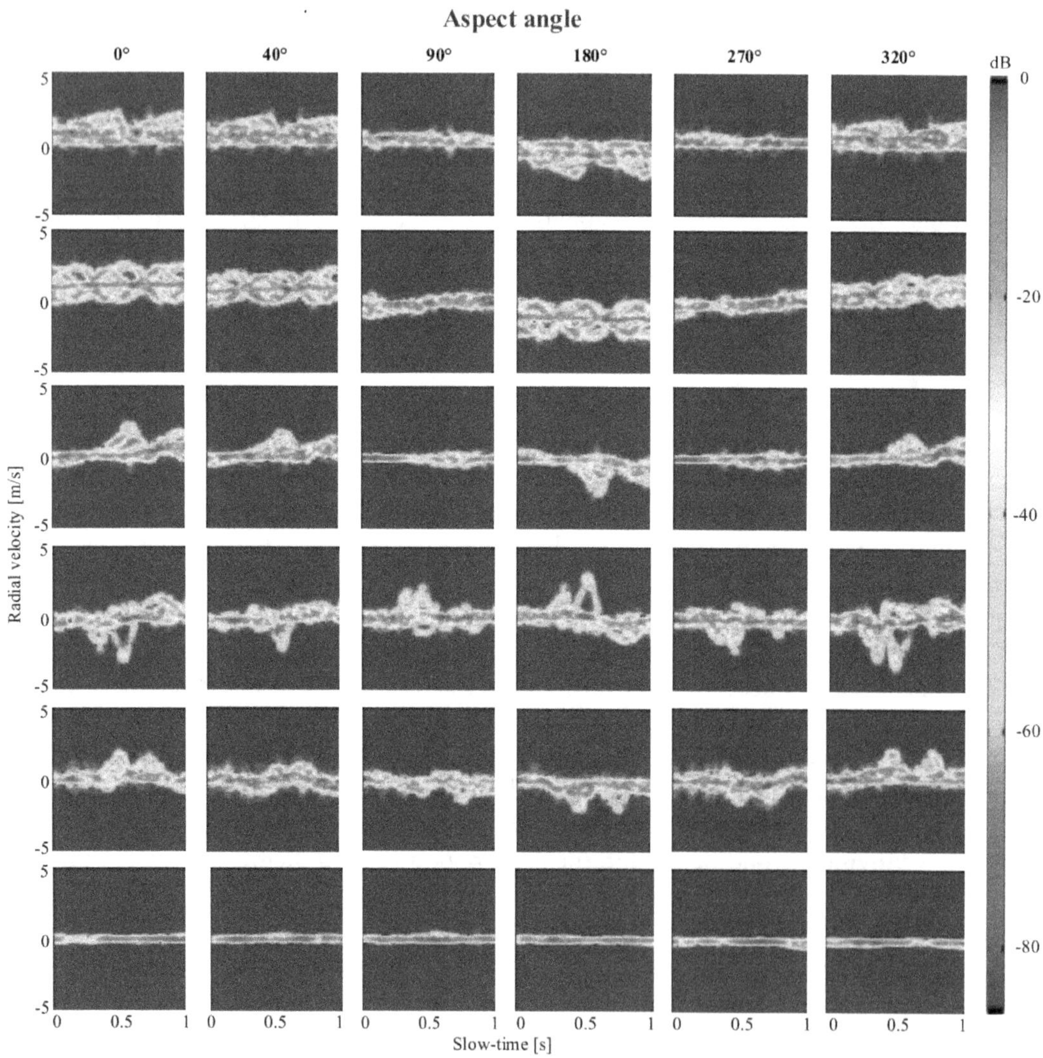

Figure 2.8 Simulated time-Doppler spectrograms of human motions with varying aspect angles. From top to bottom: walking, running, jumping, boxing, creeping, crawling. (© 2022 IEEE. Reprinted, with permission, from [28].)

images are becoming more prevalent. Notably, they are useful for imaging clutter, which collapses to a one-dimensional curve in Range/Doppler images.

- Doppler/azimuth images are not commonly used for surveillance, but can be useful for detecting interference and measuring the velocity of the radar [29].
- Another approach is to collapse a range-Doppler map by summing the powers across each angle while retaining the full data cube in memory. For those range-Doppler cells that exceed a threshold, the full column of angular data is retrieved and the points in this vector that exceed a second threshold are presented as points, making a "point cloud" representation of the scene. All of these representations provide insight to those using the radar data products, and the choice of which one(s) to utilize is an open area of optimization for autonomous vehicles today.

References

[1] Hasch, J., et al., "Millimeter-Wave Technology for Automotive Radar Sensors in the 77 GHz Frequency Band," *IEEE Trans. Microw. Theory Tech.*, Vol. 60, No. 3, 2012, pp. 845–860.

[2] C. Waldschmidt, Hasch J., and W. Menzel, "Automotive Radar — From First Efforts to Future Systems," in *IEEE Journal of Microwaves*, Vol. 1, No. 1, January 2021, pp. 135-148, doi: 10.1109/JMW.2020.3033616.

[3] Texas Instruments Inc., "Moving from Legacy 24 GHz to State-of-the-Art 77 GHz Radar," https://www.ti.com/lit/wp/spry312/spry312.pdf, October 2017.

[4] European Telecommunications Standards Institute (ETSI), "System Reference Document (SRdoc); Short Range Devices (SRD) Using Ultra Wide Band (UWB); Transmission Characteristics; Technical Characteristics for SRD Equipment Using Ultra Wide Band Technology (UWB); Radiodetermination Application Within the Frequency Range 120 GHz to 260 GHz," *ETSI Tech. Rep. 103 498 V1.1.1 (2019-02)*, 2019.

[5] Banerjee, A., et al., "Millimeter-wave Transceivers for Wireless Communication, Radar, and Sensing," in *Proceedings IEEE Custom Integrated Circuits Conference*, Austin, TX, April, 2019.

[6] Visweswaran, A., et al., "A 145 GHz FMCW-Radar Transceiver in 28 nm CMOS," in *Proceedings IEEE Int. Solid-State Circuits Conf.*, San Francisco, CA, February 2019.

[7] Sun, S., Petropulu, A. P., and H. V. Poor, "MIMO Radar for Advanced Driver-Assistance Systems and Autonomous Driving: Advantages and Challenges," *IEEE Signal Process. Mag.*, Vol. 37, No. 4, 2020, pp. 98–117.

[8] Alland, S., Stark, W., Ali, M., and A. Hedge, "Interference in Automotive Radar Systems: Characteristics, Mitigation Techniques, and Future Research," *IEEE Signal Process. Mag.*, Vol. 36, No. 5, 2019, pp. 45–59.

[9] Steinbaeck, J., et al., "Next Generation Radar Sensors in Automotive Sensor Fusion Systems," in *Proceedings. Sensor Data Fusion: Trends, Solutions, Applications (SDF)*, Bonn, Germany, October 2017.

[10] Rohling, H., and M.-M. Meinecke, "Waveform Design Principles for Automotive Radar Systems," in *Proceedings CIE Intl. Conference on Radar*, Beijing, China, October 2001.

[11] Meinl, F., Stolz, M., Kunert, M., and H. Blume, "An Experimental High Performance Radar System for Highly Automated Driving," in *Proceedings Intl. Conference on Microwaves for Intelligent Mobility (ICMIM)*, Nagoya, Japan, March 2017.

[12] Hacker, P., and B. Yang, "Single Snapshot DOA Estimation," *Advances in Radio Science*, Vol. 8, 2010, pp. 251–256.

[13] Schmidt, R., "Multiple Emitter Location and Signal Parameter Estimation," *IEEE Trans. Antennas Propag.*, Vol. 34, No. 3, 1986, pp. 276–280.

[14] Roy, R., and T. Kailath, "ESPRIT—Estimation of Signal Parameters via Rotation Invariance Techniques," *IEEE Trans. Acoust., Speech, Signal Process.*, Vol. 17, No. 7, 1989, pp. 984–995.

[15] Shan, T., Wax, M., and T. Kailath, "On Spatial Smoothing for Direction-of-Arrival Estimation of Coherent Signals," *IEEE Trans. Acoust., Speech, Signal Process.*, Vol. 33, No. 4, 1985, pp. 506–811.

[16] Pillai, S. U., and B. H. Kwon, "Forward/Backward Spatial Smoothing Techniques for Coherent Signal Identification," *IEEE Trans. Acoust., Speech, Signal Process*, Vol. 37, No. 1, 1989, pp. 8–15, 1989.

[17] Neri, F., *Introduction to Electronic Defense Systems*. Raleigh, NC: Scitech Publishing, Inc., 2006.

[18] Richards, M. A., *Fundamentals of Radar Signal Processing*, Second Edition, New York: McGraw-Hill, 2014.

[19] Brodeski, D., Bilik, I., and R. Giryes, "Deep Radar Detector," in *Proceedings IEEE Radar Conference*, Boston, MA, April 2019.

[20] Ayhan, S., Scherr, S., Bhutani, A., Fischbach, B., Pauli, M., and T. Zwick, "Impact of Frequency Ramp Nonlinearity, Phase Noise, and SNR on FMCW Radar Accuracy," *IEEE Trans. Microw. Theory Tech.*, Vol. 64, No. 10, 2016, pp. 3290–3301.

[21] Ramasubramanian, K., and B. Ginsburg, "AWR1243 Sensor: Highly Integrated 76–81 GHz Radar Front-end for Emerging ADAS Applications," *Texas Instruments, Tech. Rep.*, 2017, pp. 1–11.

[22] Wang, P., Millar, D., Parsons, K., Ma, R., and P. V. Orlik, "Range Accuracy Analysis for FMCW Systems with Source Nonlinearity," in *Proc. Intl. Conference on Microwaves for Intelligent Mobility (ICMIM)*, Detroit, MI, April 2019.

[23] Lin, K., Wang, Y., Pao, C., and Y. Shih, "A Ka-Band FMCW Radar Front-End with Adaptive Leakage Cancellation," *IEEE Trans. Microw. Theory Tech.*, Vol. 54, No. 12, 2006, pp. 4041–4048.

[24] Stove, A. G., "Linear FMCW Radar Techniques," *IEE Proceedings F-Radar and Signal Processing*, Vol. 139, No. 5, 1992, pp. 343–350.

[25] Jung, S., Kim, S., Choi, W., Kim, H., Kim, H., and Y. Eo, "High Dynamic Range Ku Band CMOS Transceiver IC for FMCW Radar Application," in *Proc. IEEE MTT-S Intl. Microwave Symposium (IMS)*, Honololu, HI, June 2017.

[26] Adela, B. B., "Antennas for Silicon-based mm-Wave FMCW Radars: Antenna Integration and MIMO System Design," PhD Dissertation, Technische Universiteit Eindhoven (TU/e), Eindhoven, Netherlands, 2019.

[27] Gurgel, K.-W., and T. Schlick, "Remarks on Signal Processing in HF Radars Using FMCW Modulation," in *Proceedings of the International Radar Symposium (IRS)*, Hamburg, Germany, September 9–11, 2009.

[28] Yang. Y., Hou, C., Lang Y., Sakamato, T., He. Y. and Xiang, W., "Omnidirectional Motion Classification With Monostatic Radar System Using Micro-Doppler Signatures," *IEEE Transactions on Geoscience and Remote Sensing*, Vol. 58, No. 5, 2020, pp. 3574–3587.

[29] Smith, L., and T. Campbell, "Calculating velocity of an autonomous vehicle using radar technology," U.S. Patent No. 10,890,919 (priority date September 22, 2017).

CHAPTER 3

Digital Code Modulation

Ralf Reuter, Arunesh Roy, Murtaza Ali, and Yaohui Liu

3.1 Introduction

The most common modulation method(s) for automotive radar use to date has been frequency modulated continuous wave (FMCW) [1] or fast chirp modulation (FCM). The main difference between FMCW and FCM is the order of the chirp time. In classical FMCW, radar up- and down-chirp in the order of milliseconds are used. In contrast, FCM radar is typically only using up- or down-chirps in the order of ten microseconds. FMCW radar sensors have been commonly used due to the low requirements on semiconductor technologies (e.g., the chirp time is much larger so slower phase-locked loop (PLLs) and ADCs can be used). Disadvantages in terms of target separation, number of trackable objects, and the availability of modern technologies (SiGe and CMOS) have made the FCM radar more attractive. Today (in 2022) nearly all radar sensors on the market are using fast chirp modulation.

Another type of radar, which has been used for military applications, but until recently was too expensive and too power hungry for high volume use cases like automotive, is phase modulated continuous wave (PMCW) [2]. For a typical commercial application, the phase is modulated by a digital code (Figure 3.1) and the carrier (center) frequency is kept constant.

In contrast to FCM radar systems where during a chirp the frequency is modulated linearly, a PMCW system is transmitting chirps, generated by a code generator.

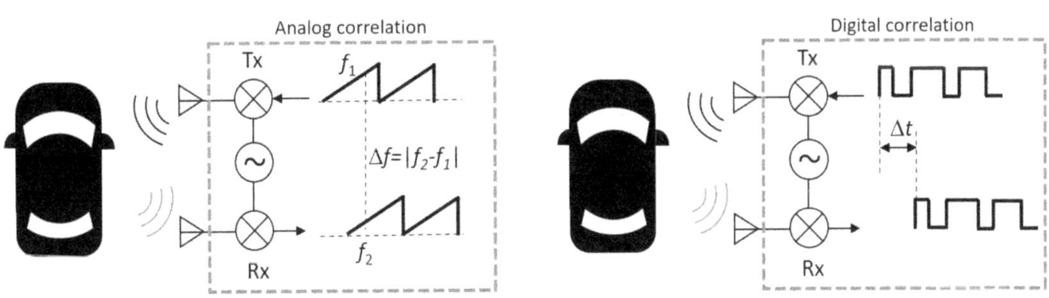

Figure 3.1 Simplified modulation principle: (a) classical frequency modulated continuous modulation and (b) phase modulated continuous wave.

On the receive site, FCM systems are identifying targets by a beat frequency (DC up to a few megahertz) in the analog domain (analog correlation), where the PMCW system is using a correlation and accumulation in the digital domain (digital correlation) as shown in Figure 3.1. To emphasize the nature of the code modulation in a PMCW system, this type of modulation is called digital code modulation (DCM).

Table 3.1 Comparison of FCM Versus DCM Radar

Modulation Type	*FCM*	*DCM*
Analog processing	Complex: Extremely linear chirp and low intermodulation products in the ADC are required.	Simple: Smaller analog down-conversion.
Signal processing	Simple: Partially inherent in the requirements of analog blocks (like chirp generator). Reduced requirements on signal processing.	High: Mostly done in digital domain that allows the use of advanced signal processing algorithms. Direct sampling that allows correction of imperfect analog chain in the digital processing chain.
Data rate	Low: Low speed ADC with approximately 50 MSps. in-phase and quadrature (IQ)-baseband not required but beneficial to increase SNR. Data rates are moderate (below GB/s).	High: High speed direct sampling ADC with Gbit/s data rate. Typically, IQ-baseband up to a couple of GHz required. Extremely high data rate requires special hardware solutions.
Range response	Good: Broader range response due the FFT and windowing that may hide small objects in close proximity of a large object. Limited discrimination between small targets near to large targets.	Better: Sharp thumbtacklike range response due to near ideal autocorrelation function of long spreading codes providing better separation of small radar cross section (RCS) objects closer to large radar cross section(RCS) objects.
Near-far problem	Good: Solved by high-pass filters in the baseband chain. Near objects are highly suppressed.	Complex: Requires special methods, code and spillover cancellation in the digital domain, usually at the cost of SNR.
Power consumption	Moderate: Driven by output power and noise figure in the analog domain. Digital processing requires less special hardware acceleration blocks.	Higher: Driven by output power and noise figure in the analog domain. High data rate processing requires advanced CMOS notes to achieve power targets.
MIMO support/ angular resolution	Moderate: Multiple receive channels have been challenging for FCM. Typical time division multiplex (TDM) is used, requiring solutions to Doppler ambiguities.	Excellent: Native (CDM code division multiplex) support for large number of simultaneous channels.
Interference	Moderate: Interference is highly dependent on the probability of sensors having the same chirp parameters used by the interfering and victim radars and may result in a strong generation of ghosts in range and/or Doppler.	Good: Due to the nature of nearly orthogonal codes, DCM will have higher robustness in high density radar environments. Increase of the noise floor will in general happen if the identical resources in time and frequency are shared.

Due to the advancements of CMOS technology and advanced signal processing, we are now able to provide a cost-effective DCM radar system that meets the increasing performance demands of the automotive industry. A comparison of the traditional FCM versus DCM radar is enumerated in Table 3.1.

3.2 FCM Versus DCM Architecture

The radar architectures of FCM and DCM are similar but differ significantly in terms of implementation and especially processing requirements. Show here are the typical implementation of a FCM radar transceiver (Figure 3.2) and a DCM radar transceiver (Figure 3.3).

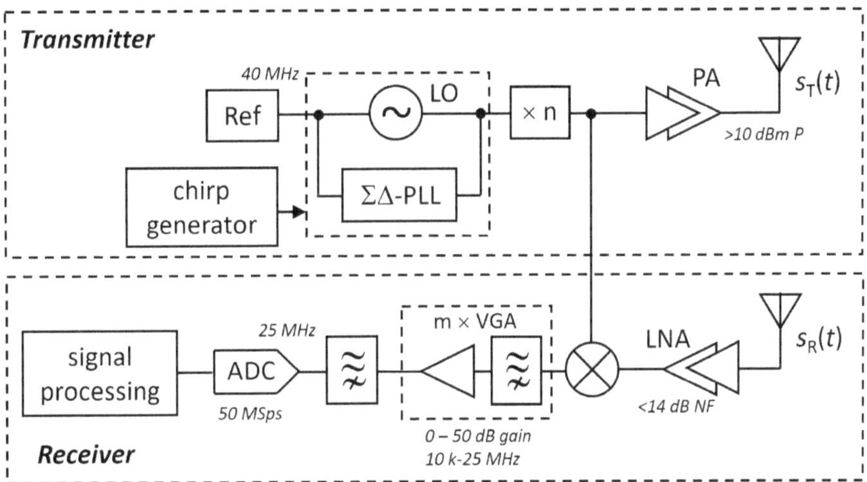

Figure 3.2 Typical implementation of an FCM radar transceiver.

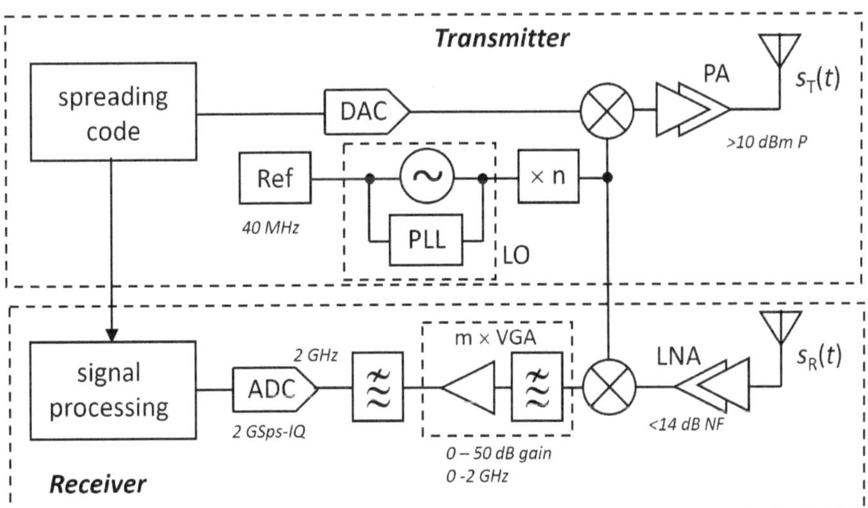

Figure 3.3 Typical implementation of a DCM radar transceiver.

In terms of the link budget [3] the output power of the power amplifier and the noise figure of the low noise amplifier (LNA) are the key contributors, and the design of these requirements is not different for FCM and DCM radar. The major difference in the analog baseband design are the requirements in terms of baseband frequency range and AC versus DC coupled circuits. Major differences, shown in Table 3.2, are visible in terms of chip/chirp generation and the requirements on ADC and data processing due to a much higher receiver baseband bandwidth. The drastically increased processing requirements have made a commercial usage of DCM radar in the past not possible, and due to improvements in CMOS and package

Table 3.2 Key Differences between FCM and DCM Radar Receive and Transmit Architectures

Modulation Type	*FCM*	*DCM*
Chirp/code generation	Chirp generator (typically a programmable sigma-delta modulator in combination with a PLL is used). Extremely linear chirps must be generated. Nonlinear chirps cause a degradation of the effective phase noise.	Spreading code generator. Architecture similar to communication system that allows to leverage similarity in terms of waveforms and implementation.
Frequency generation	Extremely linear chirp.	Defines the carrier frequency.
ADC	Approximately 50 MSps, DC to approximately 25 MHz IF bandwidth, I- or IQ architecture (complex architecture is optional and might increase SNR). Effective number of bites (ENOB) typically up to 12 bit. Intermodulation products are critical.	Approximately 2 GSps, DC to approximately 2 GHz IF bandwidth, IQ architecture (complex data required to correctly analyze phase information). ENOB determined by technology note and power consumption targets. ENOB typically up to 8 bit.
Baseband	Approximately 10 kHz–50 MHz, with high-pass filter to suppress DC and close in targets.	Approximately DC-2 GHz, no high-pass filter, need to find solution to handle near-far problem.
Data rate	50 MSps × number of Rx	2 GSps × number of Rx
Signal processing	Few hardware accelerated functions required.	Special dedicated hardware accelerated function required.
Range processing	FFT with windowing	Correlation
Doppler procession	FFT based	FFT based
Beam forming	FFT/DFT based	FFT/DFT based
TX (Transmitter)	Up-conversion modulator based on classical mixer architecture.	Digital-to-analog converter (DAC) and modulator to up-convert spreading code and to control spectrum mask.
RX (Receiver)	Controllable baseband gain, approximately 20–50 dB control range, extremely high requirements on linearity and intermodulation. High-pass filter to suppress DC, typically 10 kHz up to 25 MHz.	Controllable baseband gain, approximately 20–50 dB control range. Full DC to approximately 2 GHz. Typically, IQ architecture (complex data required).

technologies, the required processing power at low power consumption and low cost have become feasible. In a first order the FCM performance is strongly correlated to the properties of the chirp generation (e.g., phase noise, linearity) and baseband (e.g., intermodulation products). In contrast, the DCM radar performance is strongly correlated to spreading code properties (e.g., correlation properties).

3.3 Basics of DCM Radar

In DCM radar systems [4, 5], the baseband phase during a given time period (called a *chip period* or *chip duration*) is typically one of a finite number of possible phases. A spreading code consisting of a sequence of chips (e.g., +1, +1, −1, +1, −1, ...) is mapped (e.g., $+1 \rightarrow 0$, $-1 \rightarrow \pi$) into a sequence of phases (e.g., 0, 0, π, 0,π, ...) that is used to modulate the phase of the RF sinusoidal signal. The spreading code could be a periodic sequence, or it could be a pseudorandom sequence with a very large period that makes it appear to be a nearly random sequence. The resulting signal has a bandwidth that is proportional to the rate at which the phases change, called the *chip rate*, which is the inverse of the chip duration. By comparing the return signal with the transmitted signal, the receiver can determine the range and the velocity of reflected objects. This can be done using a matched filter or a bank of correlators.

As previously mentioned, phase modulated continuous wave radar transmits a phase modulated signal. The signal is of the form:

$$s(t) = \sqrt{2P}\cos\left(2\pi f_c t + \phi(t)\right) \tag{3.1}$$

where P is the power of the signal, f_C is the carrier frequency, and $\phi(t)$ is the phase of the transmitted signal. While the phase can vary continuously to minimize bandwidth, here we consider the phase function to be a sequence of phases that stay constant for T_c seconds and are called *chips*. T_c is known as the *chip duration*. That is, $\phi(t) = \phi_l$ for $lT_c < t \leq (l+1)T_c$, $l = 0,1, \ldots$.

In the case of *binary* phase modulation, the phase is either 0 or π radians (180 degrees). The signal can then be written as $s(t) = \sqrt{2P}a(t)\cos(2\pi f_c t)$ where $a(t)$ is either +1 or −1 depending on if $\phi(t)$ is 0 or π respectively. Figure 3.4 shows an example of the signal $a(t)$. Such a sequence of chips causes the spectrum of the signal to spread the energy over a bandwidth proportional to $1.T_c$. The signal $a(t)$ is a baseband signal, which is modulated up to a carrier frequency by a mixer before

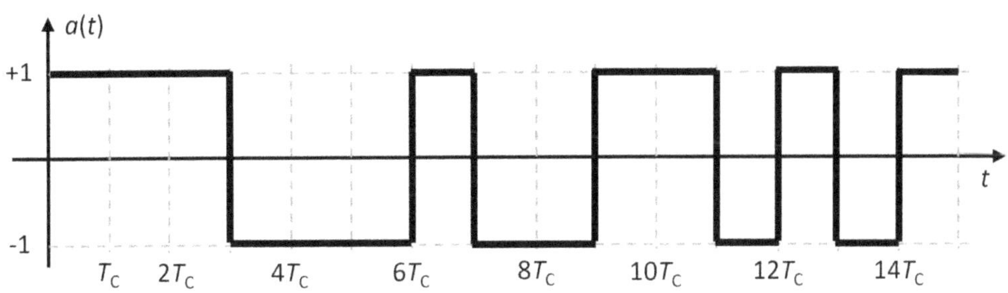

Figure 3.4 Binary modulation signal.

being transmitted. This signal is also sometimes called a *direct sequence spread spectrum signal.*

The transmitter, conceptually, is shown in Figure 3.3. The spreading waveform $a(t)$ modulates the signal from an oscillator to generate a phase modulated continuous waveform signal $s(t)$.

The transmitted signal from a single antenna is reflected by objects in the environment. The reflected signals are then received at the receiving antenna and processed. A simple model for the received signal in the absence of noise and other interfering signals is

$$r(t) = \sum_{l=1}^{L} \alpha_l s(t - \tau_l) \tag{3.2}$$

Here, there are L objects that reflect the transmitted signal that are stationary relative to the transmitter. An object that reflects the transmitted signal $s(t)$ is attenuated by the factor α_l and delayed by τ_l between the transmitter, object, and receiver. The delay between the transmitted signal and the received signal is related to the distance by $R_l = c\tau_l/2$ where c is the speed of light and the factor of two is due to the round-trip time from the transmitter to the receiver. Here, it is assumed that the transmitter and receiver are essentially co-located.

At a receiver, the received signal is amplified with a LNA, mixed down to baseband using the same oscillator signal at the transmitter. The signal is then converted from an analog signal to a digital signal with an ADC.

3.3.1 Range Processing

In DCM radars, range processing is performed using a matched filter to determine the correlations of the received signal with various delays of the transmitted signal. There are two types of such correlation functions that can be used based on how the transmitted signal is generated.

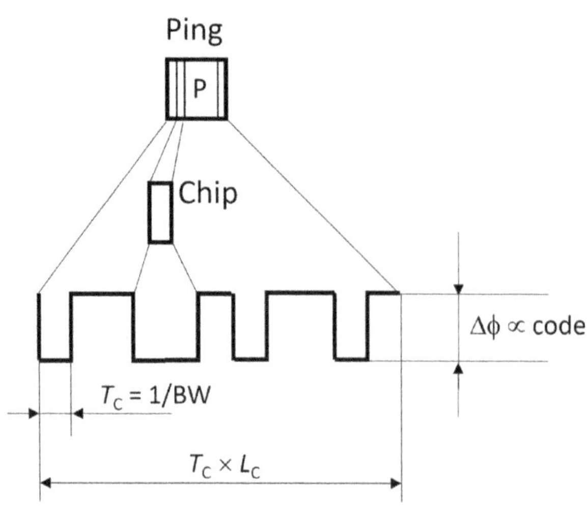

Figure 3.5 Sequence of chips forming a ping.

1. *Nonrepeating code transmission:* In this case, different codes are transmitted from one transmission time (pulse) to the next. The matched filter output is represented by the aperiodic autocorrelation function of the code.
2. *Periodic code transmission:* Here, the same spreading code is transmitted periodically. The matched filter output is represented by the periodic autocorrelation function of the code.

We illustrate these concepts through an example where there are three targets at various distances with different reflectivity (e.g., different size [RCS] objects). Consider a sequence of chips (code) that form a *ping* (see Figure 3.5). A ping consists of L_C chips with chip time T_C. The matched filter output of the ping is shown in Figure 3.6. The example shown in this figure belongs to a class of codes known as maximum-length sequences (m-sequences) [5].

A sequence of pings can be transmitted, received, and correlated to increase the coherent processing gain. A sequence of pings forms a pulse that can consist of repeating or nonrepeating chips (code). Using the same target scenario with three targets as in Figure 3.6, Figure 3.7 shows an example where multiple periods of the same spreading code (same code as the previous figure) are transmitted.

When the pulse consists of the same code sequence in the pings, the output repeats itself with a repetition period equal to the duration of the matched filter's impulse response corresponding to the length of the spreading code used. The constant distance between target replica locations is a property of the m-sequence used as the spreading code.

There are many codes that can be used depending on the objective of the radar system and the desired performance. Some codes, like Barker codes, have good aperiodic autocorrelation functions, while others have excellent periodic autocorrelation functions. Codes with good aperiodic autocorrelation are desirable if the transmitted spreading code does not repeat. When a single period of the code is transmitted, the output of the matched filter (Figure 3.6) for a maximum length sequence (known

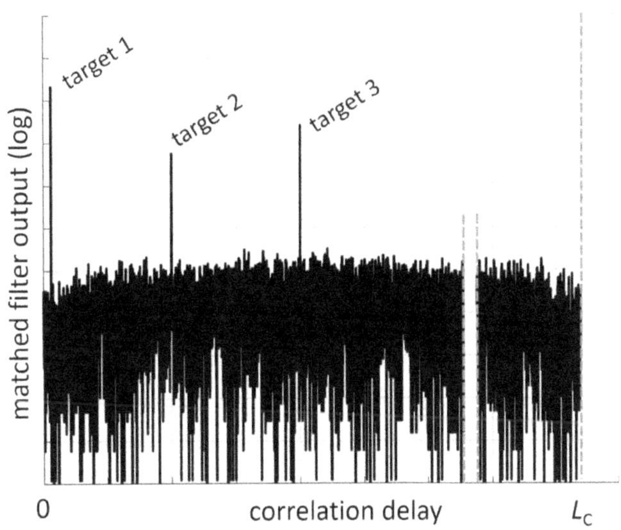

Figure 3.6 Matched filter output for a three target scenario.

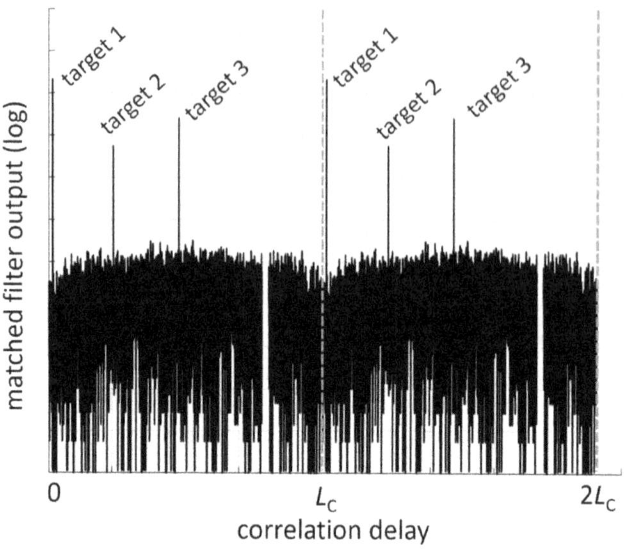

Figure 3.7 Matched filter output with periodic signal.

for good periodic autocorrelation) results in off-peak values higher than those when sent periodically (Figure 3.7). Increasing the length of the code improves the ratio of the off-peak autocorrelation function to the peak of the autocorrelation function.

The width of the mainlobe spike in the matched filter output is proportional to the chip duration. That is, the width of the spike is inversely proportional to the chip rate or the signal bandwidth. Thus, larger bandwidths yield smaller width spikes and better range resolution. The horizontal axes of Figure 3.6 and Figure 3.7 represent correlation delay in chip scale, and thus, the width of the spike in terms of range will change proportionately to chip duration.

The range resolution R_r for a DCM system is the width of the spike in the matched filter output. Two targets that are closer than the corresponding time interval of the width of the spike will be indistinguishable. Thus, the range resolution is

$$R_r = \left(\frac{T_c}{2}\right)c_0 \tag{3.3}$$

where T_c is the chip duration and c_0 is the speed of light. Since chip duration is inversely proportional to the bandwidth of the signal B, that is, $T_c = 1/B$, the above range resolution is equivalent to that in FCM radars.

$$R_r = \frac{c_0}{2B} \tag{3.4}$$

For the case of periodic code as seen in Figure 3.7, if a target is at a distance larger than $c_0(L_cT_c)/2$, where L_C is the number of chips in the spreading code before repeating, the target will produce an output that will appear like a target at a distance between 0 and $c_0(L_cT_c)/2$. For objects at a distance smaller than $c_0(L_cT_c)/2$, the range will be correctly determined. This is the maximum unambiguous range R_u

$$R_u = c_0\frac{T_cL_c}{2} \tag{3.5}$$

Two targets separated by the maximum unambiguous range will appear to the radar system as being at the same range. This is called *range aliasing*. If the chip duration, T_c, is decreased, then the range resolution would improve proportionally. As an example, if T_c is 1.33 ns corresponding to a chip rate of 750 Mchips/s, then the range resolution would be limited to 20 cm. For a periodic spreading code of length 1023, the maximum ambiguous range would be 204.6m. For a periodic spreading code of length 4095, the maximum ambiguous range would be about 818m. The unambiguous range for DCM can be increased by increasing the period of the code for a given chip duration. In the case of an aperiodic code, there is no range aliasing and thus ambiguity in range is nonexistent.

3.3.2 Velocity Processing

Another goal of an automotive radar system is to estimate the differential velocity between the radar system and a target through the Doppler effect. If the differential radial velocity (toward or away from the radar) of a target relative to a radar is Δ_v, and the carrier frequency is f_c, then the Doppler frequency shift of the signal reflected from the target is

$$f_d = 2\Delta_v \frac{f_c}{c_0} \tag{3.6}$$

Or equivalently

$$\Delta_v = f_d \left(\frac{c_0}{2f_c} \right) \tag{3.7}$$

where the factor of two comes from the fact that there is a Doppler shift going both ways: radar to target and target to radar.

In FCM radar systems chirps (linearly modulated frequency sweep) are repeated N times. In DCM systems the spreading code (same code for periodic code but different codes for aperiodic code) is repeated N times or $N\ L_C$ represents a much larger spreading code. The received signal is filtered N times. The N samples of the matched filter output, when using both an in-phase and quadrature-phase downconversion, produce complex numbers that will rotate around the circle at a rate proportional to the relative radial velocity. A plot of the Q-component versus the I-component for the first few samples of the matched filter output is shown in Figure 3.8. The rate that the samples rotate around the circle is related to the Doppler shift of the received signal, and thus the radial velocity of the target relative to the radar system. The direction of rotation around the circle is an indication of whether the target is moving toward the radar system or away from it.

One approach to estimate the Doppler frequency, and thus the velocity, is to perform a fast Fourier transform (FFT) on the N complex samples. There is a maximum relative velocity that can be estimated without ambiguity. If the points in Figure 3.8 rotate completely around the circle in time, $L_C T_C$, then it would appear the object is not actually moving. A complete rotation represents a change of 2π in phase corresponding to a relative movement of the objects by one wavelength

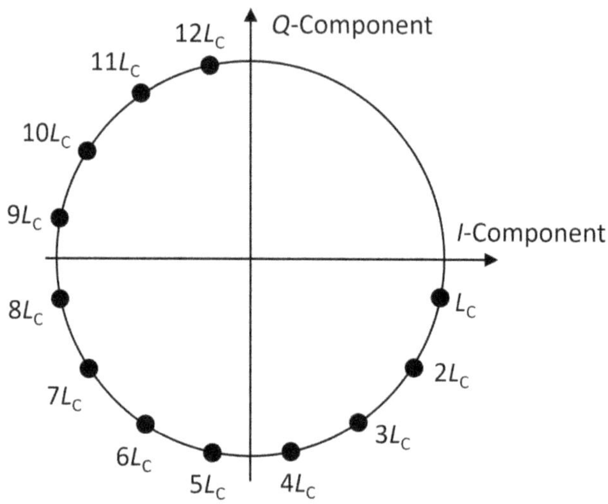

Figure 3.8 Samples of matched filter for Doppler.

or $c_0/(2f_c)$ within a time period of $L_C T_C$. As such, in a radar, the velocity can be estimated unambiguously only over a certain range. That is, two different objects separated sufficiently may appear to have the same velocity. The possible range of *unambiguous* velocities, v_u, that can thus be estimated, is given by

$$-\frac{c_0}{4f_c T_c L_c} < v_u < \frac{c_0}{4f_c T_c L_c} \tag{3.8}$$

If there are N complex samples of the matched filter output, the velocity resolution possible, V_r is then the unambiguous range of previous velocities divided by the number of samples, that is,

$$V_r = \frac{c_0}{2f_c N T_c L_c} \tag{3.9}$$

The equation for unambiguous velocity for FCM and DCM radar systems are nearly identical. In FCM systems, the factor $T_C L_C$ is replaced by the chirp time. The chirp time is limited by the PLL response to achieve a linear frequency chirp over the required modulation bandwidth, the maximum sampling frequency of the ADC, and the number of range gates. In DCM systems, the unambiguous velocity is decoupled from such design constraints and determined by the number of chips L_C. Comparing the typical chirp times in FCM system (order of 10 us) with the ping time in DCM (order of us), the DCM system has higher maximum unambiguous Doppler range.

3.3.3 Angle Processing

In modern automotive radar, the angles of arrivals from targets are estimated using multiple transmit antennas or multiple receive antennas or both. A set of angles for which a radar can detect an object is called the FOV, as illustrated Figure 3.9.

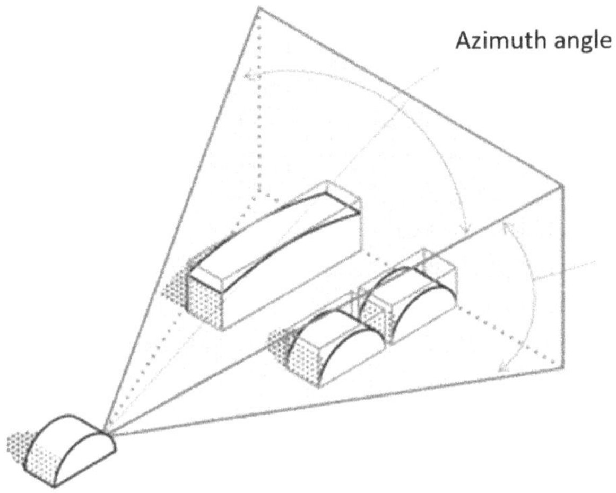

Figure 3.9 Azimuth and elevation angle measurement for a target.

The number of angles that can be determined depends on the number of transmit and receive antennas. Thus, for a given number of antennas, a large FOV results in less angular accuracy and coarser resolution, while a narrow FOV provides better angular accuracy and finer resolution.

There are several ways that multiple transmit antennas can be employed as listed in Table 3.3. One approach is to use various phase shifts of a single signal for each transmit antenna. This is called a *transmit-side phased-array* or *beam steering approach.* The radar focuses the transmitted signal energy in a certain direction. Another approach, when multiple antennas are used in both transmit and receive, includes the use of the MIMO technique described earlier, increasing the virtual aperture, and hence, providing better angular resolution. Although the basic MIMO processing is similar for both FMCW and DCM radars, they differ on how signals are transmitted on the multiple transmit antennas.

Table 3.3 provides a simplified overview of the three basic multiplexing schemes. Typically, in automotive radar TDM is used due to the reduced complexity in system and chip design.

The radar can time multiplex the transmission on different antennas (TDM MIMO). In this approach, the radar transmits using a single antenna for some given time and then switches to another antenna. The interference between antennas is thus mitigated or eliminated. This is the most common implementation used in FCM radars where the transmission is switched from antenna to antenna on a chirp-by-chirp basis. This approach has two major drawbacks. First, the maximum unambiguous velocity is reduced by a factor equal to the number of transmit antennas. Second, the scan time increases by the number of transmit antennas since each transmitter (Tx) must have enough time to transmit within its own window. It is also feasible to use FDM or frequency division multiplexing. This method is typically not used since it increases the signal bandwidth at the transmitter and the receiver proportional to the number of transmitters. Doppler division multiplex (DDM) MIMO radar systems have been introduced to the market [6]. DDM waveforms must be used to ensure the orthogonality of the transmit signal and avoid some of the problems in the application of frequency division multiplexed (FDM) and code division multiple access (CDMA) for the actual radar system. A classical approach is here to change the phase of the transit signals on a chirp time basis for FCM radar systems. CDM has been introduced in automotive radar in the past 10 years, typically the transmit phase is modulated on a chirp basis timing. To achieve highest power on target, all transmit channels must transmit simultaneously, which can be achieved by using code division multiplexing (CDM MIMO); a unique code is sent from each TX which is then separated within the RX path to achieve a true MIMO radar system. This approach requires additional digital processing but allows an improved unambiguous velocity window as well as shorter scan time, which means higher update rates are possible.

A radar system with N_T transmitters and N_R receivers has $N_T \times N_R$ transmitter-receiver pairs or virtual receivers. Using these $N_T \times N_R$ virtual receivers and the estimated range to the targets from multiple transmitters/receivers, a radar system can use the receive side beamforming approach to determine the azimuth or elevation angle or both, depending on the geometry of the antennas.

DCM naturally uses CDM MIMO where unique spreading codes are employed for different transmitters. At each receiver antenna, a filter matched to each of the N_T transmitter signals can be used. Spreading codes that are orthogonal to each other can be used when transmitting to reduce the self-interference (interference from other transmitters in the same radar), but at the expense of larger autocorrelation sidelobes. There is a tradeoff between the autocorrelation properties of spreading

Table 3.3 Multiplexing Options for MIMO Systems

Multiplexing Scheme	*Impact on Radar Image*
TDM MIMO	Time division multiplexed MIMO: Only one TX is active at a time; multiple RX are active. In dynamic environments, the multiplexing scheme can cause a blurring of the radar image because the scenario has changed between the different illumination phases. The effect can be reduced by movement compensation but is difficult in complex and highly dynamic scenarios.
CDM MIMO	Code division multiplexed MIMO: Multiple TX are active at the same time within the same frequency band; multiple RX are active. Classical modulation method in communication systems but difficult to apply to commercial automotive radar systems due to the high requirements on ADC and processing bandwidth that are typically associated with higher costs. CDM in FCM systems are typically applied on chirp basis, but this is reducing the maximum Doppler unambiguous velocity.
FDM MIMO	Frequency division multiplexed MIMO: Multiple TX are active at the same time occupying different frequency bands; multiple RX are active. Due to extremely complex implementation in hardware, the method is not really used in any commercial radar system.
DDM MIMO	Doppler division multiplexed MIMO: Multiple TX are active at the same time occupying different Doppler frequency bands; multiple RX are active. Requires special effort in hardware implementation (e.g., a phase rotator in the TX chain). Typically, in FCM systems the phase is rotated on chirp-to-chirp basis, which again reduces the maximum Doppler unambiguous velocity.

codes and the cross-correlation of spreading codes. The cross-correlation indicates the amount of interference from different antennas using different spreading codes, while the autocorrelation indicates the self-interference from one spreading code. Generally, the better the cross-correlation properties, the worse the autocorrelation properties.

Figure 3.10 shows the averaged output of all the matched filters on one of the receivers where eight transmitters are active at the same time but use different code sequences. We again use m-sequences with a three-target scenario for this illustration.

The filter output for a given receiver matched to a given transmitter spreading code now has contributions (due to cross-correlations between the codes) from the other transmitters. In this illustration, for a given receiver, we have a total of eight matched filter outputs corresponding to the eight transmitter codes. The MIMO processing for angle estimation performs a coherent summation in multiple directions (angles), and thus, effectively averages the multiple matched filter outputs.

If we compare Figure 3.10 with Figure 3.6, we see an increase in the off-peak signal energy in the MIMO case. This signal energy from the other transmitters spilling into the matched filter output for a given transmitter's code, known as *the channel isolation*, is a major metric for MIMO operations. This channel separation can only be improved by using longer spreading codes (larger L_C).

Some modern FMCW/FCM radars also use this idea to encode the transmitted chirps (sweeps) using spreading codes across multiple transmitters by adding a chirp-to-chirp pseudorandom binary sequence (PRBS) phase code for each TX antenna. Channel isolation, as determined by the code length, is, however, much shorter than DCM (chirp rate being slower than the chip rate). For example, a scan time of 5 ms with a chirp duration of 10 μs would provide use of 500 chirps. On the other hand, a DCM system with a scan time of 5 ms and a chip duration of 1 ns (for 1 GHz bandwidth) would have 5,000,000 chips in a scan, allowing many more transmit antennas to operate at the same time compared to phase coded FCM.

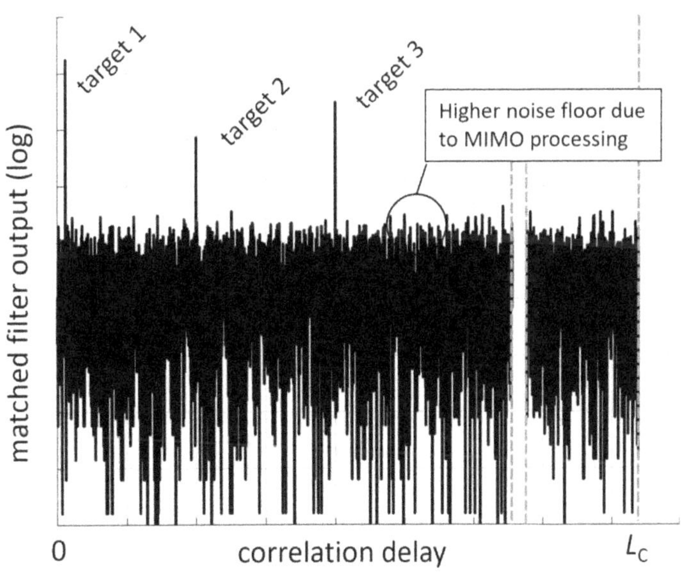

Figure 3.10 Averaged matched filter output with eight transmitters active.

The ability to use many more antennas, along with very high channel separation, allows DCM radars to have very high contrast ratio performance in angle.

3.4 DCM Radar Attributes

3.4.1 High Contrast Distance: Matched Filter

One appealing feature of DCM radars is the sharpness of the peak (often referred to as thumbtacklike response). This is due to the use of the autocorrelation function as range response compared to using the FFT in FCM radars. The FFT results in much slower drop in range response around a target. FCM is using typically using coding on chirp basis that lowers refresh rate. Due to FFT and the corresponding windowing the response is less sharp, which lead to disadvantages in distinguishing between bright and dark objects close to each other (e.g., a child standing behind a truck).

With PMCW modulation the near-in and far-out range sidelobes are of similar amplitude, removing the need for a mainlobe-broadening window function (as is used in FCM) and providing a superior capability to detect small differences in range, allowing precision detection and earlier tracking. Figure 3.11 shows a comparison of FCM versus DCM modulation, where for FCM a classical range FFT with and without window function and for DCM the correlation was used. For the simulations, the first target was set at 100m and a second target at 101m (single additional range gate) with 20 dB lower RCS was used. The DCM correlation results in a sharp response and both targets can be separated. In case of the FCM modulation the separation of the targets might become difficult.

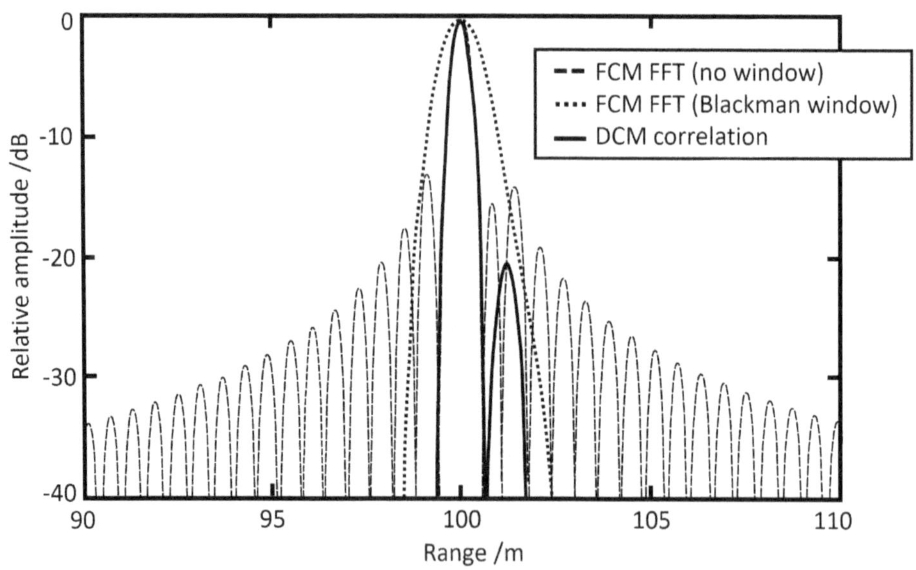

Figure 3.11 Example of FCM versus DCM range response. Two targets (100m and 101m), separated by one range gate with 20 dB difference in RCS. (Max. range = 300m; modulation bandwidth = 250 MHz; PRI time = chirp time = 20 μs, fc = 76.5 GHz, number of PRI N = 512 = number of Doppler bins.)

3.4.2 High Contrast Resolution

Traditionally, radar performance on resolution as previously defined, in any measurement dimension (range, velocity, and angle), is based on targets of the same reflectivity, or radar cross section (RCS). However, in automotive applications, this definition is not sufficient. In such applications, it is also important to measure the radar's ability to discriminate small targets in close proximity to large targets, especially in range (e.g., a small child in front of a truck) and in angle (e.g., pedestrian beside a guard rail).

The DCM radar offers unique codes for each TX with extremely high synchronization and fully matched scans between channels. DCM offers a unique ability to detect discrete objects next or above/below each other which results in a clear radar image.

3.4.3 CDM MIMO (Higher Power on Target)

A radar system with multiple transmit and receive antennas is referred to as a MIMO radar system. In the example shown in Figure 3.12, the MIMO radar system includes four transmitting and four receiving antennas. Each receiving antenna receives the waveforms for each of the four transmit channels. The signal received at receiver antenna 1 (RX1) can be compared to all the transmitted signals from antenna 1 (TX1) up to the signal transmitted from antenna 4 (TX4). Each receiver needs to be aware of the timing of the signal from TX1 up to TX4. Thus, the radar system creates a 16 virtual receiver system (i.e., 4(TX) × 4(RX) = 16 virtual receivers).

CDM MIMO systems deliver the highest power on target because all transmitter channels are illuminating the target at the same time. In comparison to a TDM

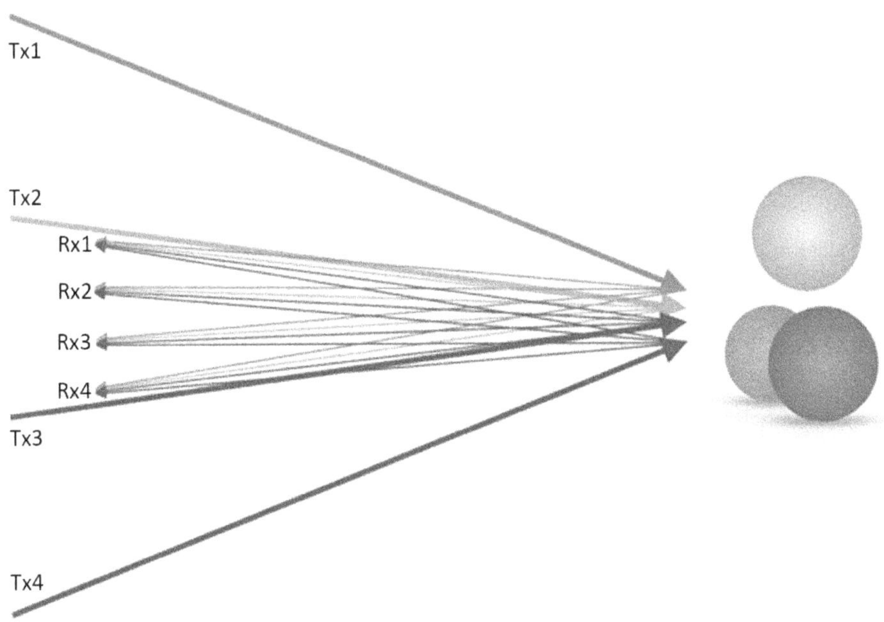

Figure 3.12 Radar system with multiple transmit and receive antennas.

system where, for example, only half of the transmitter channels are active at the same time, the CDM MIMO system has 3 dB higher SNR.

A MIMO radar system inherently allows for estimates of the angles (azimuth, elevation, or both) of an object in the environment. In general, a larger number of antennas (virtual receivers) allows radar systems to be designed with wider aperture, and thereby, higher angular resolution.

3.4.4 Interference Robustness and Interference Mitigation

Due to the increasing numbers of automotive and industrial radar sensors sharing the same frequency band or working in adjacent frequency bands, interference between different sensor modules is becoming increasingly important to radar design [7, 8]. While in the early years of automotive radar the probability for interference was low due to the low installation rates of such sensors, it is today to be one of the most urgent problems that needs to be addressed. Advanced mobile communication is already using digital modulation and allows many users to simultaneously operate in the same vicinity. Frequency modulated radar has limited intrinsic mechanisms for mitigation against multiple users, relying primarily on a narrow intermediate frequency bandwidth and effective usage of the frequency spectrum (single frequency at a time).

DCM radar is by the nature of the modulation a CDMA-based system, which offers unique, orthogonal codes and full channel identification proven for highly dense communication scenarios.

Interference robustness and mitigation techniques are an extremely wide area and the Table 3.4 should provide and initial guideline in terms of some key differences in handling interference.

Table 3.4 Introduction to Differences in Terms of Interference Robustness and Mitigation

Interference Robustness
Independent from the modulation type, the linearity of the receiver input stage (input compression point) determines the maximum range for linear operation.
FCM radar sensor are typically able to handle larger interferer power, due to the typically higher number of bits in the ADC in FCM radar sensor architectures (lower speed ADC enables larger number of bits).
FCM radar sensors are operating at narrow baseband bandwidths which helps to only see nearly in-band interference. In DCM systems large (full modulation bandwidth) baseband is used, where disturbances spread out, less strong in certain range gates but noise floor is increased over the whole baseband.
Avoidance/Mitigation
Pseudorandomization of chirp parameters (slope, frequency, sampling time) in FCM system and in DCM systems sampling time and potentially different codes can lower the probability of interference.
Zero-padding and spectral regrowth are classical methods used in narrow baseband receivers but can be applied to DCM radar systems.
Due to the wideband ADC, DCM system enables channel sensing and functions like "listen before talk." This can help to measure interference levels to increase the confidence level of the sensor SNR. Additionally, the sensor can assign the transmit power to a less occupied frequency band.

3.4.5 Cascading: Coherent and Quasi-coherent Sensors and Networks

If an extremely high angular accuracy and/or more processing power is required typically multiple chips are cascaded or individual sensors are combined via dedicated wired signal exchange. In terms of synchronization of chips and sensors, the individual ADCs inside the receive chains must be synchronized down to picosecond levels to avoid any jitter and degradation of the radar image. A picosecond synchronization is not practicable via classical dedicated wired signals due to the limitation of signal/clock speed of interface pads/pins. The solution is to directly use the transmitted signals, the unique codes per channel enables signal identification and synchronization across various chips and/or sensors. In this case, the digital code modulation enables an operation of multiple chips and/or sensors in a coherent (the LO signal is shared between chips and sensors) and quasi-coherent mode (the LO signal is not shared but lower frequency signals can be used to synchronize the carrier frequency).

3.4.6 Code Design

In Section 3.3 we already have mentioned two different types of correlation functions: periodic and aperiodic correlation (in terms of correlation we refer to auto- and cross-correlation) functions. For the definition of those functions, we refer to [9]. Additionally, to the theoretical functionality as described in [4], the methods described in this chapter take also into account a real implementation and trying to keep the flexibility in applying it to different codes and applications.

In a pulse with repeating code, for each ping, a sequence of L_c chips and is repeated times within a pulse (see Figure 3.13). If an m-sequence is used as the code,

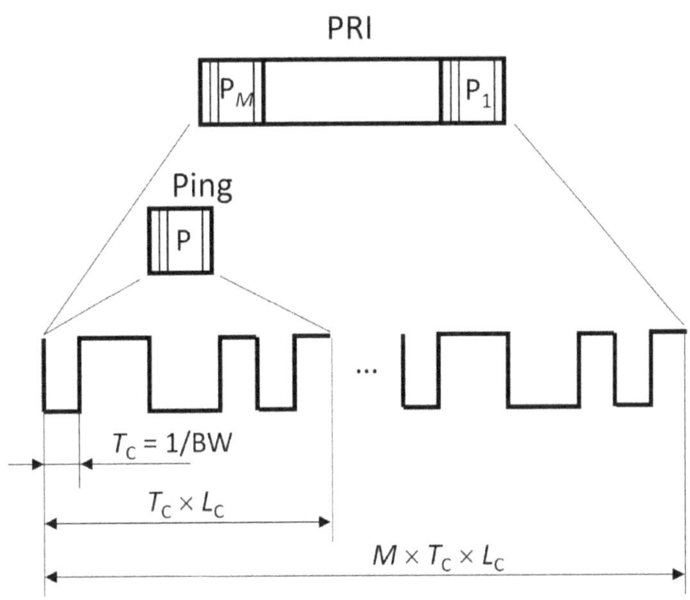

Figure 3.13 Coherent sequence of pings.

the length of the sequence determines the off-peak value of the correlation. The autocorrelation of an m-sequence guarantees that there is only one peak, and all other values are inversely proportional to the length of the sequence. This means that one must use quite long sequences to guarantee sufficient suppression of off-peak values. Alternately, one can use more advanced code families like almost perfect autocorrelation sequence (APAS) codes [10] which have very good off-peak values (almost infinity) but do not show only one peak in their autocorrelation. APAS codes have two peaks, with the second one occurring at half the length of the code length. This renders only half of the code length useable for unambiguous range measurements.

Figures 3.14 to 3.16 show the typical Doppler processing chain of a DCM system. Within a pulse, several pings are correlated and coherently summed to increase the processing gain. One can fit as many pings as possible within a pulse if the pulse repetition interval can meet the unambiguous Doppler requirements.

Once all the pings in a PRI have been coherently summed and all the pulses have been received, a bank of Doppler filters can be used to process the data to further increase the processing gain and separate out targets according to their Doppler velocities.

DCM offers an additional freedom in the design of suitable applications and properties or radar and/or communication systems: the code design.

The key parameters of a communication system are determined by the properties of the code. In case of a radar system some of the key code parameters are:

- *Orthogonality:* Determined by the length of the codes. Typically, longer codes have lower cross-correlation responses, but some codes are specially designed with this property (e.g., complementary codes).
- *Robustness against Doppler:* The Doppler shift is increasing with higher relative velocities. Even in such cases the code must be robust enough to not degrade sensor performance.
- *Applicable to flexible use cases (static and moving objects):* Especially the automotive application contains a huge variety of challenging scenario. From high speed in low dense scenarios up to low speed in extremely dense scenario.
- *Interference robustness:* The install rate of radar sensor is increasing; the probability of interference is continuously increasing but still the sensor SNR should maintain to enable a safety and reliable operation.
- *Compliance to regulations:* To allow the commercial usage of radar system in various countries certain regulations in spectrum mask, spurious emissions, and output power must be fulfilled. In the United States, the regulations are defined by the Federal Communications Commission (FCC).
- *Complexity in terms of code generation:* Especially in automotive radar sensors it is not affordable to implement power hungry code generation methods.

Nevertheless, the usage of special codes to improve the radar and/or communication system for certain applications and environments is one of the new research fields [11–13].

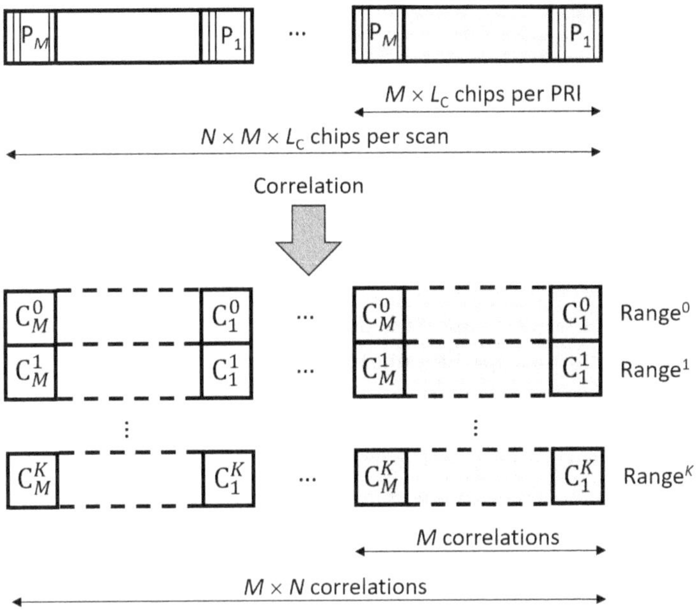

Figure 3.14 Coherent processing interval.

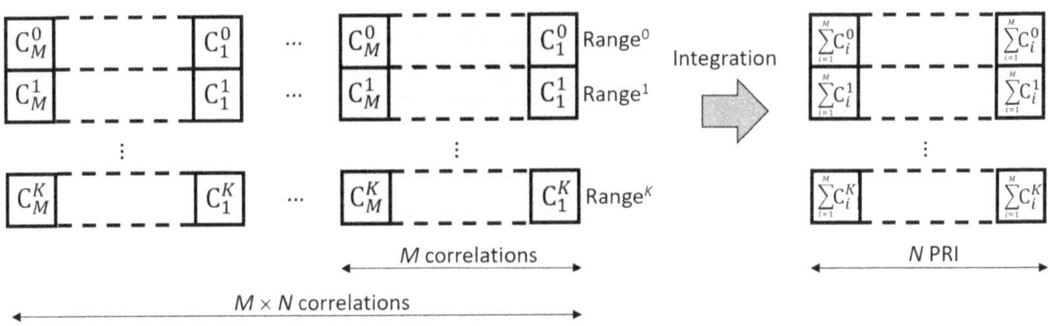

Figure 3.15 Accumulation of correlations in each PRI.

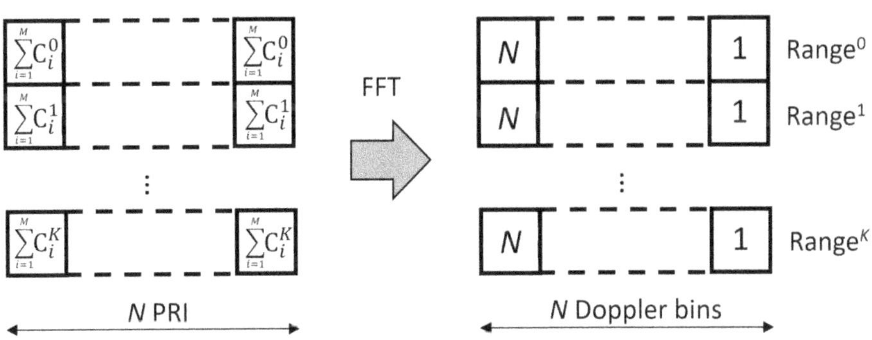

Figure 3.16 Doppler processing.

3.5 DCM Radar Implementation

Automotive radar systems have traditionally employed FCM types of radar signals. With FCM, some of the signal processing is done in analog circuitry (mixers, filters). In DCM radar systems, the analog circuitry is used only to recover a bandlimited baseband version of the received signal, and the signal processing (e.g., matched filtering) is performed primarily digitally. The DCM radars thus trade-off the simpler analog processing of FCM for much more advanced digital processing. This trade-off necessitates that DCM radars implement analog-to-digital conversion at a much higher rate (in the range of GHz) than is needed for FCM (few tens of MHz rate). This has been one of the main impediments for bringing DCM technology, in a cost-effective way, to automotive applications.

Traditionally, FCM radars based on Silicon-Germanium (SiGe) technologies have dominated the markets in the past. Millimeter-wave CMOS technologies are also becoming more popular for implementing the RF circuitry and ADC, whereas the digital processing is done in separate processors or micro-controllers. Advances in CMOS technology now allows the implementation of RF circuitry and consequent integration of digital processing, along with RF/analog processing, on a single chip. Today nearly all chip manufacturers have developed monolithically integrated CMOS-based FCM solutions for automotive radar applications. These solutions do not fully benefit from the advanced CMOS developments for high-speed low-power digital processing notes. In a monolithic integrated circuit compromises in terms of mmw-wave performance, power consumption, die-size have to be made. CMOS technologies well suitable for radar applications at 77 GHz cannot be applied for highly efficient, low-power, high-speed digital design [14]. A 77 GHz radar design with CMOS has already been demonstrated, and a low-power high speed (≥ 2G samples/sec) ADC, as needed by DCM radars, is feasible in CMOS technology [15–17]. These capabilities, coupled with a cost and power efficient digital design with intelligent hardware-software partition, makes DCM radars (Figure 3.17) a reality for automotive applications [18–22]. Flexible special hardware accelerators can be used to process high data rate digital baseband signals and reduce the rate of relevant data extracted from the raw baseband data. General purpose processors can then operate on the extracted data to perform application specific operations.

The digital hardware can also simplify the analog processing challenges. As such, a joint design of the digital and analog circuitry can achieve better performance than separate analog and digital designs typical in the automotive semiconductor industry. This integrated, flexible chip design allows for software to modify the radar performance to meet customer and application requirements, as opposed to the long and costly approach of designing custom chips.

The solution shown in Figure 3.17 and Figure 3.18 offers physically 12 transmitter channels (TX) and 8 receiver channels (RX) which are multiplex to 16 input ports. In combination with programmable code generators a huge variety of codes are supported and can be upconverted to the 76–81 GHz range. Full real time 96 virtual receiver channels (VRx) and with the usage of TDM (time domain division multiplex) 192 VRx are directly supported by one chip. High number of VRx can

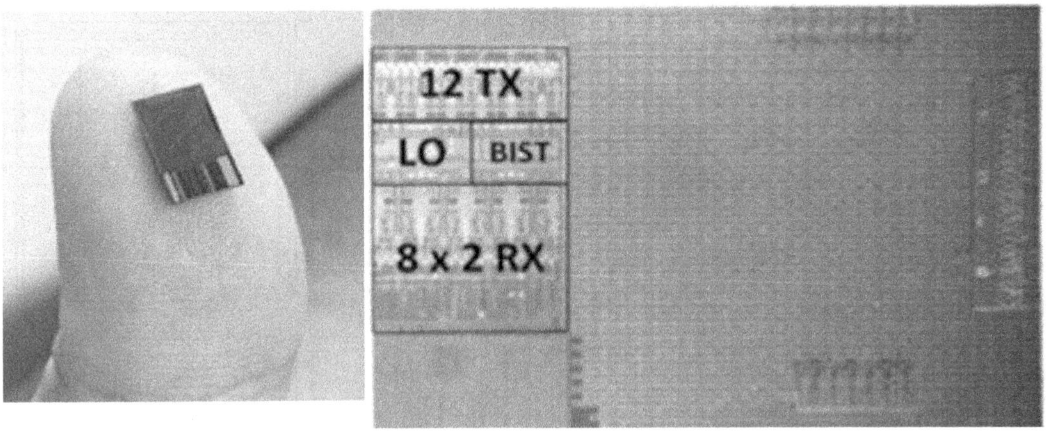

Figure 3.17 Simplified block diagram of a modern single-chip DCM radar implementation, addressing automotive and industrial radar markets (example courtesy Uhnder, Inc.).

be achieved by cascading multiple of the chips. The typical functionality of the circuit blocks shown in Figure 3.18 are shown in Table 3.5. Signal coherence over the full cascaded solution is guaranteed by special synchronization signals. The digital modulation enables cognitive sensing (adjustment transmit and receive properties accordingly to the environment), with massive MIMO and beam steering.

Table 3.5 Block Diagram Descriptions

System Blocks/ Sub-Modules	*Description*
Analog section	Receiver and transmitter with corresponding PA, LNA, baseband (VGA), and high-speed ADC, including clock generation (LO) and options to extend the number of channels by cascading the chip.
HW accelerated signal processing	To handle the extremely high data rate from the high-speed ADC (2 × 8 RX × 2 GSps ADC (10 bit) = 16 × 2 × 10 Gb/s = 40 GB/s) with sufficient bandwidth at low power consumption, programmable hardware accelerated functions are implemented on chip.
System and compute platform	General processing platform, typically using multiple arm cores to enable customers to implement feature level software.
Safety platform	ASIL or SIL, safety integrity level for automotive and industrial applications requires an independent processing platform to perform a transition into a safe state after a safety violation have been detected.
Hardware security platform	Boot security (certification of software) and cybersecurity (secured communication, for example, over-the-air updates) require a trusted platform independent from other on chip solutions.
DSP platform	The digital signal processing platform enables customers to develop and run software on classical processing solutions (e.g., tracker).
Memory	Addressing on chip SRAM, external flash, and external DRAM memory.
Connectivity and networking	The classical interface support as: JTAG, GPIO up to FlexCan and addressing multiple ethernet ports up to 1 Gb/s.

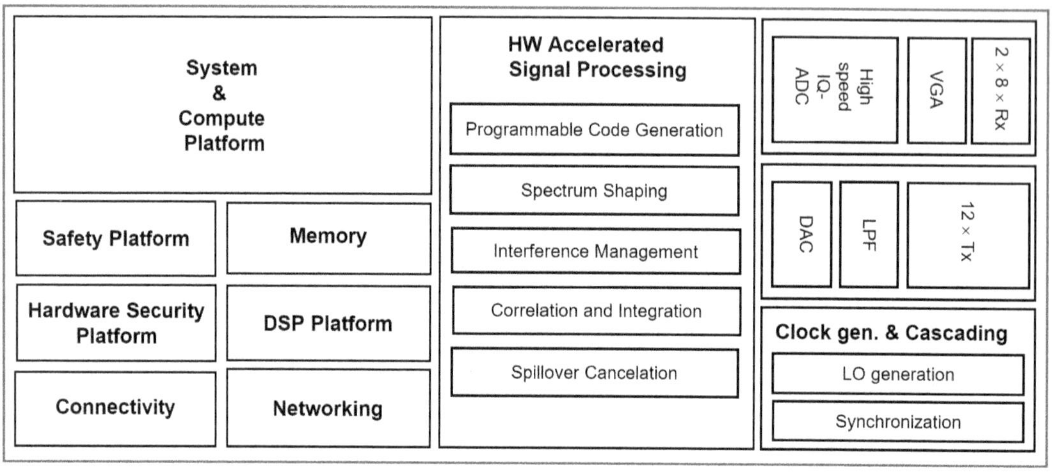

Figure 3.18 DCM radar chip to enable high-performance 4-D radar (example courtesy Uhnder, Inc.).

References

[1] Brooker, G. M., "Understanding Millimetre Wave FMCW Radars," *1st International Conference on Sensing Technology*, Palmerston North, New Zealand, November 21–23, 2005.

[2] Pearce, T. H., et al., "The Application of Digital Receiver Technology to HF Radar," *1991 Fifth International Conference on HF Radio Systems and Techniques*, July 22–25, 1991.

[3] Skolnik, K., *Introduction to Radar Systems*, Third Edition, New York: McGraw-Hill, 2001.

[4] Garcia, O. F., "Signal Processing for mmWave MIMO Radar," Master Thesis, University of Gävle, http://www.diva-portal.se/smash/get/diva2:826028/FULLTEXT01.pdf, June 2015.

[5] Sarwate, D. V., and M. B. Pursley, "Cross Correlation Properties of Pseudorandom and Related Sequences," *Proceedings of the IEEE*, Vol. 68, No. 5, 1980, pp. 593–619.

[6] Sun, H., Brigui, F., and M. Lesturgie, "Analysis and Comparison of MIMO Radar Waveforms," *2014 International Radar Conference*, Lille, France, October 13–17, 2014.

[7] Tanis, S., "Automotive Radar and Congested Spectrum: Potential Urban Electronic Battlefield," *Microwave Journal*, January 2019, pp. 48–58.

[8] Alland, S., et al., "Interference in Automotive Radar Systems," *IEEE Signal Processing Magazine*, September 2019, pp. 45–59.

[9] Chen, H.-H., "The Next Generation CDMA Technologies," West Sussex, England: John Wiley & Sons, Ltd., 2007.

[10] Wolfmann, J., "Almost Perfect Autocorrelation Sequences," *IEEE Transactions on Information Theory*, Vol. 38, No. 4, July 1992, pp. 1412–1418.

[11] Spano, E. et al., "Sequences of Complementary Codes for the Optimum Decoding of Truncated Ranges and High Sidelobe Suppression Factors for ST/MST Radar Systems," *IEEE Transactions on Geoscience and Remote Sensing*, Vol. 34, No. 2, March 1996, pp. 330–345.

[12] Morita, T. and T. Kishigami, Patent US9513365B2, "Radar System," https://patentscope.wipo.int/search/en/detail.jsf?docId=WO2014129142.

[13] Van Thillo, W., Gioffré, P., Giannini, V., Guermandi, D., Brebels, S., and A. Bourdoux, "Almost Perfect Auto-Correlation Sequences for Binary Phase-Modulated Continuous Wave Radar," *2013 European Microwave Conference*, October 6–10, 2013.

[14] Giannini, V., et al., "A 192-Virtual-Receiver 77/79GHz GMSK Code-Domain MIMO Radar System-on-Chip," *ISSCC*, 2019, pp. 164–166.

[15] International Technology Roadmap for Semiconductors (ITRS), http://www.itrs2.net/.

[16] Ginsburg, B., et al., "A Multimode 76-to-81 GHz Automotive Radar Transceiver with Autonomous Monitoring," *ISSCC*, 2018, pp. 158–160.

[17] Farley, B., et al., "An All-Programmable 16-nm RFSoC for Digital-RF Communications," *IEEE Micro*, Vol. 38, No. 2, March/April, 2018, pp. 61–71.

[18] Reuter, R., "Digital Modulation: A Breakthrough in Radar Enabling High Performance at Lowest Costs," in Workshop: Sensor and Connectivity Enabling Autonomous Cars, *IMS*, Boston, June 2–7, 2019.

[19] Reuter, R., "Beyond MIMO: 5G Radar = Digital Modulation Radar (DMR)," *16th European Microwave Week (EuMW)*, in "Automotive Forum," Paris, France, September 29, 2019.

[20] Reuter, R., "Digital Modulation Radar: A Dream Became True, Highest Performance at Lowest Costs Due to Modern Technologies and Processing Strategies," *16th European Radar Conference (EuRAD)*, Paris, France, October 2–4, 2019.

[21] Roy, A., "Let's Get Real About Imaging Radar," *17th European Microwave Week (EuMW 2020)*, in "Workshop: Automotive Radar with Increased Angular Resolution," Utrecht, Netherlands, January 10–15, 2020.

[22] Giannini, V., et. al., "Digital Code Modulation MIMO Radar Improves Automotive Safety," *Microwave Journal*, Vol. 62, No. 8, August, 2019.

CHAPTER 4

Automotive MIMO Radar

Shunqiao Sun

Automotive radar with a small number of antennas has been used for ADAS purposes since the late 1990s. These early automotive radars mostly provided target detection and velocity information. However, current generation of automotive radar for ADAS has rather limited ability to resolve closely spaced targets. Lidar systems have better angular resolution (less than 1 degree) and have been introduced in Level 4 and Level 5 autonomous driving systems. Lidar can provide point clouds. Via use of deep neural networks, such as PointNet [1] and PointNet++ [2], the point clouds can lead to target identification. However, due to its use of light spectrum wavelength, Lidar is susceptible to bad weather conditions, such as fog, rain, snow and dust in the air. In addition, the cost of lidar is high. On the other hand, automotive radar with millimeter waveform technology has the potential to provide point clouds at much lower cost than lidar, and with more robustness to weather conditions. Such radar is referred to as a "high end radar," or *imaging radar* [3]. Computer vision techniques [1, 2] that were previously reserved for high resolution camera sensors and lidar systems, can be applied to imaging radar data to identify targets. For example, a car can be identified based on 2-D radar points of an imaging radar using PointNet [4]. Imaging radars have been attracting the interest of those developing fully autonomous vehicles, major Tier 1 suppliers, and automotive radar startups.

In addition to sensitivity, the important requirements for automotive radar are high resolution, low hardware cost, and small size. MIMO radar technology has been receiving considerable attention by the automotive radar community because it can achieve high angular resolution with relatively small numbers of antennas and receivers. For that ability, it has been exploited in current generation automotive radar for ADAS as well as in next generation high resolution imaging radar for autonomous driving. MIMO radar is a good candidate for high resolution imaging radar for autonomous driving. In MIMO radar the targets are first distinguished in range and Doppler domains. Then, large virtual arrays with hundreds of elements can be synthesized to provide high resolution in both azimuth and elevation. As a result, point clouds with similar performance as lidar can be generated at a much lower cost.

In this chapter, we introduce the concept of imaging radars using MIMO technology, present some examples for synthesizing hundreds of virtual array elements by cascading multiple radar transceivers with each supporting a small number of antennas, and discuss design challenges.

4.1 Virtual Array Synthesis via MIMO Radar

In state-of-the-art automotive FMCW radar, the range and Doppler parameters of targets can be estimated by using single receive antenna. However, to estimate the angle parameter of targets, a receive antenna array is needed. In MIMO radar, the antennas transmit waveforms in a way that guarantees their orthogonality. At each receive antenna, the contribution of each transmit antenna is extracted by exploiting waveform orthogonality. For a uniform transmit antenna array with M_t transmit antennas and element spacing of d_t and a uniform receive antenna array with M_r receive antennas and element spacing of d_r, a virtual array with M_tM_r elements can be synthesized. The array response of the *synthesized* array, that is, the array corresponding to a MIMO radar can be expressed as $\mathbf{a}_t(\theta_t) \otimes \mathbf{a}_r(\theta_r)$. Here, $\otimes$ denotes the Kronecker product, and $\mathbf{a}_t(\theta_t)$ and $\mathbf{a}_r(\theta_r)$ are respectively the transmit and receive array steering vectors, corresponding to direction of departure (DOD) θ_t and direction of arrival (DOA) θ_r, defined below.

$$\mathbf{a}_t(\theta_t) = \left[1, e^{j2\pi\frac{d_t}{\lambda}\sin(\theta_t)}, \cdots, e^{j2\pi\frac{(M_t-1)d_t}{\lambda}\sin(\theta_t)}\right]^T \tag{4.1}$$

$$\mathbf{a}_r(\theta_r) = \left[1, e^{j2\pi\frac{d_r}{\lambda}\sin(\theta_r)}, \cdots, e^{j2\pi\frac{(M_r-1)d_r}{\lambda}\sin(\theta_r)}\right]^T \tag{4.2}$$

Figure 4.1 shows two MIMO radar configurations with $M_t = 2$ transmit and M_r receive antennas [5]. One is in interleaved mode with $d_r = 2d_t$ and $d_t = \lambda/2$ and the other is in stacked mode with $d_t = M_rd_r$ and $d_r = \lambda/2$. Here, $\lambda = c/f_c$ is the wavelength. It can be verified that for the both array configurations, the synthetic virtual arrays are equivalent to uniform linear arrays (ULAs) with M_tM_r elements and spacing of d_r when $\theta_t = \theta_r$.

It should be noted that because MIMO radar transmits orthogonality waveform, when isotropic array elements are used, the array beampattern—also referred to as the MIMO radar array facto—is omnidirectional. Thus, MIMO radar loses the coherent array processing gain advantage enjoyed by traditional phased-array radar systems, which is 10log$N$ for a phased-array with N elements [6]; the SNR of the array response at a given angular direction is less than that of phased-array radar with transmit beamforming. Still, in the automotive application scenario, the high resolution angle finding ability of MIMO radar coupled with its low cost are viewed as more important than the loss of coherent processing gain.

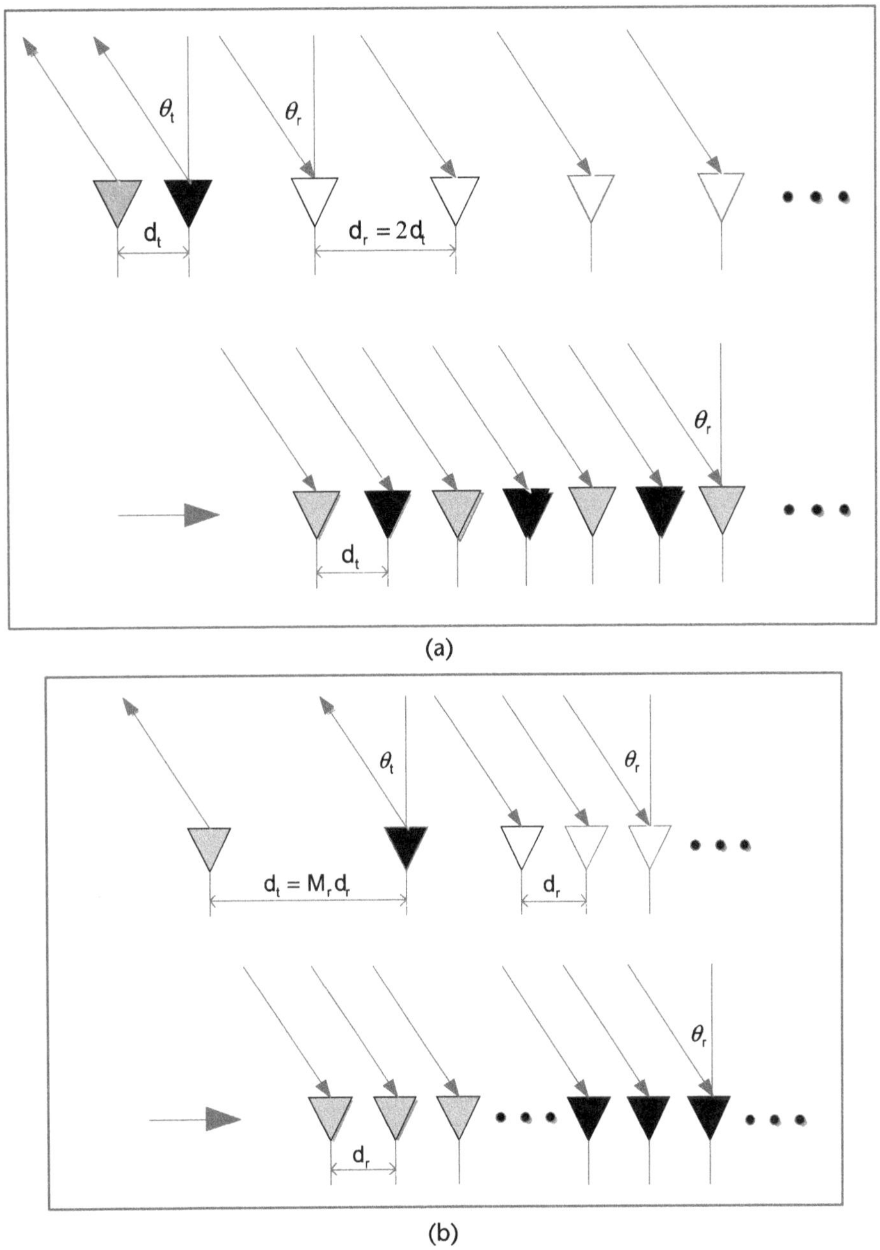

Figure 4.1 Illustration of different MIMO radar virtual array configurations [5] using TDM or DDM scheme with $M_t = 2$ transmit antennas and M_r receive antennas: (a) interleaved with $d_r = 2d_t$ and $d_t = \lambda/2$; (b) stacked with $d_t = M_r d_r$ and $d_r = \lambda/2$. Different shades indicate the transmit antenna either transmits different time slots or codes. (© 2022 IEEE. Reprinted, with permission, from [5].)

4.2 Waveform Orthogonality Strategies in Automotive MIMO Radar

Virtual array synthesis in automotive radar using MIMO radar technology relies on the separability of the transmit signals of the different antennas. The separation is easier when the transmit signals of different antennas are orthogonal. In the following, we review techniques to achieve waveform orthogonality while transmitting FMCW, such as TDM, DDM, and FDM.

4.2.1 Waveform Orthogonality via TDM

In TDM MIMO radar [7–11], only one transmit antenna is scheduled to transmit at each time slot. In Figure 4.2, a signal processing example of a MIMO radar in TDM is given by NXP Semiconductors [12], where $M_t = 2$ transmit antennas emit FMCW chirps alternatively. The switch delay between transmit antennas is $\Delta t = T_{\mathrm{PRI}}$. At each receive antenna, range FFTs of length N_r are conducted for each chirp and the FFT outputs of $2N_d$ chirps are assembled in two matrices corresponding to odd and even chirp sequences, respectively. The receive array corresponding to odd and even chirp sequences form two subarrays, which can be used to synthesize a virtual array according to interleaved or stacked configurations.

For a moving target with velocity of v, the switching delays of transmit antennas introduce a target phase migration from chirp to chirp, which is defined as $\phi = 2\pi f_D \Delta t = (4\pi/\lambda)v\Delta t$. As a result, the virtual array pattern would be distorted [9]. In Figure 4.2, the phase difference between corresponding columns in the two matrices is $\phi = (4\pi/\lambda)vT_{\mathrm{PRI}}$. If $v = -0.5v_{\max}$ and $v = -v_{\max}$, where $v_{\max}$ is the maximum unambiguous detectable radial speed and $v_{\max} = c/4f_cT_{\mathrm{PRI}}$, the phase shifts are $\phi = -\pi/2$ and $\phi = -\pi$, respectively. The array beampattern distortion is demonstrated in Figure 4.3 for a moving target with range of 35m and azimuth angle of $\theta = 0°$ [9]. Here, MIMO radar with $M_t = 2$, $M_r = 8$ operates in TDM fashion. As stated in [9], and can also be seen in Figure 4.3, for interleaved MIMO array configuration, when the target velocity increases, the grating lobes at the edge of FOV show up, while the peak at the target direction decreases and totally disappears when $v = -v_{\max}$.

For stacked MIMO array configuration, as the target velocity increases, the peak is slightly off the boresight with a mirror grating lobe at the opposite direction.

The phase migration introduced by every moving target in the virtual array response needs to be compensated for before angle finding. The phase shift estimate $\hat{\phi}$, can be obtained after each target velocity has been estimated based on the 2-D–FFT of a single receive antenna, or noncoherent 2-D–FFT integration of the same subarray. For example, in the example of Figure 4.2, the phase in the beam vector of the subarray obtained from the even chirps needs to be compensated by multiplying them with $e^{-j\hat{\phi}}$, while the phase in the beam vector of the subarray corresponding to odd chirps is kept unchanged.

It should be noted that, in TDM MIMO radar, the pulse repetition interval is enlarged by the transmit antenna number M_t. As a result, the maximum unambiguous detectable velocity will be reduced by a factor of M_t [13].

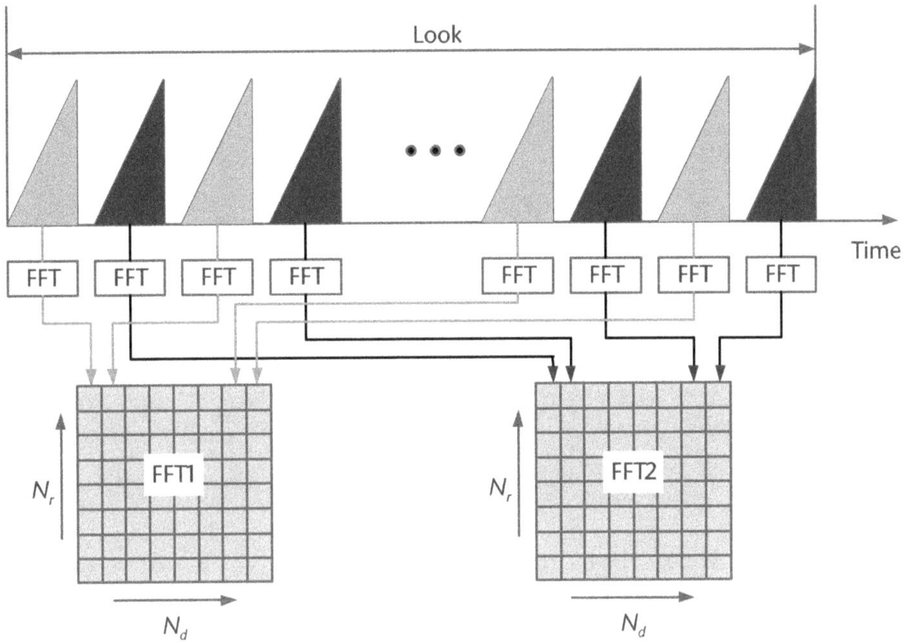

Figure 4.2 Example of radar signal processing with TDM scheme [12], where $M_t = 2$ transmit antennas transmit FMCW chirp sequences alternatively. The gray and dark color denote the odd and even echo chirp sequences, respectively. The range FFTs are conducted for each chirp and the FFT outputs are stored in two matrices corresponding to odd and even sequences, respectively for further processing. (© 2022 IEEE. Reprinted, with permission, from [5].)

4.2.2 Waveform Orthogonality via DDM

In one look, a total of N chirps (i.e., pulses) are transmitted sequentially, with pulse repetition interval T_{PRI}. All transmit antennas transmit simultaneously the same FMCW waveform, after multiplying it with a phase code that is different for each antenna, and changes between pulses, that is, $x_m(n) = e^{j2\pi\alpha_m(n)}$, $m = 1, \cdots, M_t$, $n = 1, \cdots, N$ [14]. To separate the h-th transmit signal at the l-th receiver, after range FFT, a slow time Doppler demodulation, is applied to all range bins corresponding to the same chirp. The Doppler demodulated outputs of N chirps are assembled into a vector $\mathbf{s}_l^h$. Then, the Doppler FFT is applied on the vector $\mathbf{s}_l^h$. To separate the transmit signals in the Doppler domain, one for the two methods described here can be applied.

The first approach is to design phase codes such that the Doppler FFT of the interference $e^{j2\pi(\alpha_m(n)-\alpha_h(n))}$ is shifted to a frequency that is higher than the maximum detectable Doppler frequency $f_D^{\max}$. Therefore, a low pass filter (LPF) can be applied to remove the interference [14]. One example of such phase codes is

$$\alpha_m(n) = \alpha_m n, m = 1, \ldots M_t, n = 1, \ldots N \tag{4.3}$$

where the starting phase α_m is linear across different transmit antennas, that is, $\alpha_m = a_0 m$. Figure 4.4(a) shows the range and Doppler spectrum of a target with a

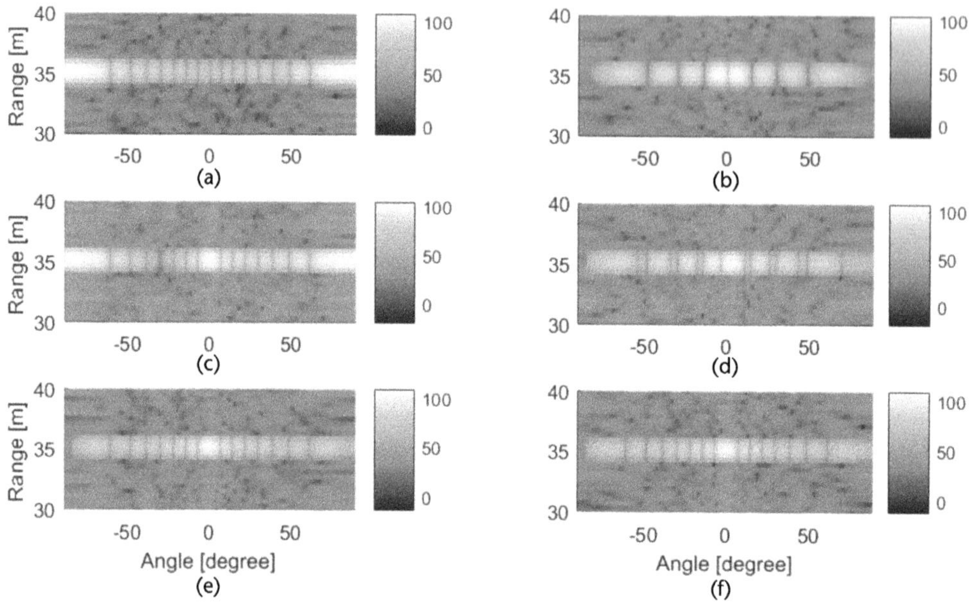

Figure 4.3 Examples of MIMO radar range and azimuth images using TDM [9] with $M_t = 2, M_r = 8$ for a single target with range of 35m, azimuth angle of $\theta = 0°$. Two MIMO array configurations, interleaved (left column) and stacked (right column) are considered. The radial velocity of (a–b) $v = -v_{\max}$, (c–d) $v = -0.5v_{\max}$, and (e–f) $v = 0$. (© 2022 IEEE. Reprinted, with permission, from [5].)

range of 75m and velocity of 10 m/s. The automotive MIMO radar has two transmit antennas and transmit phase codes given in (4.3) with $a_0 = 1$ and $N = 512$. It can be seen that signals from different transmit antennas are shifted to higher Doppler spectrum, which can be removed via a LPF in Doppler domain. Under this approach, the radar pulse repetition frequency f_{PRF} should be larger than $M_t f_D^{\max}$ [15]. Thus, if the f_{PRF} is kept unchanged, the maximum detectable unambiguous Doppler frequency is reduced by a factor of M_t. In practice, a Doppler unfolding, or de-aliasing algorithm needs to be developed with different f_{PRF} in different looks.

The second approach is to design phase codes so that the Doppler FFT of the interference can be distributed into the entire Doppler spectrum as pseudonoise. It is desired to minimize the peak interference residual (PIR) in the Doppler spectrum [16] calculated using the discrete-time Fourier transform (DTFT) for $m = 1, \cdots, M_t$, that is,

$$\mathrm{PIR} = \max_{f, m \neq h} \left| \sum_{n=1}^{N} e^{j2\pi(\alpha_m(n) - \alpha_h(n))} e^{j2\pi f n} \right| \tag{4.4}$$

where $f \in [-\tfrac{1}{2} f_{\mathrm{PRF}}, \tfrac{1}{2} f_{\mathrm{PRF}}]$. Following (4.4), the cross-correlation of the spectra of two codes needs to be flat [16], since the Fourier transform of multiplication of two codes in the time domain is equivalent to convolution of spectrum of one code with time reversed and complex conjugate of the other. The maximum autocorrelation value of an unimodular sequence of length N is N. The ideal cross-correlation of two unimodular sequences of length N has magnitude of $\sqrt{N}$. Thus, in the ideal case, according to [16] the maximum power gain of the currently transmitted signal

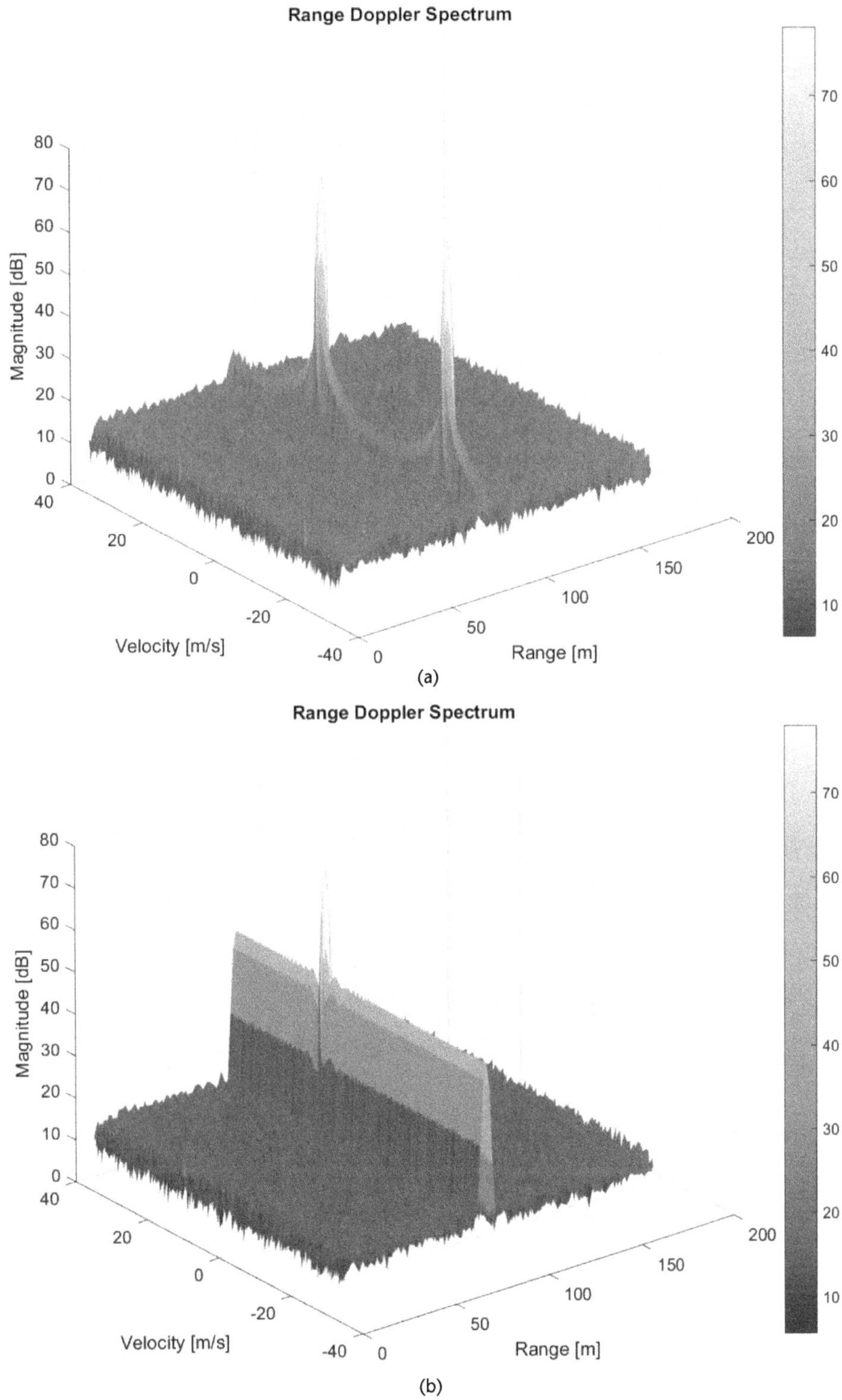

Figure 4.4 The range and Doppler spectra of a target with range of 75m and velocity of 10 m/s. The automotive MIMO radar has two transmit antennas and slow time phase coding of length $N = 512$ is applied for DDM: (a) phase shift codes defined in (4.3); (b) two Chu sequences. (© 2022 IEEE. Reprinted, with permission, from [5].)

over other signals is $\sqrt{N}$. For example, the maximum achievable waveform attenuation is about 27.1 dB for unimodular sequence set with $N = 512$.

Constant amplitude zero autocorrelation codes are good candidates for DDM. The discrete Fourier transform (DFT) of a constant amplitude zero autocorrelation code has also constant amplitude and zero autocorrelation [17]. One of such examples is the Chu sequence [18], which is defined as $x_m(n) = e^{(j\pi/N)m(n+1)n}$, $m = 1, \ldots M_t$, $n = 1, \ldots, N$, where N is a prime number. In practice, the Chu sequence of prime length is first generated and then truncated into a length for efficient FFT. For example, we generate Chu codes of prime length 521 and truncate them to length $N = 512$. By calculation with FFT, the peak interference residual defined in (4.4) is $1.08\sqrt{N}$. Therefore, the waveform attenuation for Chu sequence of length $N = 512$ is about 26.4 dB. In Figure 4.4(b), we show the range and Doppler spectrum of a target with range of 75m and velocity of 10 m/s. The automotive radar has two transmit antennas and two Chu sequences of length $N = 512$ are applied for slow time DDM. It can be seen that the waveform attenuation is about 26 dB.

In practice, binary phase codes are used due to hardware constraint [19]. The binary phase code sequences are obtained via exhaustive search such that the peak interference residual in (4.4) is low. As the code length increases, the search time will grow exponentially.

The benefit of slow time phase coding is that the interference from other transmitters does not affect different range bins. The range resolution is only determined by the bandwdith of FMCW chirp. Therefore, it avoids the range sidelobe issue using fast time phase coding. However, the Doppler sidelobes would be high due to the residual of the slow time phase coding. As a result, targets with low RCS (e.g., pedestrians) that are close to the target with strong reflections (e.g., trucks) might be masked by the waveform residual. In other words, the waveform residual reduces the radar dynamic range. Given the code length, the number of phase codes with good correlation properties is limited, or equivalently, the number of antennas that can transmit simultaneously is limited.

4.2.3 Waveform Orthogonality via FDM

In the FDM scheme, the transmitted signals are modulated by different carrier frequencies. According to [20], the separation of multiple transmit FMCW signals is achieved by shifting the m-th transmit FMCW chirp by an offset frequency $f_{\text{off},m}$. If the differences between all $f_{\text{off},m}$ are larger than twice of the cutoff frequency of the antialiasing BPF $f_b^{\max}$, which is determined by the maximum unambiguous detectable range and Doppler, the transmitted signals can be separated at the receive end. Specifically, the received signal at each receiver is first mixed with the same starting carrier frequency f_c. The separation of transmit signals in the mixer output can be implemented by a frequency shift followed by a low pass filter with cutoff frequency $f_b^{\max}$ [20]. Each receiver needs to carry out such frequency shift and filtering operation M_t times. As a result, high range resolution can be realized using a typical FMCW chirp with large bandwidth. Meanwhile, after FMCW demodulation, frequency shift and filtering operation, the FDM MIMO scheme can still utilize a low sampling rate determined by the beat signal.

Let us consider the FDM scheme in the context of the example in Section 2.6, that is, a FMCW LRR radar with maximum detectable range of 250 meters and maximum detectable velocity of 120 mph. For bandwidth $B = 150$ MHz and chirp duration $T = 50$ μs, the maximum beat frequency is $f_b^{\max} = f_R^{\max} + f_D^{\max} = 5.0274$ MHz. Therefore, the frequency shift for the m-th transmit antenna in the FDM scheme can be chosen as $f_{\text{off},m} = 12(m - 1)$ MHz. The IF should have a bandwidth of $12M_t$ MHz to hold the mixer output.

4.3 Angle Finding in Automotive MIMO Radar

In automotive MIMO radar with M_t transmit and M_r receive antennas, a virtual uniform linear array of M_tM_r elements can be synthesized with interelement spacing d. The array response can be written as

$$\mathbf{y} = \mathbf{A}(\theta)\mathbf{s} + \mathbf{n} \tag{4.5}$$

where $\mathbf{A}(\theta) = [\mathbf{a}(\theta_1), \cdots, \mathbf{a}(\theta_K)]$ is the virtual array steering matrix with

$$\mathbf{a}(\theta_k) = \left[1, e^{j\frac{2\pi}{\lambda}d\sin(\theta_k)}, \cdots, e^{j\frac{2\pi}{\lambda}(M_tM_r-1)d\sin(\theta_k)}\right]^T \tag{4.6}$$

Here, $\mathbf{n}$ is a noise term and $\mathbf{s} = [\beta_1, \cdots, \beta_K]^T$, where β_k denotes the target reflection coefficient for the k-th target. The array response at a particular time instance consisting of data obtained at all the virtual receivers and corresponding to the same range-Doppler bin is defined as the *array snapshot*. In highly dynamic automotive scenarios, usually, only a small number of array snapshots, or even a single snapshot in the worst case is available [21].

In automotive MIMO radar with virtual ULA, angle finding can be done with digital beamforming (DBF) [22–24] by performing FFTs on snapshots taken across the array elements, that is, y in (4.5) (see Figure 4.5). DBF can be implemented efficiently in an embedded DSP with a single snapshot. However, DBF is not a high resolution angle finding method. Higher resolution angle finding can be achieved

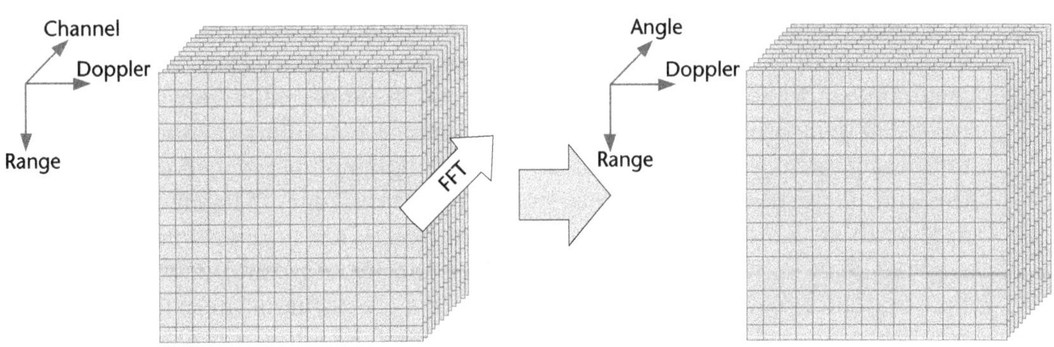

Figure 4.5 Digital beamforming (DBF). (© 2022 IEEE. Reprinted, with permission, from [5].)

with subspace-based methods, such as MUSIC [25] and ESPRIT [26–29], sparse sensing-based methods [30–38], or the iterative adaptive approach (IAA) of [39] and [40]. The performance of subspace-based angle finding methods relies on accurate estimation of the array covariance matrix with multiple snapshots, which is a challenging task in the highly nonstationary automotive radar scenarios. In such context, spatial smoothing [41] is applied for introducing virtual snapshots for array covariance matrix estimation. While sparse sensing-based methods and IAA have high computational cost, they yield angle estimates based on a single snapshot, which is important for snapshot-limited automotive radar.

Achieving high angular resolution for Level 4 and Level 5 autonomous driving requirement using ULA with $d = \lambda/2$ is expensive. According to [42], the 3dB beamwidth of antenna array with aperture size D is $\Delta\theta = 2\arcsin(1.4\lambda/\pi D)$. To achieve 3dB beamwidth of 1 degree, the antenna array aperture should be about $D \approx 51\lambda$. If the antenna array is ULA with interelement spacing as half wavelength, it should be composed of about 100 array elements. Even with the help of MIMO radar technology, the cost of synthesizing such a large virtual ULA with half wavelength element spacing is very high. One way to further reduce the cost without sacrificing the high angular resolution is via the use of nonuniform, or sparse linear arrays (SLAs) [43–48], synthesized with MIMO radar technology. In that context, selecting the locations of array elements and carrying out angle finding with the virtual sparse array are key problems.

4.3.1 High Resolution Angle Finding with ULA

4.3.1.1 Subspace Methods with Spatial Smoothing

The performance of subspace-based angle finding methods requires an estimate of the array covariance matrix. Such estimate is typically obtained based on multiple snapshots. However, in highly dynamic automotive environment, it is not possible to obtain enough snapshots before the model of (4.5) changes. In such scenarios, spatial smoothing [41] can introduce virtual snapshots for array covariance matrix estimation. In spatial smoothing, the array snapshot, $\mathbf{y}$, is divided into overlapped sub-arrays of length L and a new sampled array covariance matrix $\mathbf{R} \in \mathbb{C}^{L\times L}$ is obtained based on the sub-array snapshots.

Eigenvalue decomposition of $\mathbf{R}$, along with the Akaike information criteria (AIC) metric [49], or the minimum description length (MDL) metric [50], can be used to identify the number of targets. It should be noted, however, that many ideal assumptions in the deduction of these criteria, including additive white Gaussian noise that is uncorrelated with source signal, and the availability of enough snapshots for an accurate covariance matrix estimation, might not be satisfied in practice. To this end, forward/backward spatial smoothing techniques [51] have been widely applied to estimate the subarray covariance matrix for subspace methods–based angle estimation. When the subarray length is shorter, more subarrays are available for covariance matrix calculation, and thus a more accurate subspace estimation would be expected. However, when subarray length is shorter, the DOA estimation resolution is reduced, so a tradeoff between obtaining the number of subarrays and length of subarray should be considered [52].

The target angles can be found by identifying the locations of peaks of the MUSIC pseudospectrum [25], $P(\theta_i)$, computed at all possible θ_i's, that is,

$$P(\theta_i) = \frac{1}{\mathbf{a}_L^H(\theta_i)\mathbf{U}_n\mathbf{U}_n^H\mathbf{a}_L(\theta_i)} \tag{4.7}$$

where $\mathbf{U}_n$ is the noise subspace of $\mathbf{R}$ and $\mathbf{a}_L(\theta_i)$ is the array steering vector of length L corresponding to search direction θ_i. The computation cost of the MUSIC algorithm is high due to the angle search process. Alternatively, the ESPRIT algorithm could be used for angle estimation [26]. ESPRIT is also a subspace method, which exploits the array shift invariance properties, and has been widely used in practice. It has lower complexity than MUSIC, which, however, comes at the cost of reduced angular resolution. ESPRIT requires $2L$ sensors with $L > K$. To achieve the same angular resolution as MUSIC, ESPRIT needs twice as many sensors as MUSIC. As 2-D arrays are needed in automotive radar in order to estimate both azimuth and elevation angles, 2-D ESPRIT algorithms [29] can be applied if the array element spacing is uniform rectangular.

4.3.1.2 Compressive Sensing

MIMO radars equipped with millimeter wave technology offer wide bandwidth, thus achieving high range resolution. As a result, there are only a small number of targets that fall in the same range-Doppler bin, and thus, the targets are sparse in the DOA space. This property can be exploited by sparse sensing based high resolution methods for target angle estimation. To apply compressive sensing for DOA estimation, the whole DOA field of view is discretized into a fine grid. Assume that the DOA space is discretized on a fine grid with N points and there are K targets on the grid. The array response in (4.5) can be rewritten as

$$\mathbf{y} = \mathbf{A}\mathbf{x} + \mathbf{n} \tag{4.8}$$

where $\mathbf{A} = [\mathbf{a}(\theta_1), \cdots, \mathbf{a}(\theta_N)]$ is the basis matrix with $\mathbf{a}(\theta_i)$ denoting the array steering vector corresponding to the i-th grid point and $\mathbf{x} = [\beta_1, \beta_2, \cdots, \beta_N]^T$ is a sparse vector with K nonzero elements. The value of β_i is nonzero if there is a target at the i-th grid point. The coherence of the basis matrix, defined as

$$\mu \triangleq \max_{i \neq l} \frac{\left|\mathbf{a}^H(\theta_i)\mathbf{a}(\theta_l)\right|}{\left\|\mathbf{a}(\theta_i)\right\|_{\ell_2}\left\|\mathbf{a}(\theta_l)\right\|_{\ell_2}} \tag{4.9}$$

needs to be low for obtaining uniform recovery guarantees [53]. When meeting the required coherence conditions, the DOA can be found by solving a ℓ_1 norm optimization problem, such as Dantzig selector [54] defined here:

$$\begin{aligned} &\min \ \|\mathbf{x}\|_{\ell_1} \\ &\text{s.t.} \ \left\|\mathbf{A}^H(\mathbf{y} - \mathbf{A}\mathbf{x})\right\|_{\ell_\infty} < \eta \end{aligned} \tag{4.10}$$

or greedy methods, such as orthogonal matching pursuit (OMP) [55].

In the previous formulation, targets are assumed to be on the grid, which is not always possible in practice. While one can make the grid finer in order to capture the targets, the coherence of matrix **A** would increase, which would make the ℓ_1 norm solution invalid [56]. Thus, the performance of compressive sensing–based methods is sensitive to targets appearing off the grid [57]. Sparse sensing– and matrix completion–based methods [37, 38] can avoid grid issues without sacrificing the high resolution performance.

4.3.1.3 IAA

The covariance matrix of M array snapshots $\mathbf{y}_l$, $l = 1, \cdots, M$, can be written as $\mathbf{R} = \mathbf{A}(\theta)\mathbf{P}\mathbf{A}^H(\theta)$, where $\mathbf{P}$ is a $K \times K$ diagonal matrix whose diagonal elements contains the power of target reflections. Angle finding in the IAA algorithm [39, 40] is carried out by iteratively estimating the reflection coefficient β_k. The estimate is found by minimizing the weighted least square (WLS) cost function $\sum_{l=1}^{M} \|\mathbf{y}_l - \beta_k(l)\mathbf{a}(\theta_k)\|^2 \mathbf{Q}^{-1}(\theta_k)$, where $\|x\|^2 \mathbf{Q}^{-1}(\theta_k) = \mathbf{x}^H\mathbf{Q}^{-1}(\theta_k)\mathbf{x}$ and the interference and noise covariance matrix $\mathbf{Q}(\theta_k) = \mathbf{R} - \hat{P}_k\, \mathbf{a}(\theta_k)\mathbf{a}^H(\theta_k)$. The solution is given by [39]

$$\hat{\beta}_k(l) = \frac{\mathbf{a}^H(\theta_k)\mathbf{R}^{-1}\mathbf{y}_l}{\mathbf{a}^H(\theta_k)\mathbf{R}^{-1}\mathbf{a}(\theta_k)} \tag{4.11}$$

Then matrix $\mathbf{P}$ can be updated as $\hat{P}_k = \frac{1}{M}\sum_{l=1}^{M}\left|\hat{\beta}_k(l)\right|^2$. In IAA algorithm implementation, the DOA space is discretized into a fine grid of N points and steering matrix $\mathbf{A}$ is constructed in the same way as in compressive sensing. In addition, a standard delay and sum (DAS) beamformer is used to initialize $\mathbf{P}$,

$$\hat{P}_k = \frac{\sum_{l=1}^{M}\left|\mathbf{a}^H(\theta_k)\mathbf{y}_l\right|^2}{M\left|\mathbf{a}^H(\theta_k)\mathbf{a}(\theta_k)\right|^2} \tag{4.12}$$

4.3.2 High Resolution Angle Finding with SLA

As previously stated, the cost of synthesizing a large virtual ULA of D elements with half wavelength element spacing is very high. One way to further reduce the cost without sacrificing the high angular resolution is via the use of nonuniform, or sparse linear arrays [43, 44, 58]. With MIMO radar technology, $M_tM_r < D$ virtual array elements can be synthesized. To make sparse linear array (SLA) aperture the same as the ULA, two virtual array elements should be deployed at the edge locations of the ULA. For the remaining virtual array elements, there are multiple possibilities to deploy. The main issue with SLA is that the grating lobes may introduce ambiguity in angle finding. In that context, the key problems are how to select the locations of array elements such that the peak sidelobe level (PSL) of the virtual SLA beampattern is low, and how to carry out angle finding. There is no analytical solution to determining the antenna locations that achieve a minimum PSL for a given number of antennas [59]. Optimal sparse array design requires global optimization techniques, such as particle swarm optimization [44, 60, 61].

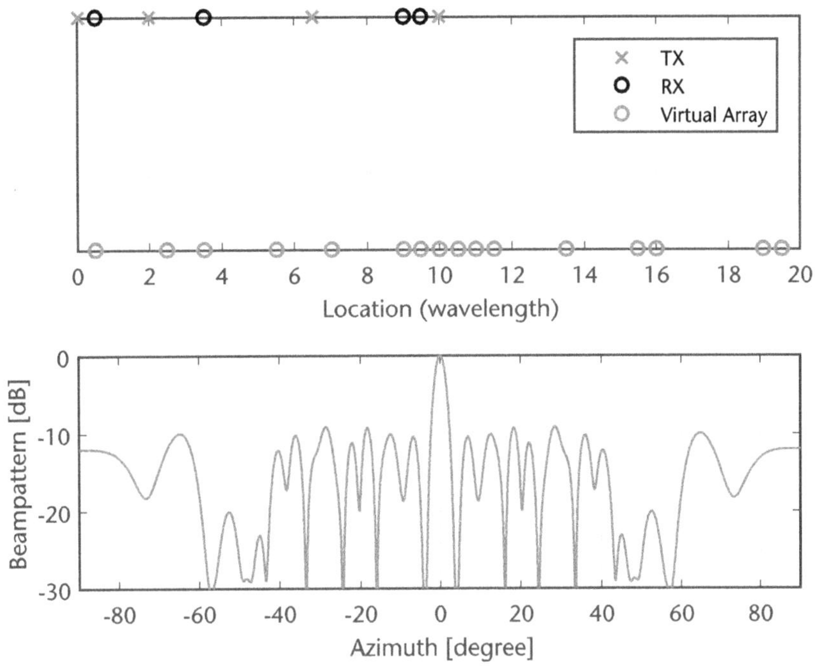

Figure 4.6 Simulated example of an SLA synthesized with MIMO radar technology using four transmit and four receive antennas. The physical limitation of array aperture is 10λ. We fix the locations of first and fourth transmit/receive antennas at $0\lambda 0.5\lambda$ and $10\lambda 9.5\lambda$, respectively such that a maximum virtual array aperture of 19λ is achieved. The remaining transmit/receive antennas are chosen such that the PSL of the synthetic virtual array beampattern is –9.1 dB. (© 2022 IEEE. Reprinted, with permission, from [5].)

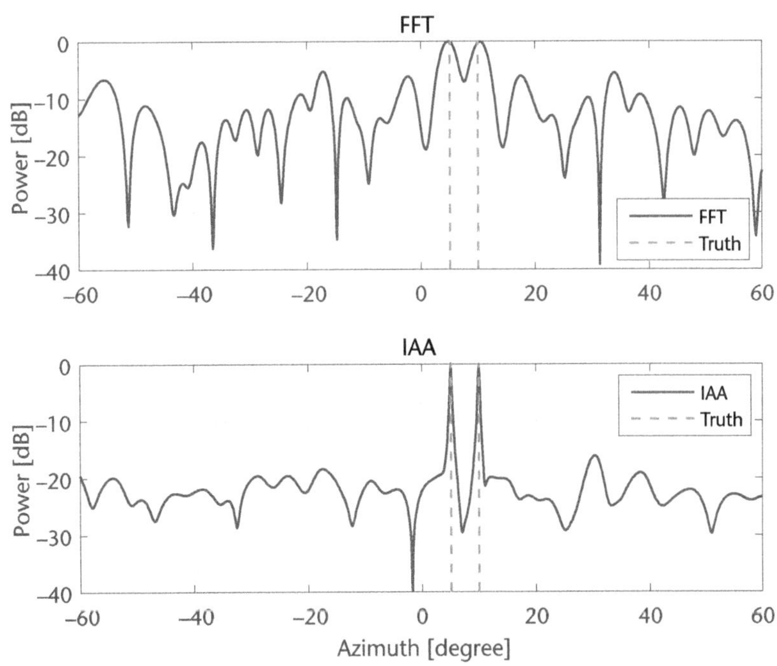

Figure 4.7 Angle finding via the IAA method and the FFT spectrum using the SLA of Figure 4.6, synthesized by MIMO radar technology. There are two targets with azimuth of 5° and 10°. The SNR is set to 30 dB. (© 2022 IEEE. Reprinted, with permission, from [5].)

In automotive MIMO radar with virtual SLA, angle finding can still be done with conventional FFT or ESPRIT methods if the holes in the virtual SLA can be filled via interpolation or extrapolation techniques to mitigate the grating lobes [62, 63]. Alternatively, instead of filling the holes, angle finding of the sparse array can be done using spatial compressive sensing (CS) ideas [64]. In the SLA scenario, it can be easily verified that the coherence of the basis matrix (see (4.9)), is the PSL of the SLA array beampattern [43]. Therefore, the coherence, or equivalently, the PSL of a sparse array plays a key role in obtaining uniform recovery guarantees for compressive sensing [53]. If the PSL of SLAs is low, angle finding using SLAs can be done via compressive sensing or IAA.

In Figure 4.6, we give an example of a virtual SLA with aperture of 19λ, synthesized with MIMO radar technology using four transmit and four receive antennas. The first and fourth of transmit/receive antennas are deployed at the edge of the physical aperture, while the remaining antennas are chosen such that the PSL is −9.1 dB. Angle estimation via IAA when using the sparse linear array of Figure 4.6 is illustrated in Figure 4.7. The ground truth involves two targets with azimuth 5° and 10°. The SNR of the received beamvector is set to 30 dB. For comparison, the FFT spectrum is also plotted. One can observe the sharper peaks around the target azimuth angles and the more attenuated sidelobe in the IAA spectrum, as compared to the FFT spectrum.

4.4 High Resolution Imaging Radar for Autonomous Driving

Currently, many of the automotive radar transceivers designed for ADAS functionality, such as MR3003 of NXP Semiconductor, and AWR2243 of Texas Instruments, can support up to three transmit and four receive antennas. Therefore, using a single automotive radar transceiver with MIMO radar technology, only 12 virtual array elements can be synthesized. In this section, we introduce the high resolution imaging radar with hundreds of virtual antenna array elements synthesized via MIMO radar technology.

4.4.1 Cascade of Multiple Radar Transceivers

To meet the requirement for Level 4 and Level 5 autonomous driving, multiple automotive radar transceivers would need to be cascaded together, with all transceivers synchronized as a single unit. The received data from all receive antennas would be processed coherently. Cascading provides a cost effective and scalable solution to achieve high angular resolution. In [65], General Motors and Texas Instruments successfully demonstrated that up to 4 Texas Instruments AWR1243 radar chips can be cascaded together to provide 12 transmit and 16 receive antennas, enabling a synthesis of 192 virtual array elements. In [66], a prototype of cascading 5 Infineon radar chips has been built to synthesize a virtual array of 128×4 elements. Such high number of virtual array elements provides a lot of opportunities in array design. Several azimuth and elevation arrays configurations can be found in [62]. Several commercial imaging radar products are available with different array configurations, such as forward-looking full-range radar of ZF and ARS540 of Continental

[67, 68]. Usually, a trade-off of balancing angular resolution in azimuth and elevation needs to be considered.

4.4.2 Examples of Cascaded Imaging Radars

Figure 4.8 shows an imaging radar design reference board that has 12 transmit and 16 receive antennas, formed by cascading 4 Texas Instruments AWR2243 radar transceivers [69]. The azimuth FOV is [−70°, 70°]. One transceiver is selected as master and all the others as slaves for clock distribution. In this way, synchronization can be achieved among 4 transceivers, allowing coherent FMCW transmission from the 12 transmit antennas and joint data processing from the 16 receive antennas. The array configuration of cascaded imaging radar is shown in Figure 4.8. There are three transmit antennas placed along the vertical direction for elevation angle finding, and nine transmit antennas placed along the horizontal direction for azimuth angle finding. The virtual array in the horizontal direction is a dense ULA with half wavelength spacing and consists of 86 virtual array elements (the overlapped virtual array elements are not shown). The array aperture in the azimuth direction is $D_x = 42.5\lambda$. In antenna theory, the 3 dB beamwidth defines the angular resolution. According to [42], the 3 dB beamwidth of the azimuth angle is

$$\Delta\theta_{AZ} - 2\arcsin\left(\frac{1.4\lambda}{\pi D_x}\right) \approx 1.2° \tag{4.13}$$

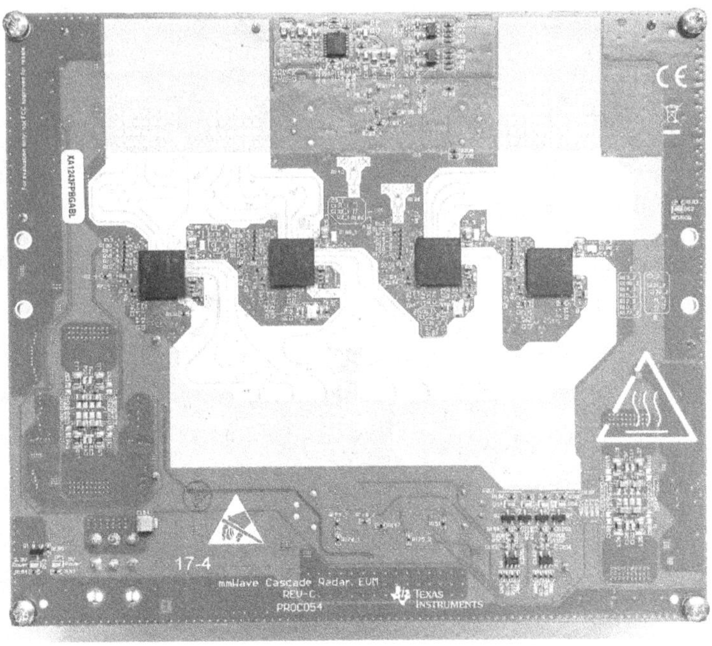

(a)

Figure 4.8 Texas Instruments AWR2243 imaging radar board (a) [69]. Four Texas Instruments AWR2243 radar transceivers are cascaded together, providing 12 transmit and 16 receive antennas (b), and enabling the synthesis of 192 virtual array elements (c). (© 2022 IEEE. Reprinted, with permission, from [5].)

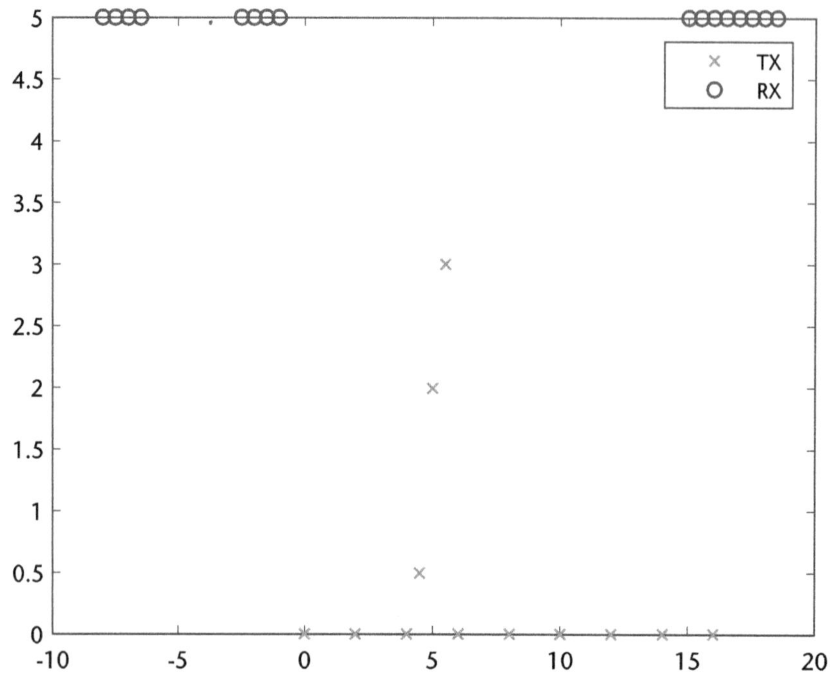

(b)

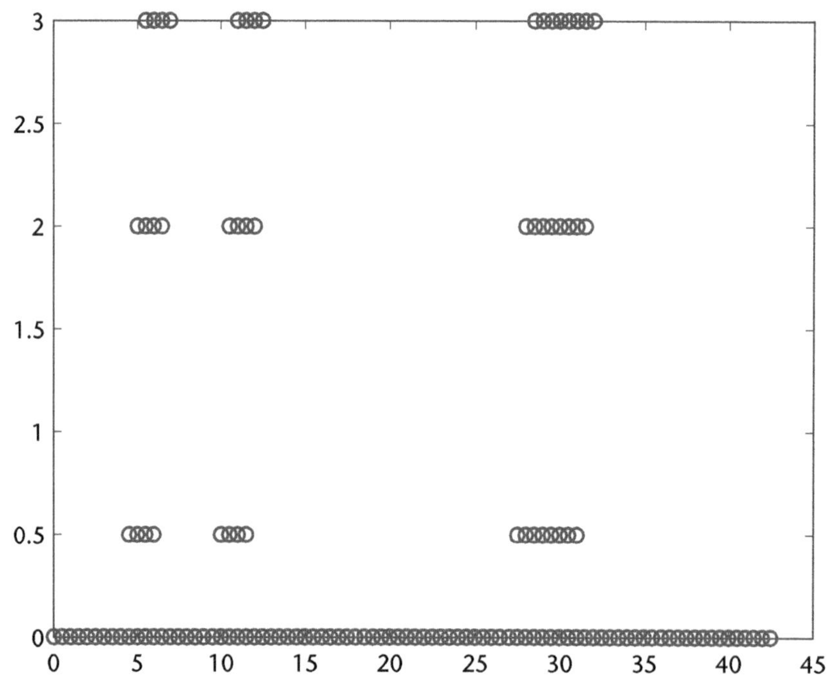

(c)

Figure 4.8 *(continued)*

In the vertical direction, the antennas in three elevation positions form multiple minimum redundancy arrays (MRAs) [70] along the horizontal direction. The angle finding in MRA requires multiple snapshots. These MRAs along the horizontal direction can be used as snapshots for elevation angle finding. The elevation array aperture is $D_y = 3\lambda$ and the 3 dB beamwidth of elevation is

$$\Delta\theta_{\mathrm{EL}} - 2\arcsin\left(\frac{1.4\lambda}{\pi D_y}\right) \approx 17^\circ \tag{4.14}$$

4.4.3 Design Challenges of Imaging Radar

Achieving waveform orthogonality in imaging radars using FMCW with a large number of transmit antennas is quite challenging. One strategy could be to divide the transmit antennas into several subgroups. In each subgroup, the transmit antennas would transmit simultaneously with slow-time phase coding (DDM), while antennas of different subgroups would be scheduled to transmit in different time slots (TDM).

Clock distribution among multiple cascaded transceivers is also challenging. For FMCW mixer operation, an LO is shared among master and slaves, and the LO routing from master to all slaves in the circuit should be matched. Also, the additional ADC sampling and data transmission among different transceivers needs to be synchronized. It is desirable to develop automotive radar transceiver that can incorporate a large number of transmit and receive antennas. For example, Uhnder has developed a radar system-on-chip (SoC), which has 12 transmit and 16 receive antennas, enabling synthesis of 192 virtual array elements [71] (discussed in more detail in Chapter 3). Thus, the four current automotive radar transceivers in the cascaded imaging radar shown in Figure 4.8 can be replaced with a single SoC radar chip. The radar on chip (RoC) developed by Vayyar has 48 transceivers at 76–81 GHz, which can provide synthesis for more than 2,000 virtual array elements [72].

4.5 Challenges in Automotive MIMO Radar

In this section, we discuss the design challenges in automotive MIMO radar, including angle finding in the presence of multipath reflections, waveform orthogonality, and efficient high resolution angle estimation algorithm development.

4.5.1 Angle Finding in the Presence of Multipath Reflections

Automotive radar runs in multipath scenarios [73]. In general, radio propagation in the presence of multipath occurs along four possible routes (i.e., direct/direct, direct/indirect, indirect/direct, and indirect/indirect routes).

Figure 4.9 shows a vertical multipath scenario, where the height of radar and target are h_S and h_T, respectively. The length of the direct/direct path is $d_1 = 2r_1$, the length of the direct/indirect or indirect/direct path is $d_2 = d_3 = r_1 + r_2 + r_3$, and the length of the indirect/indirect path is $d_4 = 2(r_2 + r_3)$. The received signal, having gone through the four paths can be written as

$$y_r = \sum_{i=1}^{4} \beta_i e^{j\frac{2\pi}{\lambda}d_i} \tag{4.15}$$

where the amplitude term β_i is function of the antenna gain, the path loss, the road reflection coefficient, and the target RCS. The signal from the indirect paths and that from the direct path would most probably arrive out of phase and thus add up destructively. As a result, the power of the received signal would fluctuate with distance [73] and thus angle finding at SNR nulls would be unstable.

MIMO radar with colocated transmit and receive antennas, also referred to as monostatic MIMO radar, is based on the assumption that DOD and DOA are equal. However, in the presence of multipath, that assumption does not hold and the system becomes bistatic [74, 75] (i.e., the transmit and receive antenna view the target from different aspect angles). Figure 4.10 shows a vehicle moving parallel to the guardrail, with an SRR sensor mounted at its front left corner. The length of the direct path of the radar signal is $d_{r_1} = r_1$, corresponding to $\theta_t = \theta_r = \theta_1$. There

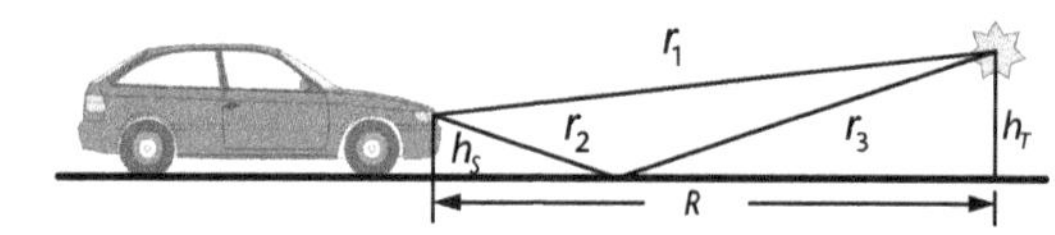

Figure 4.9 Direct path and vertical multipath [73]. (© 2022 IEEE. Reprinted, with permission, from [5].)

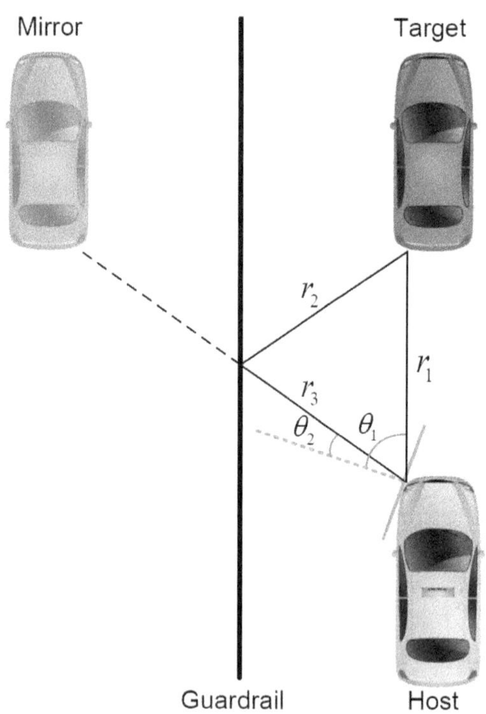

Figure 4.10 A typical multipath reflection scenario along a guardrail for a short range radar sensor mounted at the front left corner. (© 2022 IEEE. Reprinted, with permission, from [5].)

are also multipath reflections due to the guardrail. The range of the first multipath reflection is $d_{r_2} = (r_1 + r_2 + r_3)/2$, corresponding to $\theta_t = \theta_1, \theta_r = \theta_2$ or $\theta_t = \theta_2, \theta_r = \theta_1$. The range of the second multipath reflection is $d_{r_3} = r_2 + r_3$, corresponding to $\theta_t = \theta_r = \theta_2$. Compared to the direct path, multipath reflections result in longer range and smaller Doppler. For the first type of multipath, the range and Doppler bin is the same as in mirror image target detection. However, as $\theta_t \neq \theta_r$, it turns out that the phase of each virtual array element is corrupted. In other words, the monostatic MIMO radar assumption does not hold, which results in a "ghost" target whose direction is different from the mirror target.

To solve this issue, some ideas have been proposed in [74–76]. For example, joint estimation of DOD and DOA is proposed in [74], however, by ignoring the structure of the transmit array, that method cannot not enjoy the benefit of the synthesized virtual array. Polarimetric features are exploited in [75] to separate objects in a multipath scenario. However, the approach in [75] can only separate certain real target cases from their mirror targets, for example, when the real target is known and the polarization state change of the multipaths can be recognized. The method of [75] does not work when the ghost target direction is different than that of the mirror target. Doppler information can be exploited to detect moving vehicles in urban areas under multipath [76]. However, Doppler information is not always available when both objects and host vehicles are stationary. In general, there is a need for more research addressing ghost target issue in MIMO radar due to multipath.

4.5.2 Waveform Orthogonality in Automotive MIMO Radar

Different strategies, such as TDM, DDM, and FDM can be adopted in automotive FMCW radars to achieve waveform orthogonality. However, several challenges associated with each strategy need to be addressed. For example, in the TDM scheme, the scheduling delay between transmit antennas may introduce phase error for a moving target, which needs to be compensated for, otherwise, the synthesized array beampattern will be distorted. Further, the maximum unambiguous detectable velocity under TDM is reduced by a factor of M_t. In the DDM scheme, the Doppler sidelobes are high due to the residual of the phase coding. As a result, targets with small RCSs (e.g., pedestrians), that are close to the target with strong reflections (e.g., trucks) might be masked. The search time for phase codes using stochastic algorithms increases exponentially as the code length increases. Computationally efficient algorithms are needed to address this problem. In the FDM scheme, though a randomization of frequency shift among transmit antennas could reduce the range-angle coupling, a large number of transmit antennas would be needed for the improvement to be notable [15].

Recently, PMCW has been proposed for achieving orthogonality [77–80]. Each antenna transmits a sequence of phase coded pulses. Let

$$x_m = \left[x_m(1), \ldots, x_m\left(N_p\right)\right]^T \tag{4.16}$$

be the complex unimodular code sequence of the m-th transmit antenna, where $x_m(n) = e^{j\varphi_m(n)}$ is the n-th code of $\mathbf{x}_m$ and N_p is the code length. Here, the phase

$\phi_m(n)$ can be chosen arbitrarily in $[-\pi,\pi]$. The duration of a single code sequence is $T_p = N_pT_c$ with T_c the duration of a subpulse. In practice, binary code sequences have been widely used due to their simplicity. The bandwidth of PMCW is $B = 1/T_c$. The time-bandwidth product of a code sequence is $BT_p = N_p$. Since the pulses are transmitted continuously, the code sequences should have good periodic auto-correlation and cross-correlation properties [77]. The periodic cross-correlation of two code sequences $\mathbf{x}_m$ and $\mathbf{x}_l$ at lag k is defined as

$$r_{ml}^{\mathcal{P}}(k) = \sum_{n=1}^{N} x_m(n)x_l^*\left((n+k)\mathrm{mod}\left(N_p\right)\right) \tag{4.17}$$

When $m = l$, $r_{ml}^{\mathcal{P}}(k)$ becomes the periodic auto-correlation function of $\mathbf{x}_m$. Good correlation properties require the values of the periodic autocorrelation at nonzero lags, and the values of the cross-correlation at any lag to be low. The Welch lower bound on the cross-correlation between any pair of binary sequences with period of N_p in a set of M_t sequences equals [81]

$$r_{ml}^{\mathcal{P}}(k) \geq N_p\sqrt{\frac{M_t - 1}{M_tN_p - 1}} \approx \sqrt{N_p} \tag{4.18}$$

Good periodic cross-correlation properties help achieve waveform orthogonality, while good periodic auto-correlation properties make it easier to use matched filters to extract signals reflected from the range bin of interest and suppress signals reflected from other range bins.

As compared to FMCW, PMCW radar has several advantages. PMCW radar is better suited for achieving waveform orthogonality in imaging radars with a large of number of transmit antennas. PMCW radar can take advantage of existing sequences with good autocorrelation and cross-correlation properties, previously developed for code division multiple accessing (CDMA) communications, such as Gold, Kasami, and m-sequences [82–84]. Further, in PMCW radar, each automotive radar sensor can have a unique digital sequence, which may help reduce the automotive radar mutual interference. As a bonus, PMCW radar also provides certain communication capability [85], thus can be explored as a dual functional radar communication system [86].

However, PMCW radar has many implementation challenges. First, the sampling rate of ADC should satisfy the Nyquist rule (i.e., $f_s \geq 2B = 2/T_c$). The high bandwidth required for high range resolution necessitates high speed ADC and high speed processing hardware. In practice, it is required to keep the resolution of ADC as low as possible [87]. Second, according to the Welch bound of (4.18), the cross-correlation lower bound of any pair of binary sequences is of the order of $\mathcal{O}(\sqrt{N_p})$, which might not provide sufficient separation of transmit waveforms of different antennas. In practice, the autocorrelation and cross-correlation of code sequences are desired to have low sidelobes within a low correlation zone (LCZ). Furthermore, because there is no mapping relationship between range and beat signal in PMCW, it would be difficult to use high-pass analog filters to reject or attenuate ultraclose range return signals, including direct path signals from transmit antennas, reflections from radome, and vehicle bumpers. This escalates the dynamic range challenge, especially when the resolution of ADC must be kept as low as possible [87].

4.5.3 Efficient, High Resolution Angle Finding Algorithms Are Needed

A typical duration of a look in automotive radar is around 50ms, corresponding to detection update rate of 20Hz [88]. In such short duration, the current generation of automotive radar for ADAS can report a maximum of 64–200 detections. With high resolution imaging radar, the number of cells that can be selected for angle finding from the 2-D range and Doppler spectrum is around 10,000 in a single look for a typical road scenario [89]. To achieve point clouds for autonomous driving, angle finding needs to be performed thousands of times in a single look, which is a great challenge for imaging radar with hundreds of virtual array elements. Computationally efficient, high resolution angle finding algorithms are highly desirable for real-time implementation in automotive radar.

To reduce computation complexity, beamspace ESPRIT [27] and unitary ESPRIT [28] algorithms have been proposed. The idea of beamspace ESPRIT is to decompose the original ULA vector into several low-dimensional beamspaces via a transform such as the FFT. Then, if the beamspace transform matrix has the same shift invariance structure, angle finding can be carried out via ESPRIT on each beamspace in parallel, with reduced computational time [27]. The unitary ESPRIT algorithm takes advantage of the unit magnitude property of the phase factors representing the phase delays between the two subarrays and is formulated in terms of real-valued computations. As a result, it achieves great reduction of computational complexity [28].

The computation cost of each IAA iteration is $2NM^2 + NM + M^3$, where M is the number of array snapshots and N is the number of discretized grids. Fast and superfast IAA algorithms have been proposed in [90–92], respectively. The fast IAA algorithm exploits the FFT operation as well as Gohberg-Semencul (GS) representation of matrix $\mathbf{R}^{-1}$. As a result, the computation cost of each fast IAA iteration is $M^2 + 12\zeta(2M) + 3\zeta(N)$, where $\zeta(N)$ stands for the computation cost of performing FFT of size N (i.e., $O(N \log N)$) [91]. The superfast IAA uses a conjugate gradient (CG) algorithm to approximate the matrix $\mathbf{R}^{-1}$, which further reduces the computation cost.

The strengths and limitations of each DOA estimation algorithm discussed in Section 4.3 when applied to the automotive radar scenario are summarized in Table 4.1. For subspace-based high resolution DOA estimation methods, such as MUSIC and ESPRIT, the automotive radar array needs to be ULA, and multiple snapshots are required to estimate the array covariance matrix accurately. However, automotive radar operating in a highly dynamic environment typically rely on a single snapshot. While multiple snapshots can be generated via spatial smoothing or by dividing a chirp into subchirps, the associated cost is respectively reduced array aperture, or reduced SNR. SLAs have been widely used in automotive radar to further reduce the hardware cost. However, it is not straightforward to apply MUSIC or ESPRIT to SLA-based automotive radar. On the other hand, DBF and

Table 4.1 Summary of Different DOA Estimation Algorithms in Automotive Radar Scenario

Algorithms	*Resolution*	*Snapshot*	*Arrays*	*Grid Free*	*Rank Estimation*	*Robustness*	*Complexity*
DBF	Low	Single	ULA/SLA	No	No	Strong	Low
MUSIC	High	Multiple	ULA	No	Yes	Medium	High
ESPRIT	High	Multiple	ULA	Yes	Yes	Medium	Medium
OMP	High	Single	ULA/SLA	No	No	Medium	High
IAA	High	Single	ULA/SLA	No	No	Strong	High

sparsity-based high resolution methods, such as OMP and IAA, apply to SLA- as well as ULA-based automotive radar, and work with a single snapshot. In the DBF method, the number of targets can be estimated by counting the number of peaks in the DOA spectrum. DBF is not sensitive to coherent or correlated signals, which in subspace-based methods [93] need special preprocessing via spatial smoothing. It has been shown that DBF is robust to array element position errors and has low computational cost [93]. However, DBF is not a high resolution method. Also, as OMP and IAA are iterative schemes, they involve high computation cost, which limits their applicability in low-cost embedded DSPs typically used in current generation automotive radar. Further, the methods of DBF, MUSIC, OMP, and IAA assume targets are on the grid, and suffer from errors when the targets arise between grid points. In summary, more research is needed on developing computationally efficient, high resolution DOA estimation algorithms that are robust to noise and are applicable to automotive radar using SLAs with low PSL under a single snapshot.

References

[1] Qi, C. R., et al., "PointNet: Deep Learning on Point Sets for 3D Classification and Segmentation," *Proceedings of the IEEE Conference on Computer Vision and Pattern Recognition (CVPR)*, Honolulu, HI, July 2017.

[2] Qi, C. R., Yil, L., Su, H., and L. J. Guibas, "PointNet++: Deep Hierarchical Feature Learning on Point Sets in a Metric Space," *Proceedings of the 31st Conference on Neural Information Processing Systems (NIPS)*, Long Beach, CA, December 2017.

[3] Brisken, S., Ruf, F., and F. Hohne, "Recent Evolution of Automotive Imaging Radar and Its Information Content," *IET Radar, Sonar & Navigation*, Vol. 12, No. 10, 2018, pp. 1078–1081.

[4] Danzer, A., Griebel, T., Bach, M., and K. Dietmayer, "2D Car Detection in Radar Data with PointNets," *arXiv:1904.08414*, July 2019. in *Proc. IEEE Intelligent Transportation Systems Conference (ITSC)*, Auckland, New Zealand, October 2019.

[5] Sun, S., Petropulu, A. P., and H. V. Poor, "MIMO Radar for Advanced Driver-Assistance Systems and Autonomous Driving: Advantages and Challenges," *IEEE Signal Process. Mag.*, Vol. 37, No. 4, 2020, pp. 98–117.

[6] Bergin, J., and J. R. Guerci, *MIMO Radar: Theory and Application*, Norwood, MA: Artech House, 2018.

[7] Duly, A., Love, D., and J. Krogmeier, "Time-Division Beamforming for MIMO Radar Waveform Design," *IEEE Trans. Aerosp. Electron. Syst.*, Vol. 49, No. 2, 2013, pp. 1210–1223.

[8] Rambach, K., and B. Yang, "Colocated MIMO Radar: Cramer-Rao Bound and Optimal Time Division Multiplexing for DOA Estimation of Moving Targets," in *Proceedings of the IEEE 38th Int. Conf. on Acoustics, Speech, and Signal Processing (ICASSP)*, Vancouver, Canada, May 2013.

[9] Zoeke, D., and A. Ziroff, "Phase Migration Effects in Moving Target Localization Using Switched MIMO Arrays," *Proceedings of the European Radar Conference (EuRAD)*, Paris, France, September 2015.

[10] Bleh, D., and et al., "W-band Time-Domain Multiplexing FMCW MIMO Radar for Farfield 3-D Imaging," *IEEE Trans. Microw. Theory Tech.*, Vol. 65, No. 9, 2017, pp. 3474–3484.

[11] Bechter, J., Roos, F., and C. Waldschmidt, "Compensation of Motion-Induced Phase errors in TDM MIMO Radars," *IEEE Microw. Wireless Compon. Lett.*, Vol. 27, No. 12, 2017, pp. 1164–1166.

[12] Jansen, F., Filippi, A., and Z. Zivkoic, "MIMO Radar System," European Patent EP3021132A1, May 18, 2016.

[13] Roos, F., Bechter, J., Appenrodt, N., Dickmann, J., and C. Waldschmidt, "Enhancement of Doppler Unambiguity for Chirp-Sequence Modulated TDM-MIMO Radars," *Proceedings of the Intl. Conference on Microwaves for Intelligent Mobility (ICMIM)*, Munich, Germany, April 2018.

[14] Mecca, V. F., Ramakrishnan, D., and J. L. Krolik, "MIMO Radar Space-Time Adaptive Processing for Multipath Clutter Mitigation," in *Proceedings of the IEEE Workshop on Sensor Array and Multichannel Processing (SAM)*, Waltham, MA, July 2006.

[15] Sun, H., Brigui, F., and M. Lesturgie, "Analysis and Comparison of MIMO Radar Waveforms," in *Proceedings of the Intl. Radar Conference*, Lille, France, October 2014.

[16] Madsen, N., and S. Cao, "Slow-Time Waveform Design for MIMO GMTI Radar Using CAZAC Sequences," in *Proceedings of the IEEE Radar Conference*, Oklahoma City, OK, April 2018.

[17] Benedetto, J., Konstantinidis, I., and M. Rangaswamy, "Phase-Coded Waveforms and Their Design," *IEEE Signal Process. Mag.*, Vol. 26, No. 1, 2009, pp. 22–31.

[18] Chu, D., "Polyphase Codes with Good Periodic Correlation Properties," *IEEE Trans. Inf. Theory*, Vol. 18, No. 4, 1972, pp. 531–532.

[19] Feger, R., Haderer, H., and A. Stelzer, "Optimization of Codes and Weighting Functions for Binary Phase-Coded FMCW MIMO Radars," in *Proceedings of the Intl. Conference on Microwaves for Intelligent Mobility (ICMIM)*, San Diego, CA, May 2016.

[20] Feger, R., Pfeffer, C., and A. Stelzer, "A Frequency-Division MIMO FMCW Radar System Based on Delta-Sigma Modulated Transmitters," *IEEE Trans. Microw. Theory Tech.*, Vol. 62, No. 12, 2014, pp. 3572–3581.

[21] Hacker, P., and B. Yang, "Single Snapshot DOA Estimation," *Advances in Radio Science*, Vol. 8, 2010, pp. 251–256.

[22] Barton, P., "Digital Beam Forming for Radar," *IEE Proceedings F—Communications, Radar and Signal Processing*, Vol. 127, No. 4, 1980.

[23] Alland, S., and J. Searcy, "Radar System and Method of Digital Beamforming," U.S. Patent 2009/0085800), April 2, 2009.

[24] Tokoro, S., Kuroda, K., Kawakubo, A., Fujita, K., and H. Fujinami, "Electronically Scanned Millimiter-Wave Radar for Precrash Safety and Adaptive Cruise Control System," *Proceedings of the IEEE Intelligent Vehicles Symposium*, Columbus, OH, June 2003.

[25] Schmidt, R., "Multiple Emitter Location and Signal Parameter Estimation," *IEEE Trans. Antennas Propag.*, Vol. 34, No. 3, 1986, pp. 276–280.

[26] Roy, R., and T. Kailath, "ESPRIT—Estimation of Signal Parameters via Rotation Invariance Techniques," *IEEE Trans. Acoust., Speech, Signal Process.*, Vol. 17, No. 7, 1989, pp. 984–995.

[27] Xu, G., Silverstein, S. D., Roy, R. H., and T. Kailath, "Beamspace ESPRIT," *IEEE Trans. Signal Process.*, Vol. 42, No. 2, 1994, pp. 349–356.

[28] Haardt, M., and J. Nossek, "Unitary ESPRIT: How to Obtain Increased Estimation Accuracy with a Reduced Computational Burden," *IEEE Trans. Signal Process.*, Vol. 43, No. 5, 1995, pp. 1232–1242.

[29] Zoltowski, M. D., Haardt, M., and C. Mathews, "Closed-Form 2-D Angle Estimation with Rectangular Arrays in Element Space or Beamspace via Unitary ESPRIT," *IEEE Trans. Signal Process.*, Vol. 44, No. 2, 1996, pp. 316–328.

[30] Strohmer, T., and B. Friedlander, "Compressed Sensing for MIMO Radar—Algorithms and Performance," *Proceedings of the 43th Annual Asilomar Conference on Signals, Systems, and Computers*, Pacific Grove, CA, November 2009.

[31] Yu, Y., Petropulu, A. P., and H. V. Poor, "MIMO Radar Using Compressive Sampling," *IEEE J. Sel. Topics Signal Process.*, Vol. 4, No. 1, 2010, pp. 146–163.

[32] Yu, Y., Petropulu, A. P., and H. V. Poor, "Measurement Matrix Design for Compressive Sensing-Based MIMO Radar," *IEEE Trans. Signal Process.*, Vol. 59, No. 11, 2011, pp. 5338–5352.

[33] Yu, Y., Petropulu, A. P., and H. V. Poor, "CSSF MIMO Radar: Compressive-Sensing and Step-Frequency Based MIMO Radar," *IEEE Trans. Aerosp. Electron. Syst.*, Vol. 48, No. 2, 2012, pp. 1490–1504.

[34] Yu, Y., Sun, S., Madan, R. N., and A. P. Petropulu, "Power Allocation and Waveform Design for the Compressive Sensing Based MIMO Radar," *IEEE Trans. Aerosp. Electron. Syst.*, Vol. 50, No. 2, 2014, pp. 898–909.

[35] Strohmer, T., and B. Friedlander, "Analysis of Sparse MIMO Radar," *Appl. Comp. Harm. Anal.*, Vol. 37, 2014, pp. 361–388.

[36] Sun, S., "MIMO Radars with Sparse Sensing," PhD Dissertation, Rutgers University, New Brunswick, NJ, 2016.

[37] Sun, S., Bajwa, W. U., and A. P. Petropulu, "MIMO-MC Radar: A MIMO Radar Approach Based on Matrix Completion," *IEEE Trans. Aerosp. Electron. Syst.*, Vol. 51, No. 3, 2015, pp. 1839–1852.

[38] Sun, S., and A. P. Petropulu, "Waveform Design for MIMO Radars with Matrix Completion," *IEEE J. Sel. Topics Signal Process.*, Vol. 9, No. 8, 2015, pp. 1400–1411.

[39] Yardibi, T., Li, J., Stoica, P., Xue, M., and A. Baggeroer, "Source Localization and Sensing: A Nonparametric Iterative Adaptive Approach Based on Weighted Least Squares," *IEEE Trans. Aerosp. Electron. Syst.*, Vol. 46, No. 1, 2010, pp. 425–443.

[40] Roberts, W., Stoica, P., Li, J., Yardibi, T., and F. Sadjadi., "Iterative Adaptive Approaches to MIMO Radar Imaging," *IEEE J. Sel. Topics Signal Process.*, Vol. 4, No. 1, 2010, pp. 5–20.

[41] Shan, T., Wax, M., and T. Kailath, "On Spatial Smoothing for Direction-of-Arrival Estimation of Coherent Signals," *IEEE Trans. Acoust., Speech, Signal Process.*, Vol. 33, No. 4, 1985, pp. 506–811.

[42] Richards, M. A., *Fundamentals of Radar Signal Processing*, Second Edition, New York: McGraw-Hill, 2014.

[43] Carin, L., Liu, D., and B. Guo, "Coherence, Compressive Sensing, and Random Sensor Arrays," *IEEE Antennas Propag. Mag.*, Vol. 53, No. 4, 2011, pp. 28–39.

[44] Schmid, C., Feger, R., Wagner, C., and A. Stelzer, "Design of a Linear Non-uniform Antenna Array for a 77-GHz MIMO FMCW Radar," *Proceedings of the IEEE MTT-S Intl. Microwave Workshop on Wireless Sensing, Local Positioning, and RFID*, Cavtat, Croatia, September 2009.

[45] Alcalde, C. A., "Radar System and Method for Virtual Antenna Signals," U.S. Patent 9 664 775 B2, May 30, 2017.

[46] Li, Z., and C. A. Alcalde, "Angle Finding for a Detector Having a Paired Staggered Array," U.S. Patent 2018/0267555 A1, September 20, 2018.

[47] Alland, S., "MIMO Antenna with Improved Grating Lobe Characteristics," U.S. Patent 2015/0253419 A1, September 10, 2015.

[48] Searcy, J., and S. Alland, "MIMO Antenna with Elevation Detection," U.S. Patent 9 541 639 B2, January 10, 2017.

[49] Akaike, H., "A New Look at the Statistical Model Identification," *IEEE Trans. Acoust., Speech, Signal Process.*, Vol. 37, No. 1, 1989, pp. 8–15.

[50] Barron, Y., Rissanen, J., and B. Yu, "The Minimum Description Length Principle in Coding and Modeling," *IEEE Trans. Inf. Theory*, Vol. 44, No. 6, 1998, pp. 2743–2760.

[51] Pillai, S. U., and B. H. Kwon, "Forward/Backward Spatial Smoothing Techniques for Coherent Signal Identification," *IEEE Trans. Acoust., Speech, Signal Process.*, Vol. 37, No. 1, 1989, pp. 8–15.

[52] Ertan, A. E., and M. Ali, "Spatial and Temporal Smoothing for Covariance Estimation in Super-Resolution Angle Estimation in Automotive Radars," *Proceedings of the IEEE Int. Conf. Acoust., Speech, Signal Process. (ICASSP)*, Barcelona, Spain, May 4–8, 2020.

[53] Candes, E. J., and J. Romberg, "Sparsity and Incoherence in Compressive Sampling," *Inverse Problems*, Vol. 23, 2007, pp. 969–985.

[54] Candes, E. J., and T. Tao, "The Dantzig Selector: Statistical Estimation When `*p* Is Much Larger than *n*," *The Annals of Statistics*, Vol. 35, No. 6, 2007, pp. 2313–2351.

[55] Cai, T. T., and L. Wang, "Orthogonal Matching Pursuit for Sparse Signal Recovery with Noise," *IEEE Trans. Inf. Theory*, Vol. 57, No. 7, 2011, pp. 4680–4688.

[56] Tang, G., Bhaskar, B. N., Shah, P., and B. Recht, "Compressed Sensing off the Grid," *IEEE Trans. Inf. Theory*, Vol. 59, No. 11, 2013, pp. 7465–7490.

[57] Chi, Y., Scharf, L. L., Pezeshki, A., and A. R. Calderbank, "Sensitivity to Basis Mismatch in Compressed Sensing," *IEEE Trans. Signal Process.*, Vol. 59, No. 5, 2011, pp. 2182–2195.

[58] Lo, Y. T., "A Mathematical Theory of Antenna Arrays with Randomly Spaced Elements," *IEEE Trans. Antennas Propag.*, Vol. 12, No. 3, 1964, pp. 257–268.

[59] Feger, R., Wagner, C., Schuster, S., Scheiblhofer, S., Jager, H., and A. Stelzer, "A 77-GHz FMCW MIMO Radar Based on an SiGe Single Chip Transceiver," *IEEE Trans. Microw. Theory Tech.*, Vol. 57, No. 5, 2009, pp. 1020–1035.

[60] Jin, N., and Y. Rahmat-Samii, "Advances in Particle Swarm Optimization for Antenna Designs: Real-Number, Binary, Single-Objective and Multiobjective Implementations," *IEEE Trans. Antennas Propag.*, Vol. 55, No. 3, 2007, pp. 556–567.

[61] Gonalez-Huici, M., Mateos-Nunez, D., Greiff, C., and R. Simoni, "Constrained Optimal Design of Automotive Radar Arrays Using the Weiss-Weinstein Bound," *Proceedings of the. Intl. Conference on Microwaves for Intelligent Mobility (ICMIM)*, Munich, Germany, April 2018.

[62] Alland, S., et al., "Virtual Radar Configuration for 2D Array," U.S. Patent 9 869 762, January 16, 2018.

[63] Spreng, T., et al., "Wideband 120 GHz to 140 GHz MIMO Radar: System Design and Imaging Results," *Proceedings of the European Microwave Conference (EuMC)*, Paris, France, September 2015.

[64] Rossi, M., Haimovich, A., and Y. Eldar, "Spatial Compressive Sensing for MIMO Radar," *IEEE Trans. Signal Process.*, Vol. 62, No. 2, 2014, pp. 419–430.

[65] Bilik, I., et al., "Automotive Multi-mode Cascaded Radar Data Processing Embedded System," in *Proc. IEEE Radar Conference*, Oklahoma City, OK, April 2018.

[66] Och, A., Pfeffer, C., Schrattenecker, J., Schuster, S., and R. Weigel, "A Scalable 77 GHz Massive MIMO FMCW Radar by Cascading Fully-Integrated Transceivers," *Proceedings of the Asia-Pacific Microwave Conference (APMC)*, Kyoto, Japan, November 6–9, 2018.

[67] https://www.zf.com/products/en/cars/products58368.html, May 2021.

[68] https://www.continental-automotive.com. Accessed May 2021.

[69] Texas Instruments Inc., "Design Guide: TIDEP-01012 Imaging Radar Using Cascaded mmWave Sensor Reference Design (REV. A)," https://www.ti.com/lit/ug/tiduen5a/tiduen5a.pdf, March 2020.

[70] Moffet, A., "Minimum-Redundancy Linear Arrays," *IEEE Trans. Antennas Propag.*, Vol. 16, No. 2, 1968, pp. 172–175.

[71] Giannini, V., et al., "A 192-Virtual-Receiver 77/79GHz GMSK Code-Domain MIMO Radar System-on-Chip," *Proceedings of the IEEE Intl. Solid-State Circuits Conference (ISSCC)*, San Francisco, CA, February 17–21, 2019.

[72] https://vayyar.com/. Accessed May 2021.

[73] Ioffe, A., Doerr, W., Yan, H., Vu, D. H., and A. H. Arage, "RCS Characteristics of Street Curbs and the Applications in Automotive Radar Classification," *Proceedings of the European Radar Conference (EuRAD)*, London, U.K., October 2016.

[74] Engels, F., Wintermantel, M., and P. Heidenreich, "Automotive MIMO Radar Angle Estimation in the Presence of Multipath," *Proceedings of the European Radar Conference (EuRAD)*, Nuremberg, Germany, October 2017.

[75] Visentin, T., Hasch, J., and T. Zwick, "Analysis of Multipath and DOA Detection Using a Fully Polarimetric Automotive Radar," *Proceedings of the European Radar Conference (EuRAD)*, Nuremberg, Germany, October 2017.

[76] Fertig, L. B., Baden, J. M., and J. R. Guerci, "Knowledge-Aided Processing for Multipath Exploitation Radar (MER)," *IEEE Aerosp. Electron. Syst. Mag.*, Vol. 32, No. 10, 2017, pp. 24–36.

[77] Thillo, W. V., et al., "Almost Perfect Auto-correlation Sequences for Binary Phase-Modulated Continuous Wave Radar," *Proceedings of the European Radar Conference (EuRAD)*, Nuremberg, Germany, October 2013.

[78] Giannini, V., et al., "A 79 GHz Phase-Modulated 4 GHz-BW CW Radar Transmitter in 28 nm CMOS," *IEEE J. Solid-State Circuits*, Vol. 49, No. 12, 2014, pp. 2925–2937.

[79] Bourdoux, A., Ahmad, U., Guermandi, D., Brebels, S., Dewilde, A., and W. V. Thillo, "PMCW Waveform and MIMO Technique for a 79 GHz CMOS Automotive Radar," *Proceedings of the IEEE Radar Conference*, Philadelphia, PA, May 2016.

[80] Guermandi, D., et al., "A 79-GHz 2x2 MIMO PMCW Radar SoC in 28-nm CMOS," *IEEE J. Solid-State Circuits*, Vol. 52, No. 10, 2017, pp. 2613–2626.

[81] Welch, L., "Lower Bounds on the Maximum Cross Correlation of Signals (corresp.)," *IEEE Trans. Inf. Theory*, Vol. 20, No. 3, 1974, pp. 397–399.

[82] Gold, R., "Optimal Binary Sequences for Spread Multiplexing (corresp.)," *IEEE Trans. Inf. Theory*, Vol. 13, No. 4, 1967, pp. 619–621.

[83] Kasami, T., "Weight Distribution Formula for Some Class of Codes," *Report*, 1966, pp. 1–32.

[84] Dinan, E. H., and B. Jabbari, "Spreading Codes for Direct Sequence CDMA and Wideband CDMA Cellular Networks," *IEEE Commun. Mag.*, Vol. 36, No. 9, 1998, pp. 48–54.

[85] Hueber, G., and A. M. Niknejad, *Millimeter-Wave Circuits for 5G and Radar*, Cambridge, U.K.: Cambridge University Press, 2019.

[86] Liu, F., Masouros, C., Petropulu, A. P., Griffiths, H., and L. Hanzo, "Joint Radar and Communication Design: Applications, State-of-the-Art, and the Road Ahead," *IEEE Trans. Commun.*, forthcoming.

[87] Thillo, W. V., Giannini, V., Guermandi, D., Brebels, S., and A. Bourdoux, "Impact of ADC Clipping and Quantization on Phase-Modulated 79 GHz CMOS Radar," *Proceedings of the European Radar Conference (EuRAD)*, Rome, Italy, October 2014.

[88] Rohling, H., and M.-M. Meinecke, "Waveform Design Principles for Automotive Radar Systems," *Proceedings of the CIE Intl. Conference on Radar*, Beijing, China, October 2001.

[89] Meinl, F., Stolz, M., Kunert, M., and H. Blume, "An Experimental High Performance Radar System for Highly Automated Driving," *Proceedings of the Intl. Conference on Microwaves for Intelligent Mobility (ICMIM)*, Nagoya, Japan, March 2017.

[90] Xue, M., Xu, L., and J. Li, "IAA Spectral Estimation: Fast Implementation Using the Gohberg-Semencul Factorization," *IEEE Trans. Signal Process.*, Vol. 59, No. 7, 2011, pp. 3251–3261.

[91] Glentis, G. O., and A. Jakobsson, "Efficient Implementation of Approach Spectral Estimation Techniques," *IEEE Trans. Signal Process.*, Vol. 59, No. 9, 2011, pp. 4154–4167.

[92] Glentis, G. O., and A. Jakobsson, "Superfast Approximative Implementation of the IAA Spectral Estimate," *IEEE Trans. Signal Process.*, Vol. 60, No. 1, 2012, pp. 472–478.

[93] Godara, L. C., "Application of Antenna Arrays to Mobile Communications. II. Beamforming and Direction-of-Arrival Considerations," *Proceedings of the IEEE*, Vol. 85, No. 8, 1997, pp. 1195–1245.

CHAPTER 5

Synthetic Aperture Radar for Automotive Applications

Antonio Puglielli, Griffin Foster, Darsh Ranjan, Jimmy Wang, and Vinayak Nagpal

5.1 Introduction

Synthetic aperture radar (SAR) takes advantage of a radar's motion to form a high-resolution image requiring only a single transmit–receive antenna pair (virtual antenna). In this chapter, we describe an automotive SAR radar that handles all types of driving conditions, runs in real-time, and uses low-cost and off-the-shelf sensors and processors. The novel SAR mode in this radar runs in parallel to and simultaneously with a traditional point cloud signal processing architecture. Taken together, these two processing modes deliver a new level of performance for automotive radar—the SAR operation gives unprecedented resolution for stationary objects while the point cloud mode excels at imaging moving objects. The SAR technique presented here can be considered a foundational building block for next-generation automotive radar.

SAR processing is first compared to traditional radar processing, and we discuss the potential applications for SAR imaging for automotive use. In Section 5.2, we detail the mathematical foundation for performing SAR. In Section 5.3, we cover the practical issues for implementing an automotive SAR system. In Sections 5.4 and 5.5, we reflect on what has been accomplished as well as identify future paths of development.

5.1.1 Historical Background

SAR was born in the 1950s out of a desire to build airborne high-resolution radar imaging and surveillance systems. The key challenge was that the air-craft size limited the antenna's physical size and therefore its resolution. In 1951, Carl Wiley at the Goodyear Aircraft Company proposed that, by collecting a series of pulses while the aircraft was moving, those pulses could be coherently processed to enhance the radar resolution [1]. Beginning with this initial concept in the 1950s, engineers continued pushing the development of airborne SAR systems to achieve greater and greater resolution [1], requiring longer and longer apertures, and consequently more sophisticated processing techniques. In the 1970s, SAR was used for the first time in spaceborne applications with the launch of the Seasat satellite [2]. Since then, SAR has been used extensively in remote sensing both on Earth and in other space sensing applications.

Today, with the increasing relevance and value of automotive radar, it is natural to explore how to adapt SAR techniques to the automotive world. This introduces both theoretical and practical challenges. The automotive environment upends many of the assumptions built into aerospace SAR systems—in terms of number of targets and motion characteristics, among others—which requires revisiting SAR theory. At the same time, an automotive radar system must be produced in very high volume and very low cost.

5.1.2 Comparison to Traditional Radar Systems

Conventional state-of-the-art automotive radar processing relies on a real array of antennas to separate targets along the range, Doppler, azimuth, and elevation dimensions, which we will refer to henceforth as point cloud processing (because the output appears as a group, or "cloud," of points). The Doppler processing step makes the assumption that all objects in the scene, including oneself, travel at constant velocity over a coherent processing interval (CPI), simplifying this processing step to just a slow-time FFT. However, this assumption limits the processing technique to a fixed *time* interval over which this assumption holds, which in turn imposes an upper bound on the achievable Doppler resolution. A typical coherent processing interval is around 20 ms.

In order to resolve multiple targets in a scene, the targets must be separated in either range, Doppler, or angle dimensions. Conventional radars excel at detecting moving objects using the Doppler dimension because these targets' significant velocities cause them to be widely separated from other objects. Figure 5.1 shows the range-Doppler spectrum from a single virtual antenna (VX) of a typical urban driving scene. Moving objects such as the oncoming vehicles (dashed box regions) are clearly isolated. But, stationary or slow-moving objects (dotted box region) are significantly clustered together in the Doppler dimension and must instead be resolved by angle-of-arrival (AoA) estimation using a phased array. This requirement drives significant hardware cost and complexity by relying on the use of many transmit and receive channels with a physically large array geometry.

Automotive SAR differs by assuming that the relative motion between the radar and a target is known. This overcomes the constant velocity assumption and permits longer integration periods and consequently improved Doppler resolution. As long as the relative motion can be estimated accurately enough, there is no upper limit to the Doppler integration period; in practice, this bound is set simply by the latency and refresh rate. Furthermore, the Doppler resolution depends solely on the extent of the relative motion between the vehicle and the target, known as the synthetic aperture.

It is quite feasible and economical to measure one's own motion ("ego-motion") precisely using low-cost sensors. In contrast, we know little about a target's motion

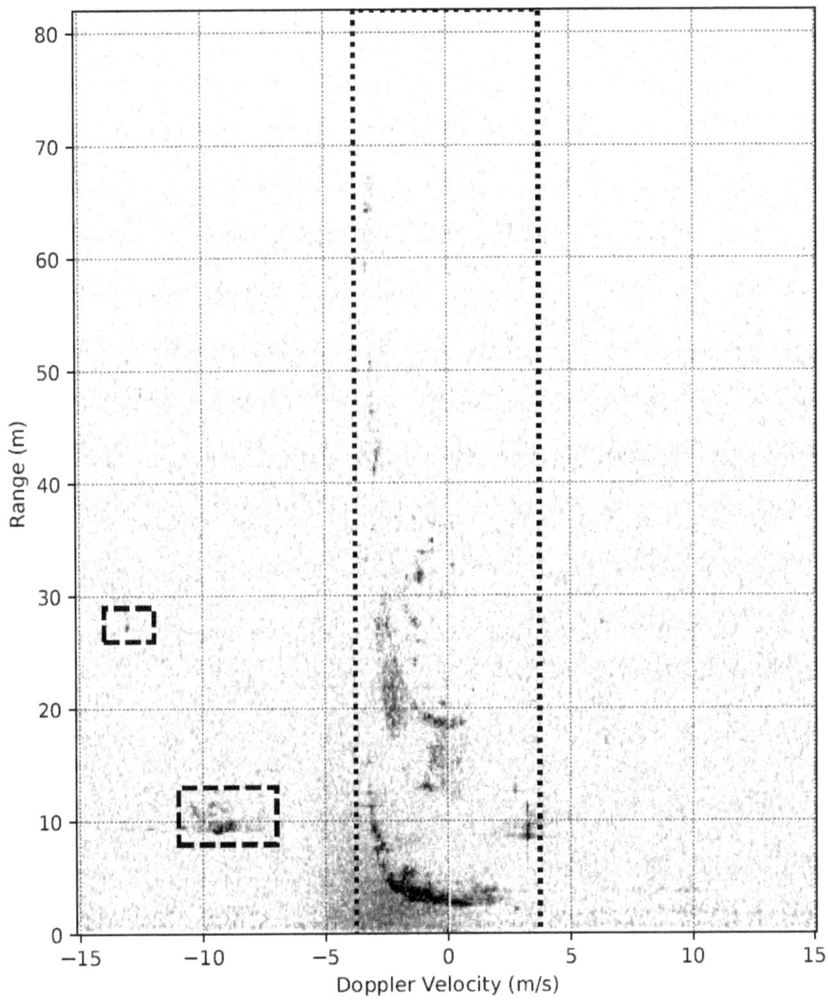

Figure 5.1 Range-Doppler spectrum of an urban driving scene for a left-facing radar. Shaded region indicates magnitude of detection. Dotted lines indicate stationary region; dashed line boxes contain signal from oncoming vehicles.

except in the case where it happens to be stationary. As a result, SAR integration has different performance characteristics for stationary and moving targets.

For stationary targets, we can coherently process an arbitrary number of pulses limited only by the latency and ego-motion quality. As such, SAR imaging performance for stationary targets has outstanding benefits:

1. The very fine Doppler resolution successfully separates almost all stationary targets along Doppler. Because a stationary target's Doppler frequency is proportional to its azimuth location, the imaging resolution is determined by the Doppler resolution and depends solely on the length of the synthetic aperture. Aperture lengths greater than 1m are quite feasible, which is an order of magnitude larger than the physical aperture of a practical phased array.

2. Since the azimuth resolution is conveyed by the synthetic aperture rather than a real phased array, very high resolution can be achieved with a single TX–RX pair, allowing the use of exceedingly simple and cost-effective hardware.
3. Operating on the same hardware and pulse stream, a SAR mode can integrate more pulses than a real-beam radar, providing significant SNR gains. For example, comparing to a baseline point cloud mode using 512 pulses, a SAR mode on the same hardware which can coherently combine 1,000 pulses gives a 3 dB improvement in SNR.

In contrast, for moving targets it is very difficult to estimate the target's motion, making SAR-style integration currently infeasible for these targets without greatly increased computational cost. This is not as significant a problem as it seems. Recall that point cloud processing is particularly good at separating *moving* targets in the Doppler dimension on account of their speed. An elegant solution is to combine the strengths of each approach by using them in parallel, with point cloud processing primarily used to sense moving objects and SAR processing primarily used for stationary objects.

The two pipelines operate on the same pulse stream and provide simultaneous and complementary information about the overall environment (Figure 5.2). A fusion step combines the information obtained from each algorithm into a common representation of the environment.

A traditional point cloud processing system relies on a large real array for high angular resolution, increasing the cost and complexity of the hardware. If targets can instead be resolved in the Doppler dimension, the burden on angle estimation can be reduced and radar hardware can be simplified. SAR processing exploits the Doppler dimension more thoroughly than previous radar architectures, so it presents a significant new tool in radar system design.

5.1.2.1 Phased Array Analogy

It is frequently useful to think of an analogy to phased-array concepts. As the vehicle moves, firing radar pulses at a regular rate, the pulse sequence mimics a physical phased array with the same extent as the synthetic aperture and its broadside oriented toward the vehicle's side. This analogy reveals some useful intuitions (especially regarding the impact of aperture length), but it has some limitations. When crafting SAR processing algorithms or studying more general scenes, we will use the Doppler perspective.

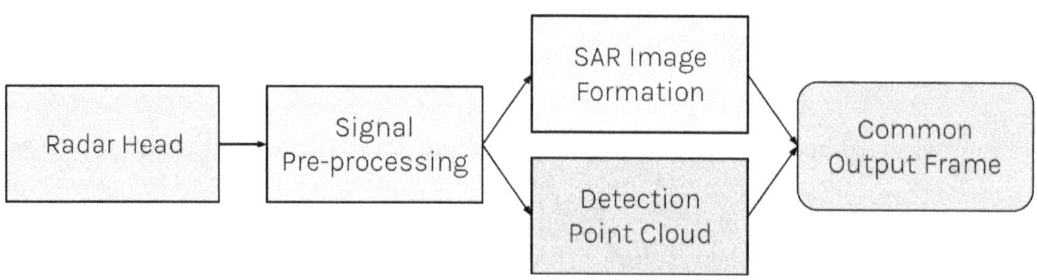

Figure 5.2 Block diagram of an automotive SAR radar, with parallel SAR and point cloud pipelines operating on the same input data.

5.1.3 SAR and Point Cloud Imaging Performance

Figure 5.3 shows a comparison between SAR, point cloud processing, and a wide-field camera for a typical urban driving scene. The range-Doppler spectrum of this scene is shown in Figure 5.1. The SAR and point cloud processed images are in the vehicle-centric coordinate frame where V is the direction of travel and U is perpendicular. In the SAR image (Figure 5.3(a)) the building facades on the left and right are apparent. Four trees can be made out 15m to the left of the vehicle, as can the curbs on either side of the road. A street sign at (–7, 16) is sufficiently bright to show azimuth sidelobes. The faint line to the right of the sign is the edge of the cement sidewalk with the grass.

The processed point cloud (Figure 5.3(c)) shows some of the same features of the SAR image but at a substantially reduced sensitivity and accuracy. The enlarged regions (Figures 5.3(b) and 5.3(d)) demonstrate how the SAR image reveals more detail and structure in both the building and the trees. At the same time, as can be seen in the camera image (Figure 5.3(e)), there are multiple moving cars (both oncoming and same-direction) that are not seen in the SAR image (dotted line regions). The oncoming cars (dot-dashed regions) can be seen as the light gray groups in the point cloud. A car and a truck which are traveling in the same direction as our car can similarly be clearly distinguished (dashed line regions).

Taken together, we can clearly see how the combination of SAR and point cloud pipelines, simultaneously operating on the same set of pulses, can provide a very high-fidelity and high-resolution radar image of the full content of a typical automotive scene.

5.1.4 Applications for Automotive Use

A primary goal of an autonomous system is the sensing of the surrounding world. Radar is widely deployed in mass-market ADAS systems to sense fast-moving objects and measure their velocities but has historically been limited to narrowly defined functions such as ACC or blind spot warning (BSW). At the same time, L4/5 robotaxis have relied heavily on lidar to perceive environment structure on account of its superior resolution compared to radar. Because most L2 passenger vehicles do not have lidars, they must rely only on cameras and radars to perceive the world. The level of autonomy that a L2 passenger car can currently achieve is often limited by the resolution of the radar sensor, so the industry is looking to radar improvements to unlock safety and convenience improvements in mass-market autos. Given the unique characteristics of SAR processing, there are a number of automotive use cases where a SAR mode provides new or better information. In this section, we discuss some of the most noteworthy applications.

5.1.4.1 Drivable-Area Detection

SAR provides a cost-effective solution to sense scene structure at high resolution using existing radar hardware. It is evident from SAR images of driving scenes that parked cars, curbs, guard rails, and dividers are clearly identifiable. For example, Figure 5.4 shows a suburban intersection where some of these features—curbs,

(a)

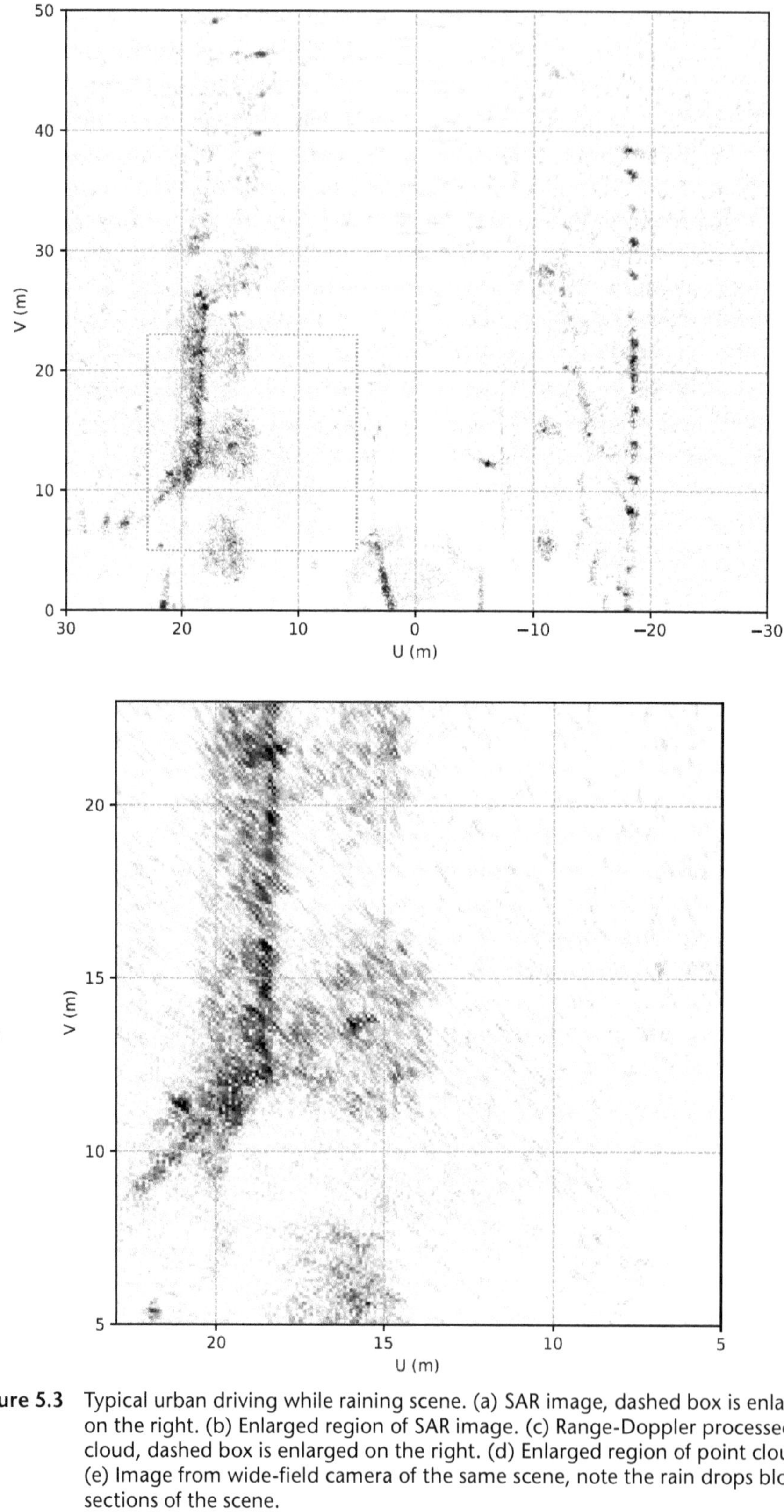

(b)

Figure 5.3 Typical urban driving while raining scene. (a) SAR image, dashed box is enlarged on the right. (b) Enlarged region of SAR image. (c) Range-Doppler processed point cloud, dashed box is enlarged on the right. (d) Enlarged region of point cloud. (e) Image from wide-field camera of the same scene, note the rain drops blocking sections of the scene.

(c)

(d)

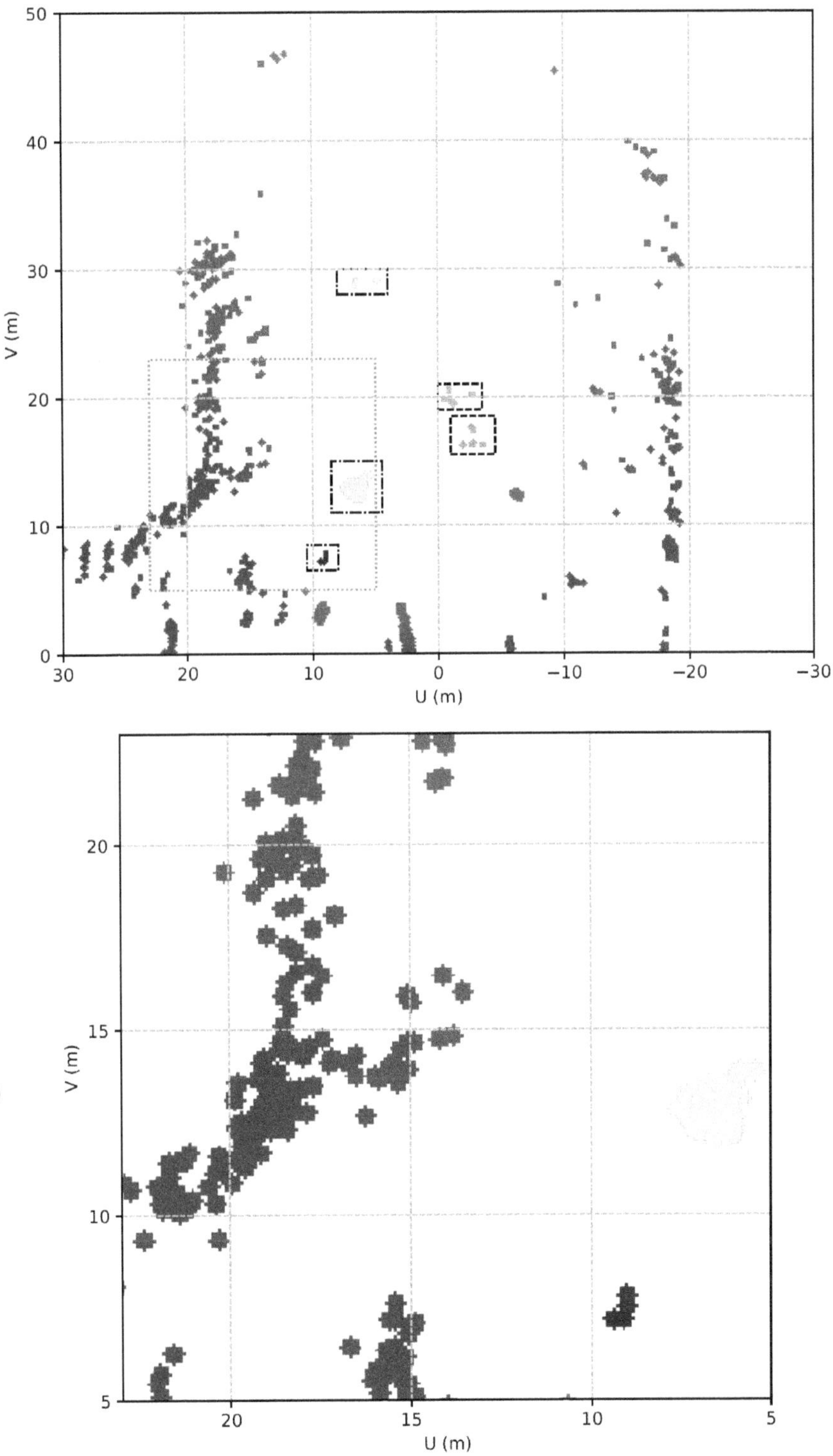

Figure 5.3 (*continued*)

(e)

Figure 5.3 *(continued)*

buildings, and sidewalks—are clearly visible using SAR. Such information can be extracted with computer vision techniques to detect drivable areas.

5.1.4.2 Radar-based Localization

To further increase the reliability of an autonomous system to sense its surroundings, many autonomous vehicle manufacturers have built detailed maps with road features such as lanes, traffic lights, traffic signs, and dividers to centimeter-level accuracy (HD Maps). The vehicle location information within the map is determined using on-board sensors such as inertial navigation system (INS), cameras, lidars, and radars. A preferred solution for L4/5 vehicles typically couples INS with lidars, where road features are extracted from the real-time lidar point cloud and matched against the HD maps. The matching is then used to calculate an accurate pose of the vehicle within the map, typically within 15 cm.

For L2 mass-market vehicles, where the deployment of lidar is limited, camera-based localization (GM Super Cruise, for example [3]) is used. Because camera depth estimation drops off exponentially with distance, this substantially limits the localization accuracy. In addition, cameras suffer from bad lighting and weather conditions. SAR provides a good complement for localization since it has similar range and azimuth resolution compared to a mid-range rotating lidar at lower cost. In addition, SAR-based localization can be implemented using image correlation, suitable for a real-time system with limited compute power.

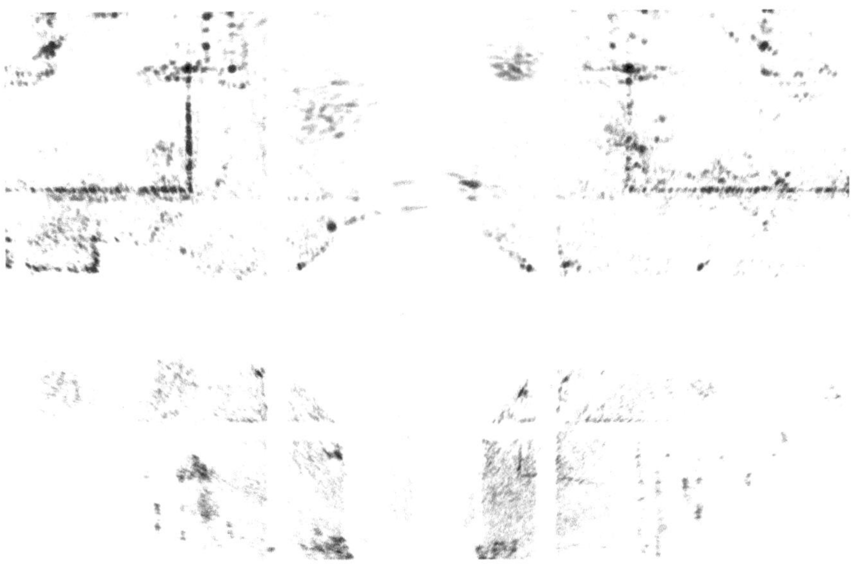

(a)

(b)

Figure 5.4 (a) SAR image of a suburban intersection. Curbs, sidewalks, buildings, trees, signs, and lamps can be clearly identified. (b) Camera view of the same scene for comparison.

5.1.4.3 Segmentation and Classification

In traditional radar, the limited Doppler resolution (on account of the shorter CPI time) caps the density and resolution of the point cloud. As a result, it can be difficult to discriminate features within the stationary environment that are relevant for driving, such as parked cars and curbs. SAR is a better approach for object classification and scene segmentation due to its high resolution for stationary scene content. Many existing computer vision techniques from the literature can be applied to SAR images as they are regularly sampled and show similar characteristics to visible images. At the same time, there are unique characteristics of SAR images that require careful consideration. In comparison to visible images SAR images have higher dynamic range but are not as dense. The brightness of objects in a scene depends on the reflectivity of the object surface at radio wavelengths, which can be different than at optical wavelengths. These factors must be considered when adapting computer vision techniques to SAR images.

5.1.4.4 Radar Point Cloud Enhancement

The SAR and point cloud layers provide complementary information about the scene and can be used to enhance each other's performance. The simplest enhancement is to segment detected points from the point cloud as either stationary or moving objects. The stationary objects can be combined with the SAR pipeline, either by fully replacing them with SAR detections or by fusing the two representations using a probabilistic model.

More advanced SAR-point cloud fusion could involve detecting moving objects in the point cloud and subtracting their Doppler contribution from the SAR image, or similarly detecting stationary objects in the SAR and subtracting them from the range-Doppler input to point cloud processing. These approaches can reduce the clutter and enhance performance of both processing pipelines by playing to their strengths.

5.2 Mathematical Foundation

Having reviewed the history and basic concepts of SAR, this section discusses the mathematical foundation of SAR imaging. The goal will be to describe the basic intuition behind SAR, with only as much mathematical rigor as is necessary.

5.2.1 Key Assumptions

For the purposes of this initial analysis, we will make a number of assumptions and simplifications (listed in Table 5.1). In Section 5.3.2, we will remove the far-field and constant velocity approximations, and in Section 5.3.4 we discuss an extension to 3-D scenes with nonzero elevation.

The Born approximation allows us to apply the superposition principle: The reflected signal from multiple illuminated objects is the sum of the signals that would be measured separately from each object in isolation. Therefore, the rest of this section is focused on analyzing the case of a hypothetical pointlike target in isolation. If by *purely linear* methods we can transform this signal into a focused response resembling a Dirac delta function, then the same method will work for the general scene.

Table 5.1 List of assumptions for general formulation of SAR

Assumption	*Explanation*
Born approximation	Targets do not interact (e.g., by multiple scattering).
Stop-and-go approximation	Neglect any motion of the vehicle during pulses (FMCW chirps). Because pulse time is approximately 30 μs, this is reasonable.
Far-field approximation	Targets are in the far field of the synthetic aperture (i.e., azimuth angle to a target does not change over one synthetic aperture).
Constant velocity approximation	The radar is traveling in a straight path at constant velocity.
Zero elevation	We ignore the elevation dimension.

5.2.2 Signal Model

Consider a radar system operating on a vehicle moving through a stationary scene. At fixed time intervals, the radar will transmit a fixed linear FMCW[1] waveform from a transmit antenna (TX), and a receive antenna (RX) will simultaneously measure the echoes from the objects in the radar's beam. The received signal is down-converted via stretch processing and IQ-demodulation, as described in other chapters of this work. The scene consists of a single stationary pointlike object at distance R_t from the transmitter and distance R_r from the receiver.

For a chirp duration T, we define a time variable t relative to the middle of the chirp, that is, $t \in [-T/2, T/2]$, known as "fast time." The FMCW transmit and receive operation is summarized as:

$$tx(t) \propto \exp\left(2\pi j\left(f_0 + \alpha t\right)t\right) \tag{5.1}$$

$$rx(t) \propto \exp\left(2\pi j\left(f_0 + \alpha\left(t - \frac{2R}{c}\right)\right)\left(t - \frac{2R}{c}\right)\right) \tag{5.2}$$

The received signal after deramp can then be described as:

$$\sigma(t) = \frac{A}{R_t R_r} W(\mathbf{u}) \exp\left(2\pi j\left(\left(f_0 + \alpha t\right)\frac{2R}{c} - \frac{\alpha}{2}\left(\frac{2R}{c}\right)^2\right)\right) \tag{5.3}$$

where we use the definitions:

- A is a target-dependent complex reflectivity;
- $R = (R_t + R_r)/2$;
- $W(\mathbf{u})$ is the angle-dependent antenna gain;
- f_0 is the chirp start carrier frequency;
- α is the FMCW ramp slope;
- c is the speed of light.

1. SAR also works with other range compression techniques, such as binary phase coding. To keep things simple, however, the analysis in this chapter will only study FMCW waveforms.

Now we can simplify (5.3) further. The material properties and distance-dependent attenuation factors are irrelevant to SAR image formation, so we set $A/R_tR_r = 1$. The phase term $-\alpha/2(2R/c)^2$ is typically compensated in signal processing prior to SAR image formation, so we will ignore it. Finally, the antenna response only becomes relevant in very particular corner cases, so we ignore it: $W(u) = 1$. With these simplifications, the signal model for a unit-reflectivity pointlike object for a single chirp is

$$\sigma(t) \approx \exp\left(2\pi j\left((f_0 + \alpha t)\frac{2R}{c}\right)\right) \tag{5.4}$$

5.2.3 Slow Time

Now as the radar moves, and we build up a synthetic aperture, the object's distance will depend on time. We introduce a second time variable s called "slow time," which is discrete and only changes from chirp to chirp. The signal model as a function of both time variables is:

$$\sigma(s,t) \approx \exp\left(2\pi j\left((f_0 + \alpha t)\frac{2R(s)}{c}\right)\right) \tag{5.5}$$

where the distance R now explicitly depends on s.

This signal model captures the impact of any arbitrary path $R(s)$. As discussed in Section 5.3.2, this model is used to perform SAR imaging in general conditions. However, here we add in the simplifying assumptions of far-field and constant velocity so as to make $R(s)$ linearly depend on s.

Choose a fixed time duration S, so that $s \in [-S/2, S/2]$. During this time interval the radar's fixed velocity is v and the synthetic aperture length is Sv. We choose the center of the aperture as the origin of our coordinate system and define the azimuth angle as the angle θ between the target's position and the aperture. Using the far-field approximation, θ does not change over one synthetic aperture. Then, the range of the object from the radar as a function of slow time is:

$$R(s) = R_0 - sv\cos(\theta) \tag{5.6}$$

where $R_0 = R(0)$ is the target range to the center of the synthetic aperture and $-v\cos(\theta)$ is its Doppler velocity. Substituting (5.6) into (5.5) results in:

$$\sigma(s,t) = \exp\left(2\pi j\frac{2f_0R_0}{c}\right)\exp\left(2\pi j\frac{2}{c}\left(\alpha tR_0 - (f_0 + \alpha t)sv\cos(\theta)\right)\right) \tag{5.7}$$

The final step is to define a coordinate transform $\beta \equiv s(f_0 + \alpha t)$, which just warps the two-dimensional time coordinates of (s, t). Then, we obtain

$$T(\beta,t) \equiv \sigma(s,t) = \exp\left(2\pi j\frac{2f_0R_0}{c}\right)\exp\left(2\pi j\frac{2}{c}\left(\alpha tR_0 - \beta v\cos(\theta)\right)\right) \tag{5.8}$$

This is the key result. Observe that the signal model is a 2-D complex exponential in variables (t, β). The fast-time frequency R_0 tells us the target's range and the slow-time frequency $-v \cos\theta$ uniquely tells us the azimuth location because of the SAR assumptions—namely, that the target is stationary and that we know the vehicle's ego-velocity. We can simply use the 2-D Fourier transform to measure these frequencies, and because the signal model is a pure complex exponential it is an impulse in the frequency domain. It follows that we know the object's position R_0,θ *with high resolution*, since this frequency-domain impulse can be distinguished from other closely spaced signals at slightly different range-Doppler frequencies. Even when we relax the constant velocity assumption and employ more sophisticated processing schemes, this same relationship will hold.

5.3 Building an Automotive SAR

Before describing how a SAR should be built, we must first determine the key specifications to achieve. The half-power beamwidth (HPBW) of a SAR radar system is approximately [2]:

$$HPBW \approx \frac{\lambda}{L\sin(\theta)} \tag{5.9}$$

where θ is the target azimuth angle relative to the vehicle direction of motion (i.e., dead ahead is $\theta = 0$) and L is the synthetic aperture length in meters [2]. (The more general formula can be found in the reference.)

The half-power beam width is plotted in Figure 5.5 for operation in the 77 GHz band. As shown there, SAR integration can achieve excellent resolution for a very wide FOV. Specifically, for a 1m aperture SAR can achieve 0.5 degree resolution over $\theta \in [11, 169]$ degrees; by increasing the aperture to 2m, we can extend it to $\theta \in [5, 175]$. The effect of aperture length is shown in Figure 5.6. Not only does the resolution improve as the aperture length is increased, but as noted earlier, the sensitivity also increases to reveal finer scene details.

To target competitiveness with lidar for static object imaging, it is commonly desired to achieve a SAR image resolution of 0.1 degrees. This can be achieved with a synthetic aperture length of 1m, resulting in excellent SAR imaging performance for an extremely wide FOV of 160 degrees.

Recall that (Section 5.2.3) with SAR processing for stationary targets, the angle-of-arrival information comes from the Doppler dimension. This tells us that the SAR mode imposes essentially no requirements on the radar hardware and in fact can be activated on any radar sensor. All that is needed to implement a SAR mode is a positioning solution that estimates the vehicle's ego-motion and a processor for fusing the two data streams.

5.3.1 Measuring Ego-Motion

In order to coherently process a 1m aperture of radar pulses, it is necessary to measure the vehicle's ego-motion sufficiently accurately over that aperture. In aerospace

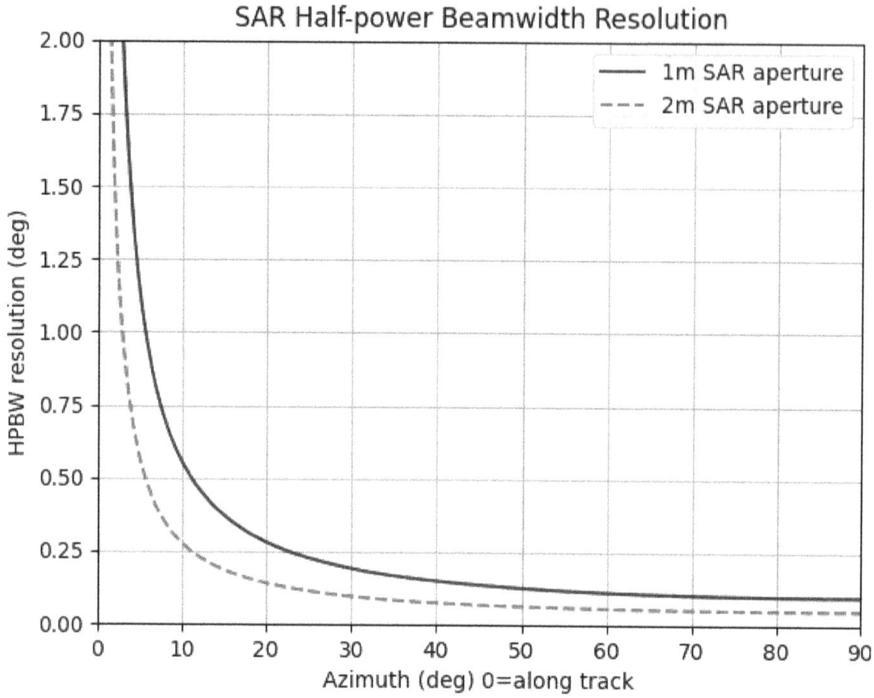

Figure 5.5 SAR resolution versus angle to vehicle direction of travel.

SAR, the motion of the platform is very steady and predictable, particularly in space. Typically, motion estimation in aerospace SAR can be approached by splitting the motion into an ideal path with small perturbations.

In contrast, automotive motion characteristics include acceleration, deceleration, turning, and other significant deviations from constant velocity motion. Therefore, high-fidelity measurement of arbitrary vehicle ego-motion, known as the *tracklog*, is a key requirement for building an automotive SAR system. Fortunately, it is common for vehicles today to be equipped with positioning sensors such as GPS, inertial measurement unit (IMU), and wheel encoders. Furthermore, improvements in technology are continually driving down the cost and improving the performance of these positioning sensors [4].

5.3.1.1 Positioning Accuracy Requirements

A common myth regarding SAR operation is that the position must be estimated to subwavelength precision, which at 77 GHz would imply $< 1\ mm$ positioning. Fortunately, SAR operation merely requires the system to know the travel distance *between* pulses, not the absolute position where those pulses were emitted. A simple analogy makes it clear: In a phased array, the baseline between antennas must be known very precisely but the absolute coordinates of the radar are only relevant for global image alignment. From this argument, we can intuit that any constant or slowly-varying error in absolute position does not affect the ability to coherently

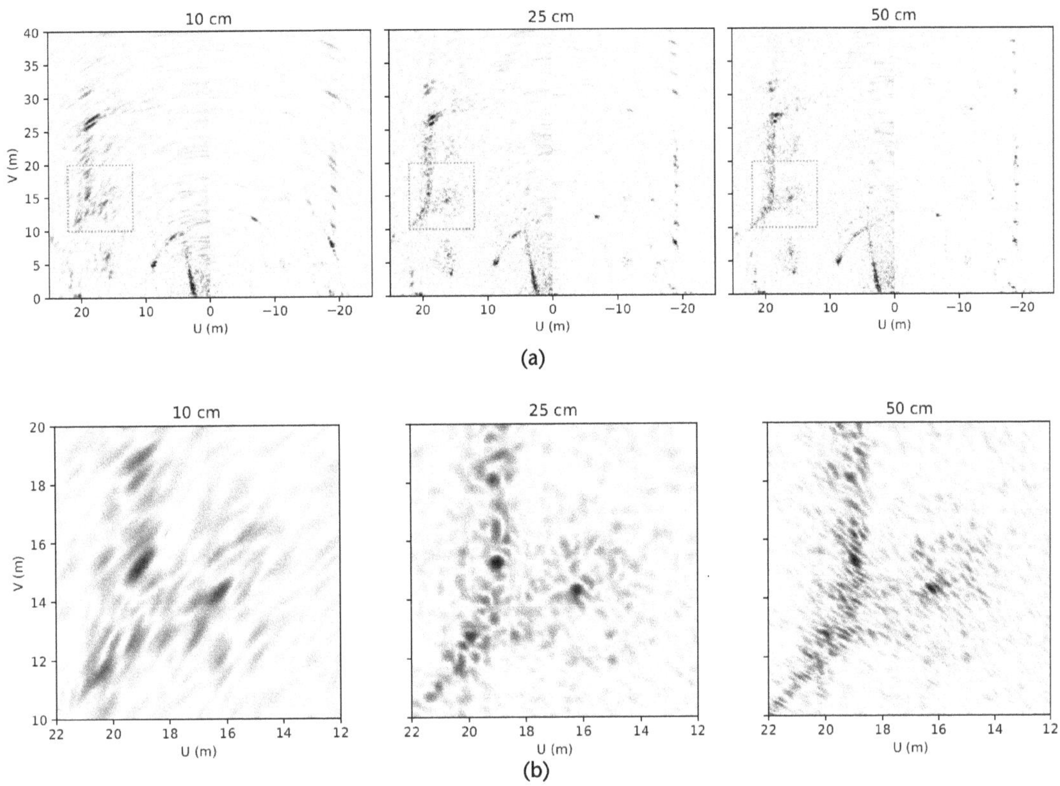

Figure 5.6 (a) SAR image of the same scene as in Figure 5.3, as a function of aperture length. (b) Enlarged view of the box region in the full image (5.6(a)).

integrate SAR. Essentially we need high-fidelity estimates of velocity and acceleration, not position—a much more feasible task.

This makes it clear that velocity error is the most significant positioning specification for a SAR radar. Velocity errors stretch or shrink the aperture and therefore the distance between every pair of chirps within the aperture. In the SAR image, this manifests as AoA warping, with the most severe errors either straight ahead or directly behind the vehicle. This effect is shown in Figure 5.7 where the lidar detections are shown for reference in the gray points. In this example, the building faces in the SAR image are warped as a function of azimuth; note that the more severe warping occurs straight ahead of the vehicle. The direction of the warping (toward or away from the vehicle track) depends on the sign of the velocity error. We have empirically found that velocity error of ±2% is sufficient for reliable imaging up to approximately ±10 degrees from the vehicle track.

One other tracklog accuracy requirement bears mentioning. Heading accuracy refers to the error, in degrees, of the vehicle heading. During linear motion, heading errors primarily result in rotation of the SAR image in global coordinates. During turns, however, the heading error plays a significant role in determining the characteristics of the curved aperture. Therefore, the heading must be measured sufficiently accurately so as to capture the vehicle trajectory during the turn.

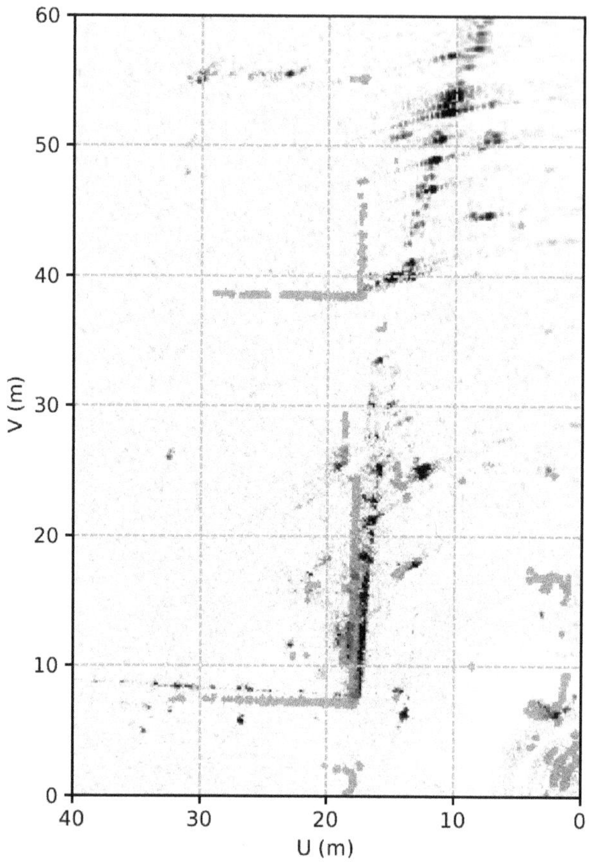

Figure 5.7 Example of a SAR image with velocity error corruption. Note the warping of the building relative to the reference lidar (gray points), with greatest angular error near the vehicle direction of travel.

SAR image formation also requires precisely synchronizing each radar pulse with the tracklog. Because the vehicle is moving, an error in times-tamping a radar pulse is equivalent to a tracklog error. Naturally, the system is least sensitive to synchronization error when traveling at constant velocity and most sensitive to it when accelerating or turning. Fortunately, it is possible to leverage a number of automotive technologies for this task including hardware timestamping abilities on many embedded processors along with multisensor synchronization technologies such as PTP (IEEE-1588).

5.3.1.2 Positioning Sensors

The three core sensor types that are available on most mass-market vehicles today (IMU, wheel encoder, and GPS) provide raw data to an Extended Kalman Filter (EKF), which performs sensor fusion to estimate the vehicle's motion in 3-D space (Figure 5.8). By knowing the accuracy, noise, and update characteristics of each sensor, the EKF can optimally combine them and also incorporate known motion models. The EKF output is used as the tracklog for SAR processing.

The usefulness of the measured tracklog for SAR imaging depends on how the positioning sensors are installed relative to the radars. It is important that the radar and positioning sensors be rigidly mounted relative to one another, which is typically accomplished by installing all sensors on the vehicle chassis or body. The translation and orientation ("extrinsics") between each radar and all positioning sensors are measured at design time and calibrated after installation. The extrinsics are used to refer the vehicle tracklog to the actual path of each radar sensor based on the dynamics of an extended rigid body. This is most crucial during turns, when each radar on the car traverses a different path through space. Finally, since the radar and positioning sensors are rigidly installed, it is possible to neglect deformation of the radar-IMU axis.

5.3.2 SAR Image Formation

In Section 5.2 we discussed the intuition for SAR imaging in an idealized scenario as the 2-D frequency spectrum of the time-domain signal. In fact, this intuition can be turned into a full-fledged SAR imaging algorithm. In the process, we should remove the far-field and constant velocity assumptions made above, since those are not realistic.

To do so, we can consider frequency-domain processing from a more general perspective. Let's revisit (5.8) and introduce some new variables to simplify the notation and discussion:

- Let $f = f_0 + \alpha t$, which is the instantaneous carrier frequency;
- Let $\dot{R} = -v\cos(\theta)$, which is the Doppler velocity of the stationary object.

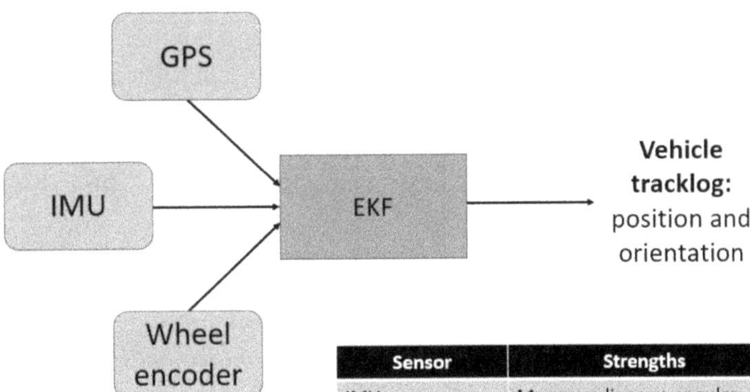

Sensor	Strengths	Weaknesses
IMU	Measures linear + angular accelerations; used for dead reckoning	Cost vs noise/bias tradeoff
GPS	Measures velocity Can calibrate IMU bias	Outages Low position accuracy
Wheel encoder	Measures velocity + turning	Wheel slipping Bias from tire inflation

Figure 5.8 Extended Kalman Filter (EKF) processes raw positioning sensors to estimate the vehicle's motion in a 3-D space.

We can rewrite (5.8) to display the dependence of the signal model on the range-Doppler coordinates explicitly now:

$$\rho\left(\beta,f \mid R_0,\dot{R}\right) = T(\beta,t) = \exp\left(2\pi j\frac{2}{c}\left(f R_0 - \beta\dot{R}\right)\right) \tag{5.10}$$

The total received signal will be denoted by $x(\beta, f)$, and it corresponds to the summed contribution from every object in the scene.

Our goal in SAR image formation is to recover the target's position in 2-D space. In the simplified preceding description, we observed that we could do this simply by looking at the two-dimensional frequency spectrum of the signal. To do this, we choose the range-Doppler coordinates $(\hat{R}_0, \hat{\dot{R}})$ we want to look at, and then evaluate which is a finite summation in practical terms. The first equation is simply a restatement of the inverse Fourier transform for the simplified, straight-path case.

$$I\left(\hat{R}_0,\hat{\dot{R}}\right) = \iint x(\beta,f)\exp\left(2\pi j\frac{2}{c}\left(f\hat{R}_0 - \beta\hat{\dot{R}}\right)\right)df\,d\beta \tag{5.11}$$

$$= \iint x(\beta,f)\overline{\rho\left(\beta,f \mid \hat{R}_0,\hat{\dot{R}}\right)}\,df\,d\beta \tag{5.12}$$

However, the second line provides a much more general perspective. This looks like a *correlation* of the total received signal with the impulse response of a target at range-Doppler coordinates $(\hat{R}_0, \hat{\dot{R}})$ (i.e., matched filtering). Unlike (5.11), (5.12) does not imply anything about the radar's path nor the processing scheme. To use (5.12), we compute the signal model *rho* empirically based on the measured tracklog, which recalls the general description in (5.5), and use it to compute the correlation response. In fact, we can directly compute the correlation in the spatial Cartesian domain instead of range-Doppler coordinates.

This more general correlation perspective also applies when the object being imaged is near the synthetic aperture, in which case we just incorporate azimuth migration into the signal model. These considerations lead us to the *back projection* algorithms, which are attractive in the automotive setting due to their generality and especially their ability to accommodate arbitrary vehicle paths and targets locations.

5.3.2.1 Time-Domain Back Projection

The direct discretization of (5.12) gives us the *time-domain back projection* algorithm. The equation is discretized over the slow time samples and the empirical signal model is used at each slow-time pulse. While it is quite useful theoretically, its practical usefulness is limited by its computational complexity. In practical settings, the fast time, slow time, range, and azimuth axes may all have samples on the same order of magnitude, say approximately N. In this case, since the 2-D integral over fast and slow time must be evaluated for each output pixel, we see that the computational complexity of this method is $O(N^4)$. A value of N 10^3 is desirable, so this is clearly impractical.

5.3.2.2 Fast Back Projection (FBP)

A quite straightforward optimization is to consider the iterated integral in (5.12) and take apart the signal model into its fast and slow time components.

$$I\left(\hat{R}_0,\hat{\dot{R}}\right)=\int\left[\int x(\beta,f)\exp\left(-2\pi j\frac{2}{c}f\,\hat{R}_0\right)df\right]\exp\left(-2\pi j\frac{2}{c}\beta\hat{\dot{R}}_0\right)d\beta \tag{5.13}$$

The fast-time integral can be replaced by a fast-time Fourier transform into the range domain. Subsequently, slow time processing can be applied by discretizing the outer integral.

The more general non-straight path case defies analytical formulation but can empirically be shown to be a very good approximation. The key intuition is that the fast and slow time domains are almost fully decoupled. Therefore, this improved algorithm proceeds as follows. The first step is to compute the one-dimensional Fourier transform of ρ over fast time. Then the inner integral reduces to an indexing into the range profile (generally interpolated), which is constant-time. Finally, the slow-time integration is computed directly from the discretization of the outer integral in (5.13). The computational complexity of this two-step algorithm, called fast back projection (FBP), is still dominated by the second step, namely the integration, which now becomes $O(N^3)$.

5.3.2.3 Fast Factorized Back Projection (FFBP)

We can obtain a further speedup by sharing subcomputations among output samples. Recall from (5.9) that the azimuth resolution of the SAR system is directly proportional to the length of the synthetic aperture. To deal with the outer integral in (5.12), we observe that it can be split into a finite number of smaller integrals:

$$\int_{-S/2}^{S/2} f(s)\,ds=\sum_{k=0}^{M-1}\int_{s_k}^{s_{k+1}} f(s)\,ds \tag{5.14}$$

where $-S/2 = s_0 < s_1 < \cdots < s_M = S/2$. Each integral on the right-hand side is over a smaller subset of the original synthetic aperture, which we can call a *subaperture*, and thus its result will have a lower azimuth resolution than the full integral. The key here is that "resolution" really means *information content*: an image with lower azimuth resolution can be represented with fewer degrees of freedom.

This leads us to the improved algorithm called the FFBP [5]: we split the full synthetic aperture into a small number of sub-apertures of equal size and recursively evaluate the integral on each. Crucially, when we do so, we represent each output subimage with coarser azimuth resolution by using fewer pixels. This reduces the compute burden, but no information is lost. Then, we sum the subapertures to bootstrap up to the larger aperture, increasing the azimuth resolution and number of pixels with each summation step (via 2-D upsampling).

FFBP has computational complexity of $O(N^2\log(N))$, which in terms of asymptotic complexity is the best known result, to our knowledge.

5.3.2.4 Real-Time Operation

The algorithmic manipulations previously described make it possible to achieve manageable asymptotic complexity while addressing arbitrary scenes and vehicle motions. SAR images are commonly on the order of 1 mega-pixel per quadrant where we use 1,000 pixels in range along with 1,000 azimuth pixels to cover a 100-degree FOV at 0.1 degree resolution. The input data may have comparable dimensionality, using 512 range bins and around 1000 slow time samples.

With these parameters, we have estimated that computing a 1 megapixel SAR image using FFBP requires approximately 2 Gflops. If this computation is repeated for four corner radars in a vehicle, at 10 frames per second, a 360 degree SAR image stream can be computed with about 80 Gflops/sec. Fortunately, the availability of cheap, high-performance processors for automotive applications makes it possible to obtain the required number of flops within a suitable platform for automotive integration.

5.3.3 Coexistence with Point Cloud Pipeline

As we have noted, SAR processing makes essentially no demands on the radar antenna geometry and therefore a SAR mode could be added to nearly any radar system. For a SAR and point cloud mode to coexist on the same processor, certain processing steps such as time-domain and range processing are applied in common. At the appropriate point in the processing chain, the data is split into two pipelines which perform the SAR and point cloud detection algorithms, respectively.

In addition to the hardware and processing requirements, we must also ensure that SAR and point cloud modes can operate on the same pulse stream. SAR processing typically works best on a spatially uniform pulse sequence, and must be sampled at the appropriate Nyquist rate. We achieve uniformity in the sequence by operating the radar with high duty cycle (typically over 95%) so that gaps between pulses are minimized in number and quantity. As an illustrative example, we could program a radar frame with a 20 ms duration, using 512 pulses at a pulse repetition interval (PRI) of 38 μs; this gives a duty cycle of 97%. Two successive frames are combined to give a synthetic aperture of 1,024 pulses over 40 ms while only one of every two frames is used for computing a traditional point cloud at 25 Hz. Note that number of frames (and partial frames) used for a single synthetic aperture would vary based on speed in order to achieve the target 1m aperture.

The Nyquist sampling rate for the SAR mode is determined by the wavelength, FOV, and the mounting orientation relative to the direction of travel. For instance, for a 100-degree FOV radar that is mounted facing 45-degrees away from the direction of travel, the Nyquist sampling rate is approximately half-wavelength or 1.9 mm in the 77 GHz band. Note that this is a *spatial* sampling requirement; the PRI will depend on the vehicle's speed. For a PRI of 38 μs, this sampling requirement can be met up to 50 m/s (110 mph or 180 kph), which is in line with the operating parameters of a traditional point cloud radar sensor.

Another emerging consideration is the use of MIMO radar modulation formats for enhanced sensitivity and angular resolution. There are a number of MIMO modulation formats in the industry, including time-domain multiplexing (TDM) and various coding schemes, each with unique processing considerations that can be adapted to SAR operation. For instance, to process TDM-MIMO pulses with SAR, the pulses from each transmitter can be independently back-projected using the appropriate signal model, which depends on the TX-RX antenna baseline. The back projection operation knows the spatial position of the vehicle that corresponds to each TDM-MIMO pulse and therefore inherently compensates for the motion between pulses. (This same compensation is performed in point cloud processing via a Doppler-dependent phase shift.) Then, the subimages from each transmit stream can be combined to give the MIMO processing gain. Similar algorithms can be devised for other MIMO modulation formats.

In summary, we have found that SAR and point cloud modes have fully compatible modulation and pulse sequence requirements. This allows both pipelines to operate simultaneously on the same set of pulses and give complementary outputs describing the scene.

5.3.4 Elevation Information

SAR operation provides resolution only in azimuth since it is the azimuth angle that determines the Doppler spectrum. Another way of seeing this is by noting that the SAR processing previously described is invariant under rotations around the synthetic aperture, so an object at any 3-D position will appear rotated into a single horizontal plane with the height projected along the range dimension.

To recover the elevation information, the system needs to be augmented by including at least one pair of real antennas with a vertical baseline. Then a fairly standard AOA calculation or interferometry approach can be used to estimate the elevation angle of a target from the phase information of the SAR image. If the elevation content of the scene is very rich, in that multiple elevation targets lie in the same range-azimuth cell, then more than one elevation baseline may be needed to distinguish them.

5.4 Future Directions

Technology trends in radar silicon, compute chips, and positioning sensors have made it possible to build a cost-effective automotive SAR system operating in any driving conditions over a wide FOV, producing very high-resolution images of the surrounding stationary scene. Looking to the future, there are several research directions which could further enhance the utility and performance of automotive SAR.

5.4.1 Forward-Facing SAR

As described in Section 5.3, though SAR operation results in impressive resolution gains for a very wide FOV, the resolution degrades for targets within 5–10 degrees of the vehicle direction of travel. This is called "forward-facing SAR" in the literature. Some researchers have studied how to improve resolution in the forward-facing cone by combining SAR operation with a real array [6].

Alongside the fundamental resolution limits is also the question of how to practically achieve them. Recall from Section 5.3.1.1 that SAR operation is most sensitive to velocity errors for targets with $\theta \sim 0$. Empirically, we have found that reliable SAR imaging in $\theta \in [-5, 5]$ degrees requires velocity errors of around $\pm 0.3\%$. This is extremely challenging and currently cannot be achieved with cost-effective positioning technologies.

5.4.2 SAR for Moving Objects

Throughout this chapter, we have used SAR imaging for stationary objects and relied on traditional point cloud processing to sense moving objects. In order to unify the pipelines, we would need to estimate the target motion with sufficient accuracy to perform SAR integration. This is a very challenging problem, particularly for fast targets, which is why this remains an open research topic. Two approaches could be:

1. Bootstrap the SAR process by estimating target motion using a tracking operation on the point cloud output. The estimated motion can be fed into the back projection algorithm for the target's azimuth location.
2. Test motion hypotheses and use a decision criterion to choose the best fit. For instance, we could image a region at different test velocities, and then pick which velocity hypothesis gives the most focused image.

Imaging moving objects with SAR techniques could unlock longer integration times and therefore higher Doppler resolution; increased Doppler resolution lowers the burden on angular estimation techniques and could permit the use of simpler and cheaper hardware.

5.4.3 Gapped SAR

As discussed in Section 5.3.3, SAR processing works best with a uniform pulse stream which leads us to use pulse duty cycles over 95%. In order to reduce radar power consumption and reserve time for monitoring and calibration functions, it is preferred to reduce the radar duty cycle. In a SAR mode this would come at the cost of introducing gaps into the pulse stream. For instance, a radar duty cycle of 70% would result in a significant gap between two Nyquist-sampled sub-apertures.

Gapped SAR has been studied in [7] and [8]. Gaps of 20–30% of the aperture length can be handled with relatively little loss in image quality [8]; however, new processing techniques were needed to handle the spectrum estimation across the aperture gap. It would be valuable to understand both the theoretical limitations of gapped SAR as well as practical methods for achieving those limits that are suitable for automotive environments and constraints.

5.5 Conclusion

This chapter has introduced automotive SAR technology and demonstrated how it can improve radar sensing quality. By combining SAR processing side-by-side with a traditional point cloud pipeline, the SAR pipeline can provide lidarlike resolution for stationary objects and complement the strength of the point cloud processing in sensing moving objects. This approach surpasses what point cloud processing can do alone and unlocks new frontiers in radar performance. Furthermore, with recent technology trends in positioning sensors and processor technology, SAR techniques can be economically realized in practical automotive cost and form factor envelopes and are also able to deal with the wide diversity and complexity of automotive motion characteristics. Building off these developments, SAR processing can be considered a core building block for next-generation automotive radars.

References

[1] Lasswell. S. W., "History of SAR at Lockheed Martin (previously Goodyear Aerospace)," *Proc. SPIE 5788*, Radar Sensor Technology IX, 16 May 2005.

[2] Engineering and Technology History Wiki (ETHW), *Synthetic Aperture Radar*, https://ethw.org/Synthetic_Aperture_Radar, 2021.

[3] Wired, *The Laser Maps Behind Cadillac's Superb Self-Driving Skills*, 2021, https://www.wired.com/story/cadillac-super-cruise-ct6-lidar-laser-maps/.

[4] GPS World, *Using Consumer-Grade Sensors for Precise Positioning*, 2021, https://www.gpsworld.com/using-consumer-grade-sensors-for-precise-positioning/.

[5] Ulander, L. M. H., Hellsten, H., and G. Stenstrom, "Synthetic-Aperture Radar Processing Using Fast Factorized Back Projection," *IEEE Transactions on Aerospace and Electronic Systems*, Vol. 39, No. 3, 2003, pp. 760–776.

[6] Franceschetti, G., Iodice, A., and D. Riccio, "Forward-Looking Synthetic Aperture Radar (FLoSAR): The Array Approach," *IEEE Geoscience and Remote Sensing Letters*, Vol. 11, No. 1, 2014, pp. 303–307.

[7] Larsson, E. G., Stoica, P., and J. Li, "Spectral Analysis of Gapped Data," *Conference Record of the Thirty-Fourth Asilomar Conference on Signals, Systems and Computers*, Vol. 1, 2000, pp. 207–211.

[8] Salzman, J., et al., "Interrupted Synthetic perture Radar (SAR)," *Proceedings of the 2001 IEEE Radar Conference*, Atlanta, GA, 2001, pp. 117–122.

CHAPTER 6

Radar Transceiver Technologies

Mike Keaveney

6.1 Background and Introduction to Automotive Radar

Radar has been around now for more than 100 years, having first been patented in 1904 by German inventor Christian Huelsmeyer and used initially to prevent ships colliding in foggy weather on the River Rhine. Major advances were made to the technology during World War II where radar played a key role in helping to win the Battle of Britain. It was only in the last decade however that radar has started to make inroads into automotive and is now one of the most important sensors in the advanced driver assistance systems (ADAS) of today's automobiles.

Frequency modulated continuous wave (FMCW) radar [1–2] has become the dominant technology for ADAS mainly because it can achieve high resolution in range with relatively low implementation complexity and power consumption compared to other radar types. The transmit waveform is in the form of a frequency chirp that sweeps linearly over the desired RF bandwidth for a given range resolution requirement. When this chirp is reflected off an object and picked up at the receiver after the time of flight, it is mixed with the transmit chirp to produce a beat frequency at baseband. This beat frequency is proportional to the time difference between the chirps, from which the range can be calculated. If an N point FFT is performed on the digitized baseband signal from the mixer, each of the $N/2$ output bins will contain a signal if there is a reflection from an object at the corresponding range bin or noise otherwise.

This frequency chirp "pulse" is an ideal radar waveform for short range applications such as automotive radar, because its operation depends on the overlap between the transmit and receive pulses. The pulse signal itself can have very wide bandwidth, with up to 5 GHz of spectrum available in the 76–81GHz band, for best range resolution. A great advantage of FMCW is that the "de-chirping" operation at the receive mixer, known as *pulse compression*, reduces this bandwidth down to no more than the tens of MHz needed to contain the beat frequencies from reflected objects. Because of this, the receive baseband comprises low bandwidth, low power consumption circuits. The data rate out of the receiver ADC needs to be no higher the two times the beat frequency corresponding to the maximum range bin, leading to modest data bandwidths, memory, and processing requirements in the DSP.

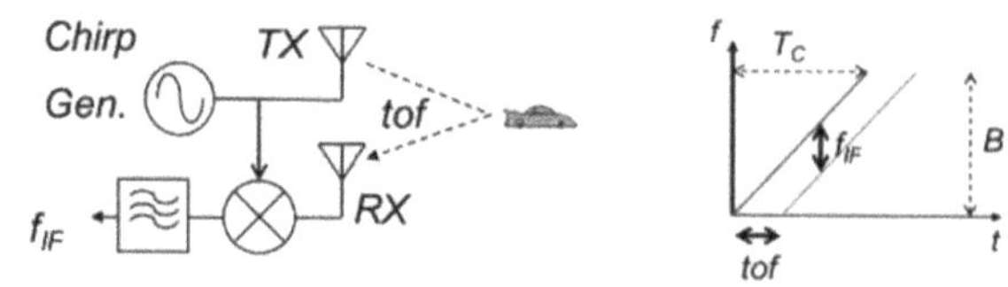

Figure 6.1 FMCW Radar signal chain, waveform, and response from a target whose return is delayed tof seconds due to its range.

Throughout this chapter, we use the abbreviations TX to represent the transmitter and transmit chain, and RX to represent the receive chain (Figure 6.1).

Early ADAS radar signal chains were implemented using a mix of device process technologies, such as CMOS for the baseband sections with the RF frontends on more exotic device technologies such as Gallium Arsenide (GaAs), which could produce transistors with both high speed and high output power [3]. As volumes increased, the frontend blocks were quickly ported on to emerging lower cost technologies on silicon such as Silicon-Germanium (SiGe) bipolar, which offered high speed transistors capable of high enough power levels. With SiGe BiCMOS (Bipolar plus CMOS on the same silicon substrate), the integration level increased, further reducing cost. The radar modules were typically configured and applied to fulfill a particular ADAS function such as blind spot Detection (BSD) or ACC. As more advanced applications emerged (e.g., AEB), better spatial resolution was required so it became necessary to increase the channel count so that larger antenna apertures and digital beamforming techniques could be used to improve the angle resolution. With three TX and four RX channels, MIMO techniques could be used to generate a wider virtual aperture that could be spatially separated into $3 \times 4 = 12$ digital beamforming sectors [4].

As Moore's Law advanced, the transistors on deep submicron technologies at 40 nm and lower became fast enough to make millimetre-wave circuits on CMOS feasible. There still remained challenges, particularly with the generation of sufficiently high output power and low oscillator phase noise on CMOS. These will be discussed in the following sections, along with the new architectures that have emerged to overcome these challenges.

These new circuit architectures now enable fully integrated radar transceiver SoC integrated circuits to be implemented on low cost CMOS process technology. The ability to integrate the complete radar signal chain from bits to 77 GHz on advanced CMOS also opens up the possibility of digitally encoded radar pulses, such as PMCW [5] or orthogonal frequency–division multiplexing (OFDM) [6] radars. These digital enabled radar waveforms enable new degrees of flexibility and adaptability for the radar that can be adapted depending on the use case for each measurement cycle. Digital encoding broadens out the bandwidth of these radar pulses and is used to uniquely identify the pulse waveform from a given TX channel at the receiver. These properties make the waveforms more resilient to interference from other radars, which enable more radars to coexist in the same environment. These more sophisticated waveforms do come at a price of full signal bandwidth being required along the complete TX and RX signal chains. This wider signal chain and processing bandwidth is costly in power consumption, particularly in the data converters that now must digitize multiple GHz bandwidth signals, but also the amount of memory and processing horsepower able to handle the data rates. This situation will continue to improve however as technology continues to scale.

Figure 6.2 compares the signal chain for a digital encoded radar waveform, such as PMCW, to the chirp sequence FMCW signal chain. The measurement cycle in radar is known as a coherent processing interval (CPI) where a range-Doppler map image is produced from a sequence of time synchronized measurements. In CS-FMCW, the digitized samples from each received chirp are processed by what is known as a fast-time FFT (i.e. *fast* because the samples are at the ADC output data rate). This is also known as a range FFT since the output frequency bins correspond to increasing range gate values or range bins. The samples in each range bin are then processed across successive chirps in what is called a slow-time FFT. This is also known as the Doppler FFT because phase changes from chirp to chirp, due to Doppler will appear in the intersecting range and doppler bins in the 2-D range-Doppler map [4]. Note that each chirp in the sequence provides one sample for the Doppler FFT, so the shorter the chirp-to-chirp time interval, the higher the Doppler sampling rate. Hence short, fast chirps are desirable for high unambiguous (i.e., without aliasing), velocity detection [4]. For example, to detect relative velocities, unambiguously in the range of +/– 180 km/h, requires a chirp-to-chirp time interval of less than 10 μs. In PMCW, the equivalent task to the range FFTs is performed by a bank of correlators to identify the range bins that correlate with the transmit PMCW bit sequence [5]. In this case, for high unambiguous Doppler, short PMCW pulse widths are desirable. This requires high transmit and receive signal chain bandwidths to avoid distorting the PMCW bit sequence within the pulse.

As illustrated in Figure 6.2, the RF frontends are essentially interchangeable between FMCW and PMCW; both need to deliver similar output power levels over the 5 GHz BW from 76–81GHz and also have broadly similar requirements on the RX frontends. The chirp generator controlling the VCO is probably the most demanding block on the FMCW signal chain. This frequency chirp forms the

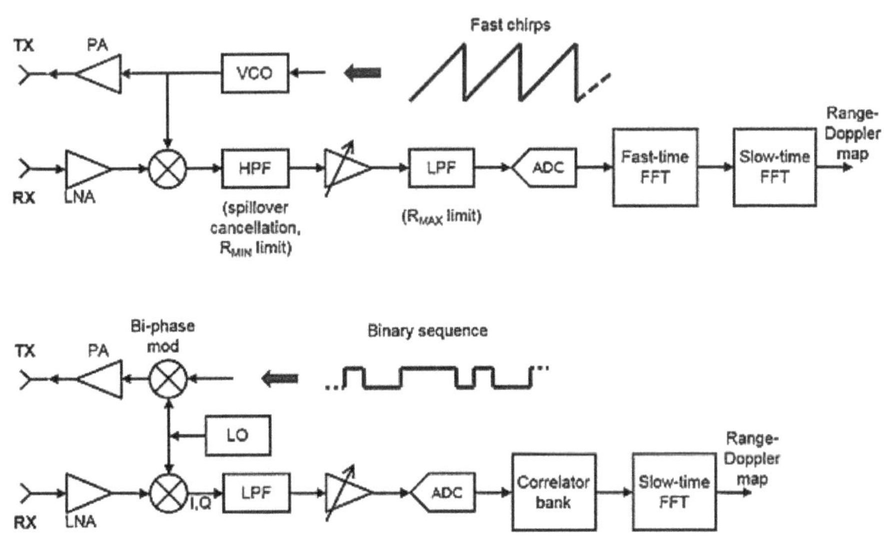

Figure 6.2 Comparison of FMCW and PMCW radar waveforms and signal chains. (© 2017 IEEE [7].)

transmit signal and is also the LO for the receive mixer. LO phase noise sidebands appear around strong reflectors and can hide weaker objects close by. Nonidealities in the chirp linearity also appear as sidebands in the range and Doppler FFTs that blurs the resolution achievable. An even worse artifact that can arise for chirp nonlinearity is a spur in the baseband spectrum, known as a *ghost target*. If this is mistakenly detected as a real target it could trigger an undesirable response such as engaging automatic emergency brakes in an ADAS system. Another weakness of FMCW is its susceptibility to interference. If the TX signal of another FMCW radar sensor is captured by the receiver, it is likely that this direct path signal will be much stronger than the reception of its own TX signal via the reflection from an object. Due to the frequency sweeping nature of the FMCW waveform, the probability of this overlap condition occurring is quite high. This will likely overload the receiver during the overlap period, or at least cause significant degradation of the noise floor of the range FFT. Various techniques can be used to mitigate this weakness such as the use of faster chirps (e.g., short 10 μs chirps), frequency hopping between chirp sequences, and time shifting between chirps.

6.2 Block Diagram Overview of an FMCW Radar Transceiver

Figure 6.3 shows a block diagram of a modern representative radar transceiver for 77 GHz FMCW radar. The 3 TX channels and 4 RX channels can work together in a MIMO configuration to provide up to 12 virtual channels. For imaging radar applications, up to 4 transceivers can be cascaded together to up to 192 virtual channels in MIMO by routing the LO out from the master transceiver back into itself and to another 3 slave devices.

Output power per transmit channel is usually targeted at 13 dBm or 20 mW over the 76–81 GHz automotive radar bands. The higher the TX output power the longer the range possible for the radar although, as can be seen from the radar equation [3], a doubling of range would call for a 16 fold increase in TX power,

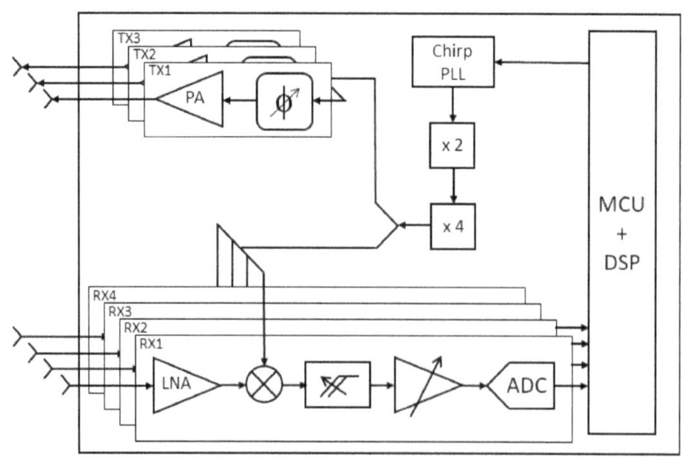

Figure 6.3 Block diagram of a 3T4R 77GHz FMCW radar transceiver.

all else being equal. Even if this could be achieved on a monolithic integration, it would be counterproductive at the system level due to the amount of self-interference it would produce, degrading its own receiver sensitivity. Thus a maximum output power of 13 dBm (20 mW into a 50 ohm antenna load) is a popular specification for the TX power amplifier in the automotive radar application. This also complies with the regulatory standards worldwide for this band [8] as well as export control regulations (e.g., [9]).

LO phase noise is one of the most important specifications in FMCW radar as the phase noise sidebands appear around the detected targets baseband spectrum and smaller objects such as a motor cycle could be buried in the phase noise sideband of a large reflector such as a truck. A phase noise level of < –100 dBc/Hz at 1 MHz offset from the carrier would be state of the art.

A receiver noise figure in the range of 12–15 dB would be typical for this application, given that the frontend gain must be limited, to avoid saturating the signal chain through amplification of signal leakage from its own transmit power amplifier, as well taking into account frontend losses through the package and on-chip passives at mm-wave frequencies.

6.3 Challenges with Deeply Scaled CMOS

CMOS process technologies have always been optimized for digital circuits as these account for, by far, the highest volume applications driving the technology advances. Consequently, analog and RF circuits have had to make do with whatever active devices (i.e., different transistor types) and passive devices (e.g., resistors and capacitors) that could be implemented on these technologies without any significant cost adders. Electrical models of these devices are needed so that they can be used for circuit design using computer aided design (CAD) tools. The standard device models available from CMOS processes would typically only be valid for operating frequencies of digital circuits, so RF extensions are needed to enable circuit design at mm-wave frequencies. The model construction is based on an ideal electrical model for the device, complemented with components to represent the nonideal electrical characteristics, known as *parasitic components*, which reflect the actual implementation of that device on the silicon substrate. The parasitic components can be further broken down into intrinsic and extrinsic parasitic components. *Intrinsic components* are part of the intrinsic device (e.g., the gate capacitance of a MOS transistor). Extrinsic components are due to effects outside of the device, for example the capacitance between the metal wiring to the transistor gate contact and the grounded substrate. Due to the complexity and low breakdown voltages of deeply scaled CMOS process technology, metal oxide semiconductor (MOS) transistor intrinsic parasitic parameter values tend to have a significant voltage and temperature dependence that complicates the creation of accurate models compared to the case of a transistor on a high resistivity GaAs substrate. However, as will be seen later in this chapter, the influence of reactive parasitic components such as interconnect inductances and capacitances becomes more dominant at mm-wave frequencies, and these values can be extracted quite accurately from the IC layout interconnect and are more stable over voltage and temperature.

A particular challenge for power amplifier design is the low breakdown voltages of deep submicron CMOS. The correspondingly low supply voltages allowed make it difficult to achieve more than 10 dBm of power at RF when using standard architectures, such as those employed on GaAs or SiGe technologies. At 77 GHz the influence of parasitics limits the maximum transistor dimensions which further limits the output power achievable from the transistor on CMOS. More exotic and thus expensive technologies, such as GaAs and even SiGe, thanks to the smaller device sizes required for a given output power, have far better parasitics with respect to mm-wave operation. This means that the layout of the transistor plays a crucial role in its performance on CMOS. Furthermore, the presence of parasitics limits the maximum gain and compromises stability.

The metal backend of standard CMOS is not optimized for RF/mm-wave performance. The lower layers in the metal stack are particularly challenging because they are thin and close to the substrate and thus have high resistive and capacitive parasitic components. However the thick upper metal layers available for power distribution can be exploited to advantage in mm-wave design; more on this later in this chapter. The strict metal coverage density rules of advanced CMOS require metal dummy filling, which in turn impedes the design of matching and tuning circuits used in RF design. Process spreads are also wider in deeply scaled CMOS with further leads to the challenges of producing a robust and high yielding design for the high volume automotive radar market.

Despite these challenges, the drive for integration, in order to reduce overall system cost, means a solution has to be found for mm-wave circuits with sufficient performance for the application on CMOS. Discussion on how this can be achieved will be the focus of the rest of this chapter.

6.4 Active Devices in CMOS

There are two main categories of active (transistor) devices on CMOS, namely N-channel MOS transistor (NMOS) and P-channel MOS (PMOS), which operates with complementary or inverse bias voltages to NMOS. The control terminal of a MOS transistor is called the *gate*, the voltage of which controls the amount of current flow between the source and drain terminals. In general, the smaller you can make the length of the gate terminal between source and drain, the faster the transistor will operate. This is in fact the most important dimension that characterizes a CMOS process technology, so much so that the technology node is called after it. For example, 28 nm CMOS means that the PMOS and NMOS transistors can be manufactured with a gate length, between the source and drain regions, as short as 28 nm. For the more than 50-year history of MOS process technology, the minimum gate length manufacturable has roughly halved every 18 months, in line with Moore's Law. Thanks to such technology scaling of the gate length, active devices with operating frequencies in the 100s of GHz range are now available in technology nodes such as 28 nm CMOS. Conventionally, mostly NMOS devices were used for RF applications due to their approximately two times higher charge carrier mobility (speed at which the current can flow) compared to PMOS. However, in the more advanced CMOS technologies, such as those with high-k metal gate,

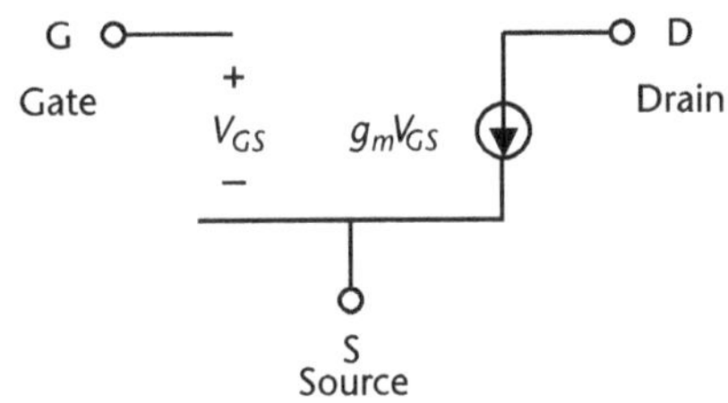

Figure 6.4 Ideal MOS transistor model.

the channel in the PMOS is "strained" to enhance its mobility up to a level similar to the NMOS, so nowadays both types can be used in RF and mm-wave circuits.

Figure 6.4 shows an electrical circuit schematic of an ideal MOS transistor model. The ideal gate terminal has an infinitely high impedance (shown as open circuit) and the connection between source and drain is an ideal current source, also infinitely high impedance. The current that flows between the source and drain terminals is modulated by the voltage between gate and source with a gain factor, known as the transistor transconductance, g_m. In this ideal model, the frequency of operation or the bandwidth is also infinite.

Figure 6.5(a) illustrates how the layout of a MOS transistor "stripe" looks like on a silicon substrate with the critical transistor dimensions of gate length and gate width identified. As you can imagine, for a given gate length, the wider the gate width, W, the more current can flow between drain and source. This is a nice feature of circuit design using MOS transistors, where the transconductance gain, g_m

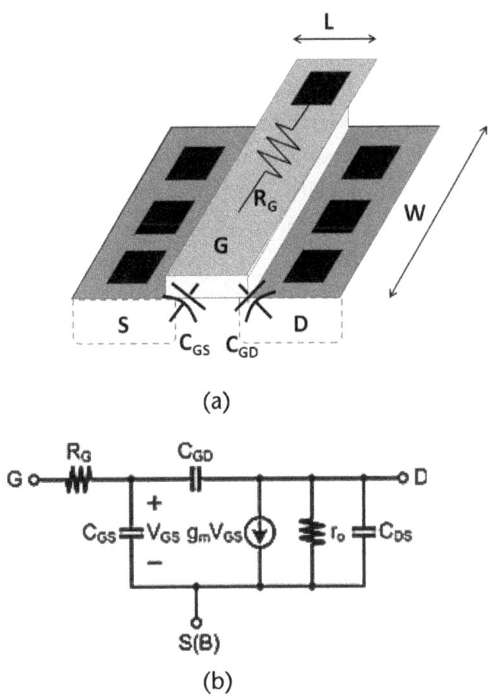

Figure 6.5 (a) MOS transistor cross-section with main parasitics. (b) MOS transistor model with parasitics.

can be set by dimensioning W. However, the intrinsic parasitic capacitance of the gate terminal to the source, drain and substrate (also known as *bulk* region) is proportional to W as well. Since these parasitic capacitances will limit the bandwidth or frequency of operation of the transistor, a more detailed model with parasitic components is needed for high frequency circuit design.

Figure 6.5(b) shows such a model of a MOS transistor with the main parasitic components, which impact high frequency operation, added. Note that the bulk and source are shorted together which is the most common configuration. The high-frequency performance of a transistor on a given technology is generally characterized by the transition frequency f_T and the maximum oscillation frequency f_{MAX}.

The transition frequency is defined as the frequency for which the current gain with the output shorted in AC is unity. Assuming that the source and bulk are both connected to ground, an approximation for the transition frequency is:

$$f_T = \frac{g_m}{2\pi\left(C_{GS} + C_{GD}\right)} \tag{6.1}$$

The parasitic resistances are considered with f_{MAX}, which is the frequency at which the power gain drops to unity, with input and outputs matched. It can be approximated by [10]:

$$f_{\text{MAX}} = \frac{f_T}{\sqrt{4R_G\left(g_{DS} + f_T C_{GD}\right)}} \tag{6.2}$$

It is clear from this equation that to achieve high f_{MAX}, the gate resistance as well as the gate to drain capacitance should be minimized. Figure 6.6 shows a multistripe transistor layout on CMOS, optimized for mm-wave operation. The multiple stripes are also known as *fingers*. The transistor gate and drain connections are routed through vias up to the thick metal layer through the shortest path possible through the thinner and more resistive lower metal layers. Routing these connections as much as possible on thick metal layers provides the lowest extrinsic parasitic resistances and capacitances. Note that the source and bulk are connected together and distributed on metals layers 1 to 3. In a common source configuration, the impedance on this network should also be kept low to avoid gain reduction through excessive degeneration. The use of the thick metal was avoided here, despite its much lower resistance, as the priority is to minimise the extrinsic C_{GS} and C_{DS} parasitics for high frequency operation.

When choosing transistor sizes in millimetre wave design, the channel length L is usually set to the minimum value for highest speed. In order to minimize R_G, the gate finger width is kept small and is contacted on both sides by metal, which further reduces it in theory by a factor of four. A transistor of width W can then be made up of an integer number of unit fingers or stripes up to a practical limit of around 40 stripes before "current crowding" into the stripes closer to the thicker metal has an impact. From the formulas for f_T and f_{MAX}, it is seen that changing the number of fingers does not affect the f_{MAX}. Also, f_T doesn't change since g_m and the capacitances are proportional to W. For a given finger width, R_G is inversely

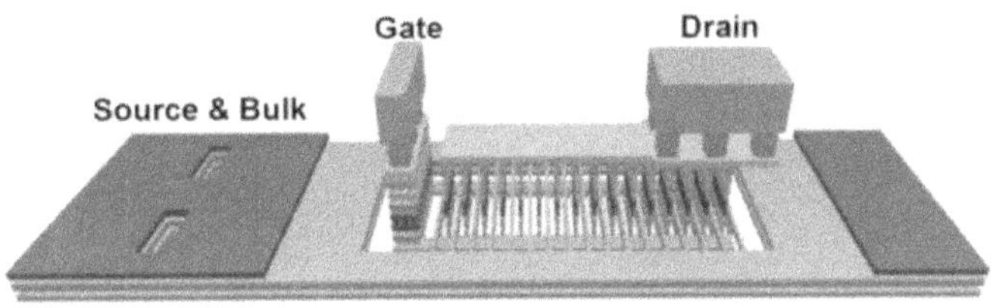

Figure 6.6 Transistor layout optimized for mm-wave. (© 2015 IEEE [11].)

proportional to the number of fingers, while g_{ds} and C_{gd} are proportional to the number of fingers.

The device models supplied with the process design kit on CMOS include the intrinsic device parasitics, such as the overlap capacitances from source and grain to the gate. The extrinsic parasitic components, due to the interconnect can be extracted from the layout using a parasitic extraction tool and included in the circuit simulations. For larger structures, it is more accurate to use a finite element based electromagnetic simulator so that inductance and transmission line effects can be included in the model.

Referring to Figure 6.6, the input signal to the gate is brought in from the left on one of the thick metal layers higher up in the stack and the drain output is brought out to the right. This simplifies top level signal path routing makes it easy to scale up the overall transistor size with multiples of this unit structure in parallel, for higher power applications while minimizing the extrinsic C_{GD} interconnect parasitic. This unit transistor structure and layout has been proven effective for mm-wave for process technologies from 40 nm planar down to 16 nm finFET [11–13]. Meeting the design rules for electromigration gets tougher as the lower metals get thinner and contacts and vias get smaller as technology scales; this ultimately sets a limit on the current per stripe that is achievable.

6.5 Passives in CMOS

While the performance of active devices gets worse at mm-wave frequencies, the situation generally gets better for passives (Figure 6.7). For the same impedance, inductors are smaller for higher frequencies and as wavelengths become smaller transformers also become smaller and less lossy. In addition, advanced CMOS processes have two or more thick metal layers available at the top of the stack, furthest away from the substrate which can be used to make high quality factor "Q" inductors and transformers and lower loss transmission lines. However, as Moore's Law advances, even these metals are getting thinner and closer to the lossy substrate.

On-chip transformers are popular at mm-wave as they serve multiple important purposes; they can act as baluns, making a balanced (differential) signal from a single-ended signal. They can be used for DC biasing, through the common-mode, center-tap point in the differential secondary inductor. Thirdly, they can be used to provide a power match between two capacitive impedances, for example between

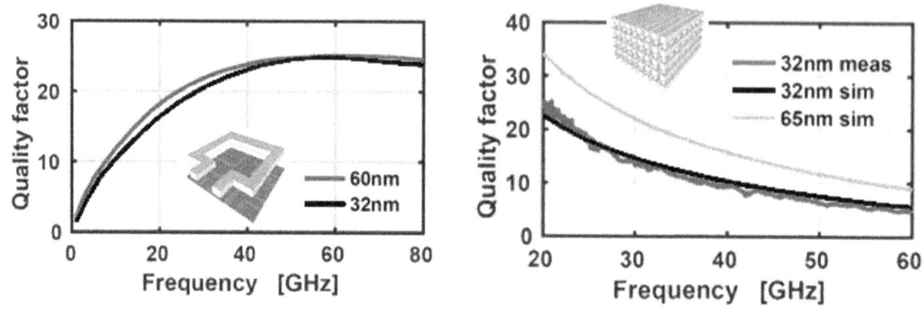

Figure 6.7 Inductor and capacitor Q versus frequency. (© 2013 IEEE [14].)

the drain capacitance of a differential transistor output stage and the gate capacitance of the next stage.

Different topologies of on-chip transmission lines are possible. They can be either single-ended or differential and in the latter case, can be implemented with or without a ground plane. The most important parameters of a transmission line are its characteristic impedance Z_0, the phase constant and the attenuation constant. These are determined by the physical dimensions of the metal tracks and the electrical characteristics of the conductors and the dielectrics in the backend metal stack. Losses are caused by conductor resistance and the conductivity in the oxides and the silicon bulk material.

The situation with capacitors is mixed. Metal on metal (MOM) caps are now commonplace on advanced CMOS, thanks to the more closely spaced metal layers and provide a high Q alternative to the MOS caps readily available on CMOS from the transistor gate capacitance. The MOM's capacitance per unit area improves with technology scaling, as the distance between metals and metal layers reduces. The downside is that the metals get thinner resulting in higher loss due the increased interconnect resistance. Capacitor Q naturally decreases first order with increasing frequency but on deeply scaled CMOS the decrease in Q is more than first order, due to thinner metals on the more advanced nodes.

6.6 Circuit Architectures Suitable for Advanced CMOS

6.6.1 The Transmit Power Amplifier

The transmit power amplifier is a particular challenge on advanced CMOS where the peak drain voltage of the high speed transistors cannot go much more than approximately 1V for reliable operation. Referring back to Figure 6.3, the TX power amplifier drives the antenna directly and this is typically a single-ended drive to a patch antenna structure on the PCB. Delivering 13 dBm or 20 mW into a 50 ohm load calls for a peak-peak voltage swing of 2.82V. A way forward is to transformer a couple number of low voltage swing, but high current output, amplifier stages to the 50 ohm output port to the TX antenna [11].

The gain stage is designed as a fully differential, neutralized, common-source amplifier as shown in the bottom right of Figure 6.8. The DC bias voltage on the

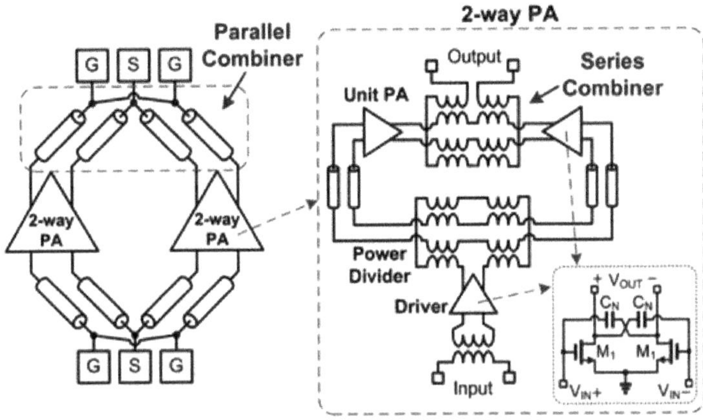

Figure 6.8 PA with parallel—series power combining. (© 2015 IEEE [11].)

input gates and output drains are set to ensure compliance with reliability requirements, with the gate bias usually set for high efficiency, class B operation of the differential pair for the high power output stages. The DC biases are applied via the common mode center tap points of the input and output transformers to each stage. The cross-coupled capacitors shown are sized to neutralize the gate to drain Miller capacitances of the amplifier, thus enhancing its gain and stability [11]. The NMOS transistors in the differential pair are each made up of the transistor structure described earlier, laid out to minimize the effect of the parasitic components at mm-wave.

A number of these optimized unit transistor cells have to be paralleled so that the output current sums up to the required value. There is a limit to how far you can go with this as the top level interconnect parasitic on the drain increases faster the output power improvement, resulting in diminishing returns as the number of paralleled units exceeds five [11]. A further problem is that the output impedance of such high current output stages will be significantly lower than 50 ohms and thus will be difficult to provide a broadband match over the 5 GHz from 76–81 GHz. A solution to both of these problems is a power-combining network that uses a combination of series and parallel combining where the transformer ratio can be kept small to provide a broadband match. As illustrated in Figure 6.8, the winding in series can be designed to match to the low impedance amplifier output and these sum up in series to provide an approximate 100 ohm impedance in the transformer secondary. These are then paralleled at the output to match to 50 ohms.

While the series/parallel transformer combines the output currents of multiple low voltage amplifier stages with good efficiency at mm-wave, the area taken by the overall power amplifier (PA) is relatively large. The power amplifier structure in [12] is a lot more compact in terms of $_{PSAT}$/mm of core area, thanks to the use of higher power cascode amplifier stages on 22 nm SOI-CMOS technology. This amplifier topology works well even at mm-wave thanks to the much lower intrinsic source/drain parasitic capacitances on SOI. On bulk CMOS, these much larger parasitics on the intermediate nodes of the stacked transistors leak a lot of the mm-wave signal current to ground, impacting the amplifier efficiency.

6.6.2 The TX Phase Shifter

An RF phase shifter that can shift the phase of one TX channel relative to another is a useful capability to have on a multichannel FMCW radar transmitter for beam pattern alignment and/or to implement phase coded MIMO schemes [4]. Ideally true time delay shifters are preferred to avoid beam squint for wide bandwidth radars [15], but these tend to be large, expensive, and lossy structures. A vector modulator based active phase shifter [16], more suitable for CMOS implementation, is an acceptable compromise over the relatively modest fractional bandwidths required over the 76–81 GHz automotive radar band. Referring to Figure 6.9, the input RF signal can be passively split into I/Q quadrature components using either lumped element or a quarter wave hybrid coupler, both of which are small at mm-wave frequencies. This can have programmable phase steps over the full 360-degree range with the digital coding mapped to appropriately scaled I and Q vectors for the Gilbert Cell based VGA's gain settings. Some sort of calibration scheme is required to maintain phase step accuracy over process and temperature below the 5 degree typically required for today's automotive radar beam patterns. Loop back schemes, [17–19], which leverage the advantage that comes from full SoC implementation, are now commonly used for this sort of calibration in phased array communications systems and radars.

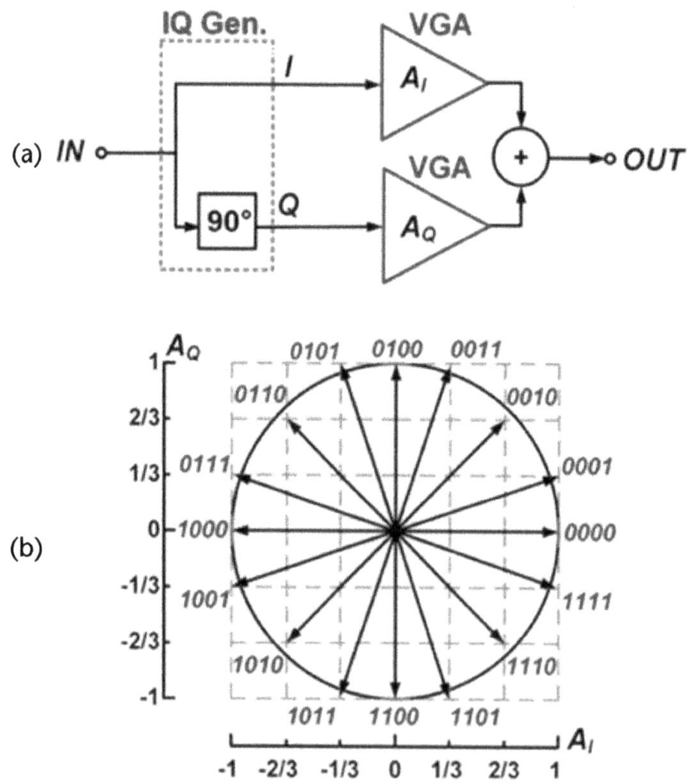

Figure 6.9 Vector modulator based phase shifter. (© 2015 IEEE [16].)

6.7 The LO/FMCW Chirp Generator

A fractional-N PLL is the most popular choice for FMCW chirp generation in today's automotive radars. As illustrated in Figure 6.10, the desired chirp shape and parameters can be applied digitally to the sigma-delta modulator that controls the feedback divider and through feedback action of the loop, onto the VCO output. The loop filter stabilizes the loop and determines overall loop dynamics, such as settling time of the output and chirp retrace time. In general, the wider the loop BW, the faster the settling time of an output frequency chirp.

Output phase noise is another important characteristic of the PLL. This is also a key performance metric in radar as weak reflections from small targets can get buried in the phase noise sidebands around strong target reflections. There are a number of noise sources in the PLL that contribute to output phase noise. The phase detector noise as well as quantization noise from the sigma delta modulator-controlled divider can be referred to the phase detector input and has a low pass transfer function to the output, with a roll-off set by the loop filter profile. Phase noise from the VCO is suppressed inside the loop bandwidth but is the dominant noise contributor at frequency offsets outside the loop bandwidth, where it follows the free running oscillator phase noise profile. As shown in Figure 6.11, the overall output phase noise profile is strongly shaped by the loop bandwidth setting. Wider loop bandwidths suppress more of the oscillator noise but pass more of the phase detector noise to the output. Narrow loop bandwidths attenuate the phase detector noise more but suppress less of the oscillator noise. The optimum loop bandwidth for lowest noise tends to be at a frequency close to the intercept of the phase detector and oscillator noise profiles. Wider loop bandwidths result in output phase noise levels significantly higher than the intrinsic noise of the oscillator. This will result in high sideband levels around strong target reflections in the receiver baseband spectrum of an FMCW radar, making smaller targets in nearby range bins more difficult to detect.

Unfortunately, the optimum loop bandwidth for phase noise is generally not wide enough to allow the PLL output frequency switch fast enough for fast chirp FMCW radar. Speed-up techniques such as dual loop [19] or loop bandwidth switching between PLL acquisition mode and ramping mode have limited success in

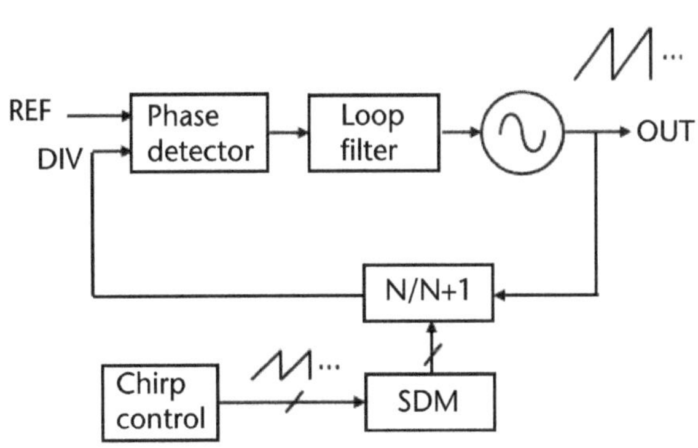

Figure 6.10 FMCW chirp generation using a fractional-N PLL.

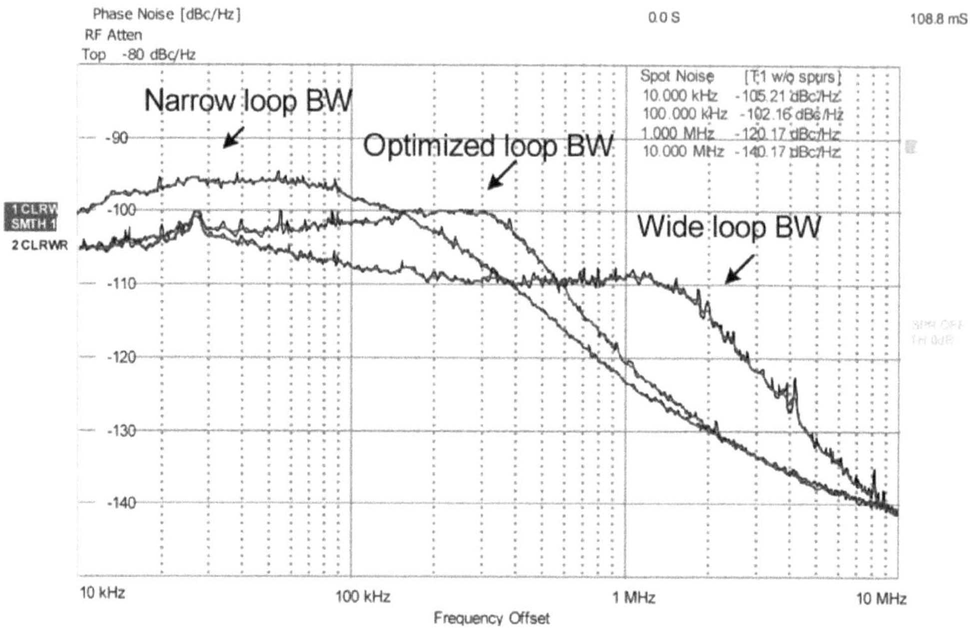

Figure 6.11 PLL phase noise profiles for various loop bandwidths. (© 2014 IEEE [21].)

charge pump PLLs with analog phase detectors and loop filters with highly variable components such as resistors and capacitors. Loop settling speed-up techniques have better success in an *all digital* PLL where the phase detector is replaced by a time to digital converter (TDC) with a well-defined gain and the loop filter is a digital filter with highly deterministic and adjustable loop dynamics parameters [20–24]. The gain of the digitally controlled oscillator, which replaces the analog tuned VCO is the only unknown, but this can be measured easily on chip [24]. The approach described in [23–24] estimates the amount of steady state error that a type II PLL maintains at its phase detector inputs while ramping and preloads the digital loop filter with this amount so that the settling transient due to the acquisition of the steady state error is minimized. This feed-forward technique works independently of the PLL loop bandwidth as shown in the results of Figure 6.12(b) compared to the conventional type II settling of Figure 6.12(a), where the chirp retrace settling time is highly dependent on the loop bandwidth setting.

With the digitally enabled speed-up scheme, as demonstrated in Figure 6.12(b), chirp settling is essentially independent of the loop bandwidth. This means that the PLL loop bandwidth can now be independently optimized to achieve minimum overall phase noise, allowed by the phase detector and oscillator phase noise contributions. The goal thereafter is to minimize both of these noise contributors to achieve lowest overall integrated phase noise.

The major limiting factor for oscillator performance at mm-wave is the quality factor of the tuning capacitors. We saw earlier in the chapter how capacitor Q decreases with frequency and degrades even further with technology scaling. This is reflected in Figure 6.13 which shows phase noise increasing more than the 6 dB/octave expected with higher oscillator frequency. This indicates that best LO phase

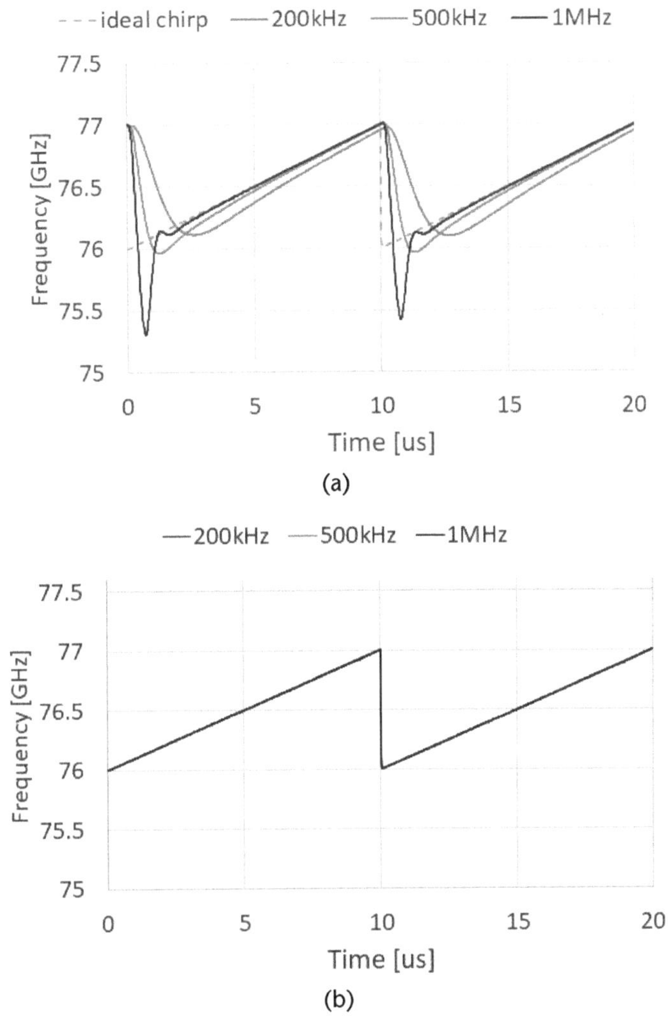

Figure 6.12 (a) Conventional chirp retrace settling transients with various PLL loop bandwidths. (b) As with Figure 6.12(a), but with the feed-forward settling technique [23–24].

noise performance can be obtained with a lower frequency oscillator followed by a frequency multiplier. This approach enables a wider oscillator tuning range also as the ratio of tuning capacitance to background capacitance is more favorable at the lower frequency. Another benefit is that it avoids the need for a mm-wave frequency divider after the oscillator, when used in a PLL.

While a digital PLL has significant advantages over an analog PLL when it comes to fast chirp generation, its major performance bottleneck is the phase detector. An analog phase detector can very precisely output a voltage or current, to an analog loop filter, that is proportional to the phase error between its reference and feedback inputs. In a digital PLL this phase error must be quantized to a digital code so that it can be input to the digital loop filter. This function can be performed by a TDC [20] or better still, a phase to digital converter (PDC), that is exactly scaled to the oscillator period [26]. This turns out to be the main performance bottleneck in a

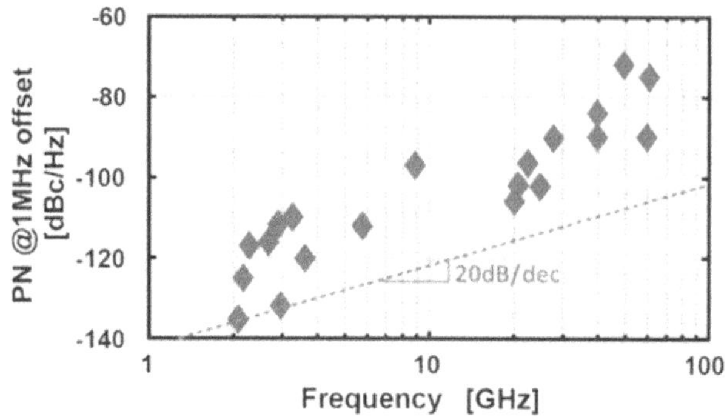

Figure 6.13 Published phase noise performance versus oscillator frequency. (© 2015 IEEE [25].)

digital PLL as the phase detector noise is limited by the quantization noise level or resolution that can be achieved in the TDC or PDC. Poor TDC resolution shows as increased phase noise and spurious in the PLL output. The rotary travelling wave oscillator (RTWO) used in [26] doubles up as the digitally controlled oscillator (DCO) and also the PDC function in the digital PLL. This is an architecture well suited for SoC implementation; good RTWO phase noise performance requires thick metals for a low loss transmission line and fast inverters to maintain sharp edges in the travelling wave, both readily available on advanced CMOS. In addition, as shown in [26], a very high resolution PDC can be implemented by placing latches, clocked by the reference, at evenly spaced taps along the RTWO transmission line. The RTWO can be designed to operate at low frequency for best oscillator phase noise and tuning range and, as shown in [27], a frequency multiplier can be implemented by combining the appropriate number of edges from phase taps spaced around the RTWO transmission line.

The LO on a PMCW radar transceiver can be greatly simplified in comparison to what is required in FMCW. With PMCW the bandwidth required for radar range resolution is provided by the wideband TX waveform. The LO is no longer ramping, thus avoiding all of the challenges with ramp linearity seen with FMCW. Since the LO now operates at a fixed frequency for the duration of the measurement cycle, a much simpler, integer-N PLL can be used for frequency synthesis and output frequency resolutions of multiples of the reference frequency are sufficient. Furthermore, since PMCW does not rely on frequency differences between TX and RX, the oscillators for the TX LO and RX LO can be different. This avoids the need for 77 GHz LO routing between the TX and RX channels, thus opens up the possibility of simpler and lower power LO distribution and good separation between TX and RX on chip.

6.8 The Receiver Signal Chain

The RX signal chain at block diagram level looks similar between FMCW and PMCW. However, after the mixer on the RF frontend, there are significant differences in the signal chain bandwidths and data processing rates between the two.

6.8.1 RX Frontend

In both FMCW and PMCW receivers, the feed from the RX antenna is typically single-ended, which is converted to differential signalling as soon as possible on chip using a balun. The balun also helps greatly with electrostatic discharge (ESD) robustness of the RX as a direct electrical connection to the LNA input is avoided and ESD diodes can be placed at the center tap of the transformer without impacting the RF signal strength. The LNA is the first active circuit and gain stage that the RX signal sees, and its purpose is to provide gain without contributing significant noise. As in any receiver, the noise figure of the RX frontend and LNA plays a dominant role in setting the overall noise figure for the radar receiver. The fully differential, neutralized common source NMOS amplifier is a popular choice for the amplifier stages in the LNA [5]. The other purpose of the RX frontend is to provide a good broadband match to the 50 ohm RX antenna. The absolute value of return loss (reflection parameter S11) is a good measure of input matching and is widely adopted in the specifying RX inputs. When $|S11| < -10$ dB more than 90% of the power is delivered to the load and the receiver is considered matched.

A down-conversion mixer is needed to realize frequency translation of the received signal at mm-wave down to baseband. Direct conversion is necessary in FMCW since the TX signal is the LO for the RX mixer. Good linearity is essential in the RX mixer as intermodulation products between two or more strong target reflections would result in ghost targets appearing in the baseband spectrum. Second order nonlinearity is to be minimized at all costs as a second order product creates a frequency difference that—for two strong targets at similar range—will appear close to DC, at a range equivalent to their range difference. Such a ghost target would risk falsely triggering automatic emergency braking. Hence a double balanced mixer topology is preferred and care is taken to keep the layout symmetric and well matched differentially.

Conflicting requirements of the RX frontend are noise figure and input compression (P1dB). Good noise figure is needed for weak reflections coming in from long range targets. High P1dB is needed to handle strong signals from targets at short range. It is difficult to achieve for both scenarios with a mm-wave receiver on scaled CMOS as the high LNA gain is required for noise figure causes compression at the LNA output/mixer input stage. A way to mitigate this problem is to use a complex down converter architecture as shown in [19]. The complex or I/Q down converter cancels the down conversion of the image band, which in FMCW radar does not contain a useful signal and thus only contributes noise. Hence this approach reduces the noise figure of the frontend by ideally up to 3 dB since only half the noise BW is down converted. This means that the LNA gain can be reduced and still maintain the existing RX noise figure while providing more headroom for strong signal reception before compression kicks in. The price for this is area and power consumption for the additional quadrature path from the RX mixer to the ADC output.

6.8.2 Radar RX Baseband

It is after the down convert mixers that the RX signal paths for FMCW and coded waveforms such as PMCW diverge. Thanks to the FMCW de-chirping operation, the baseband bandwidth just needs to cater for the beat signals between the TX and RX chirps, which is typically just in the 10s of MHz bandwidth. The PMCW baseband signal path needs to preserve the RF bandwidth that was required for range resolution. Thus baseband filtering and RX ADCs need to be multi-GHz in bandwidth and produce data to be processed at multi-Gbps rates. This inevitably results in RX chains that are significantly higher power than in FMCW. The magnitude of this drawback reduces though with Moore's Law scaling over time but with today's advanced CMOS nodes this is still a significant disadvantage.

The gain and phase matching between multiple RX chains is also important for digital beamforming. Channel filters, particularly those of high order are the most difficult to match, particularly for signal components close to the filter corner frequencies. This problem can be easily managed through the use of feedback-based active filter architectures in the analog baseband of FMCW radars. If necessary, built-in self-test (BIST) schemes using on chip test tone generators can be utilized here as well at low cost.

This is a more difficult and costly problem to solve with the multi-GHz baseband bandwidths required by digital coded waveforms such as PMCW and OFDM, particularly as the RX channel count scales upward. As well as the beamforming distortions caused by gain and phase mismatches, channel group delay mismatches result in blurring of the main lobe of target reflections, analogous to what nonlinearity in the chirp does to reflections in FMCW radar.

The baseband frequency in an FMCW receiver is directly proportional to the target range. This can be exploited by using a high-pass filter-based equalizer to suppress the strong reflections from nearby targets, thus preserving more of the ADC dynamic range for the weaker reflections from far away targets. Since the RX frontend section at mm-wave is usually fixed gain for simplicity, it falls on the baseband section to provide RX gain programmability. It makes sense to introduce this after the high pass filter (HPF) stage in the FMCW receiver as with the strong reflections from short-range targets suppressed, there is scope to amplify the residual signals in order to maximize the signal input to the baseband ADC. Hence a programmable gain amplifier follows the high pass filter (HPF) in the FMCW receiver. In many scenarios, such as adaptive cruise control, there are no short range targets in the FOV and this it can be useful to have some programmable gain in the HPF block also, as shown in the FMCW receiver of Figure 6.14. The best overall RX noise figure will be achieved for any given gain setting when the gain reductions are first made at the backend of the signal chain and sequenced toward the frontend. This

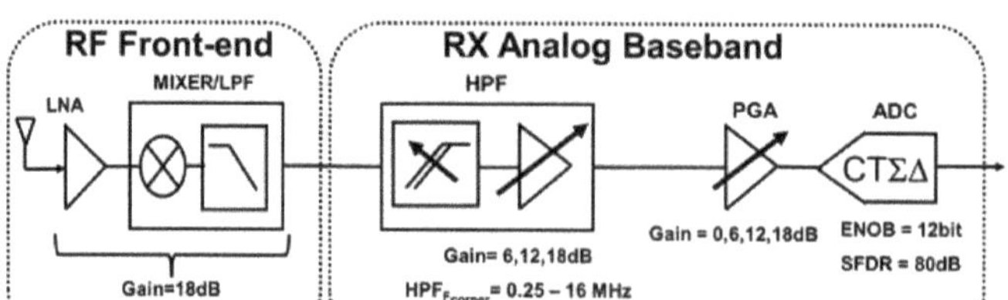

Figure 6.14 An FMCW radar receiver.

is particularly suitable in the ACC use case where lowest RX noise figure is often desired in order detect targets up to 300m.

Aliasing in the ADC of an FMCW receiver has to be avoided at all costs, as a strong reflection from a target at a range that corresponds to beyond the Nyquist frequency will alias down to a low baseband frequency and as appear as a shorter range ghost target and could trigger emergency brakes. A high order antialiasing filter prevents this and was the approach commonly used in FMCW. This approach is undesirable for a number of reasons on advanced CMOS such as area, power consumption, and group delay variation across channels with such high order filters. Hence oversampled ADC architectures are more preferred because they greatly relax the order required of the antialiasing filter. They instead place the burden on backend digital filtering to decimate the high speed but low resolution input samples down to low rate, high resolution output samples. These architectures are well suited to implementation on advanced CMOS and with the filtering mostly digital, there is perfect matching across RX channels, which is ideal for digital beamforming. A popular oversampling ADC choice is the continuous-time sigma-delta ADC architecture that already has an inherent first order low pass response in the signal transfer function [28].

6.9 Summary

Process technology scaling as well as new circuit architectures suitable for CMOS implementation have now enabled complete radar SoCs to be implemented that can meet the needs of automotive radar applications. The performance requirements of the radar transceiver have increased with operating bandwidths increased from 1 to 5 GHz while remaining within power and cost budgets of the original 76–77 GHz systems.

Phase noise, chirp settling time, and chirp linearity are the big challenges for performance of FMCW radar. Impairments in these performance metrics show up as ghost targets artifacts and spectral spreading in range-Doppler plots.

Digitally coded radar waveforms operate with a fixed LO frequency and thus avoid the complications of generating a highly linear frequency chirp. However, maintaining group delay matching between channels across multi-GHz bandwidths is now the challenge as impairments here will result in spectral spreading in range-Doppler, analogous to effects due to chirp nonlinearity in FMCW. Power consumption of the wideband data converters as well as in the processing of the much higher data rages output from them are also major challenges today but are ones that should reduce as technology continues to scale.

Interference among radar sensors is an emerging problem as more radar sensors are deployed in cars. The digital coded radar waveforms can be more resilient against this; however, their wider RX signal bandwidths are now open to interferers throughout the band. Interference received in FMCW does more damage to the desired signal, however, fast settling chirp generators can be utilized in avoidance and mitigation schemes such as fast frequency hopping and randomized time shifting of the transmit chirps.

As vehicles traverse from levels of autonomy towards Levels 4 and 5, the requirements of the radar sensor continue to advance. It is inevitable that more digital techniques will be employed in the waveform design and signal processing. Thanks to Moore's Law scaling and the continuing innovations in circuit architectures, CMOS semiconductor technology has shown it can also rise to meet the challenge.

References

[1] Barrick, D. E., "FM/CW Radar Signals and Digital Processing," *NOAA Technical Report* ERL 283-WPL 26, July 1973.

[2] Schroeder, C., and H. Rohling, "X-Band FMCW Radar System with Variable Chirp Duration," *2010 IEEE Radar Conference*, 2010, pp. 1255–1259.

[3] Kingsley N., and J. Guerci, *Radar RF Circuit Design*, Norwood, MA: Artech House, 2016.

[4] Feger, R., et al., "A 77-GHz FMCW MIMO Radar Based on an SiGe Single-Chip Transceiver," *IEEE Transactions on Microwave Theory and Techniques*, Vol. 57, No. 5, May 2009, pp. 1020–1035.

[5] Guermandi, D., et al., "19.7 A 79 GHz Binary Phase-Modulated Continuous-Wave Radar Transceiver with TX-to-RX Spillover Cancellation in 28 nm CMOS," *2015 IEEE International Solid-State Circuits Conference (ISSCC) Digest of Technical Papers*, 2015, pp. 1–3.

[6] Sturm, C., et al., "Performance Verification of Symbol-Based OFDM Radar Processing," *2010 IEEE Radar Conference*, 2010, pp. 60–63.

[7] Bourdoux, A., Parashar, K., and M. Bauduin, "Phenomenology of Mutual Interference of FMCW and PMCW Automotive Radars," *2017 IEEE Radar Conference (RadarConf)*, 2017, pp. 1709–1714.

[8] Recommendation ITU-R M.2057-1: "Systems Characteristics of Automotive Radars Operating in the Frequency Band 76–81 GHz for Intelligent Transport Systems Applications," January 2018.

[9] U.S. Department of Commerce, Export Control Classification Number 3A001 b.4.e.2.

[10] Tsividis, Y., *Operation and Modelling of the MOS Transistor*, Second Edition, Oxford, UK: Oxford University Press, p. 503, 1999.

[11] Zhao, D., and P. Reynaert, "An E-Band Power Amplifier with Broadband Parallel-Series Power Combiner in 40-nm CMOS," *IEEE Transactions on Microwave Theory and Techniques*, Vol. 63, No. 2, February 2015, pp. 683–690.

[12] Çelik, U., and P. Reynaert, "An E-Band Compact Power Amplifier for Future Array-Based Backhaul Networks in 22nm FD-SOI," *2019 IEEE Radio Frequency Integrated Circuits Symposium (RFIC)*, 2019, pp. 187–190.

[13] Philippe, B., and P. Reynaert, "24.7 A 15dBm 12.8%-PAE Compact D-Band Power Amplifier with Two-Way Power Combining in 16nm FinFET CMOS," *2020 IEEE International Solid-State Circuits Conference (ISSCC)*, 2020, pp. 374–376.

[14] Mammei, E., et al., "A 33.6-to-46.2GHz 32nm CMOS VCO with 177.5dBc/Hz Minimum Noise FOM Using Inductor Splitting for Tuning Extension," *2013 IEEE International Solid-State Circuits Conference Digest of Technical Papers*, 2013, pp. 350–351.

[15] Longbrake, M., "True Time-Delay Beamsteering for Radar," *2012 IEEE National Aerospace and Electronics Conference (NAECON)*, 2012, pp. 246–249.

[16] Pepe, D., and D. Zito, "Two mm-Wave Vector Modulator Active Phase Shifters with Novel IQ Generator in 28 nm FDSOI CMOS," *IEEE Journal of Solid-State Circuits*, Vol. 52, No. 2, February 2017, pp. 344–356.

[17] Orlandi, F., "Method for Checking an FM/CW Type Radio Altimeter, and Radio Altimeter Designed for the Implementation of this Method," U.S. Patent US6,437,730, August 2002.

[18] Ku, B., et al., "75–85 GHz Flip-Chip Phased Array RFIC with Simultaneous 8-Transmit and 8-Receive Paths for Automotive Radar Applications," *2013 IEEE Radio Frequency Integrated Circuits Symposium (RFIC)*, 2013, pp. 371–374.

[19] Ginsburg, B. P., et al., "A Multimode 76-to-81GHz Automotive Radar Transceiver with Autonomous Monitoring," *2018 IEEE International Solid-State Circuits Conference (ISSCC)*, 2018, pp. 158–160.

[20] Staszewski, R. B., et al., "All-Digital TX Frequency Synthesizer and Discrete-Time Receiver for Bluetooth Radio in 130-nm CMOS, *IEEE Journal of Solid-State Circuits*, Vol. 39, No. 12, December 2004, pp. 2278–2291.

[21] Wu, W., Staszewski, R. B., and J. R. Long, "A 56.4-to-63.4 GHz Multi-Rate All-Digital Fractional-N PLL for FMCW Radar Applications in 65 nm CMOS," *IEEE Journal of Solid-State Circuits*, Vol. 49, No. 5, May 2014, pp. 1081–1096.

[22] Chetniak, D., Samori, C., and S. Levantino, "Digitally-Intensive Fast Frequency Modulators for FMCW Radars in CMOS" (Invited Paper), *2019 IEEE Custom Integrated Circuits Conference (CICC)*, 2019, pp. 1–8.

[23] Chillara, V., Dalton, D., and P. Dato, "Fast Settling Sawtooth Ramp Generation in a Phase Locked Loop," U.S. Patent US10,340,926, July 2019.

[24] Chillara, V., Dalton, D., and P. Dato, "Fast Settling Ramp Generation Using Phase Locked Loop," U.S. Patent US10,931,290, February 2021.

[25] Jany, C. et al., "A Programmable Frequency Multiplier-by-29 Architecture for Millimeter Wave Applications," *IEEE Journal of Solid-State Circuits*, Vol. 50, No. 7, July 2015 pp. 1669–1679.

[26] Takinami, K., et al., "A Distributed Oscillator Based All-Digital PLL with a 32-Phase Embedded Phase-to-Digital Converter," *IEEE Journal of Solid-State Circuits*, Vol. 46, No. 11, November 2011, pp. 2650–2660.

[27] Shehata, M. A., et al., "A 32–42-GHz RTWO-Based Frequency Quadrupler Achieving >37 dBc Harmonic Rejection in 22-nm FD-SOI," *IEEE Solid-State Circuits Letters*, Vol. 4, 2021, pp. 72–75.

[28] Dong, Y., et al., "A Continuous-Time 0–3 MASH ADC Achieving 88 dB DR with 53 MHz BW in 28 nm CMOS," *IEEE Journal of Solid-State Circuits*, Vol. 49, No. 12, December 2014, pp. 2868–2877.

PART 2

Challenges and Solutions for the Automotive Environment

CHAPTER 7

Radar Challenges from the Automotive Scene

Gary Clayton, Tim Campbell, Ryan Fan, and Matt Markel

7.1 Introduction

The operating environment of automotive radar presents several challenges for radar designers. Automotive radars must operate in weather, in dense clutter environments, and at low grazing angles that make fading from multipath a common occurrence. This operating environment, coupled with the need for low cost and the challenging requirements associated with higher levels of autonomy, imposes many challenges for the designer of radar systems for autonomous vehicles that are different from other radar applications. For this discussion we focus on SAE Level 4 autonomy, which assumes that the sensor suite for an autonomous vehicle (AV) must provide reliable information about the world around the AV with high precision and recall.

In this chapter, we present data showing the *challenges* of the automotive environment, specifically those related to the large dynamic range in the scene and the deleterious effects of multipath reflections. In addition, the automotive radar environment presents three unique and very challenging *constraints* on the radar designer. These three constraints—which exacerbate the environmental challenges—are (1) the very long range swath (ratio of the farthest range to shortest) needed, (2) the requirement to create images of a very nonsparse clutter scene, and (3) the need for pulse widths equal to or greater than the range swath, requiring a simultaneous transmit and receive capability.

Arguably, removal of any one of these three constraints will dramatically simplify the work of the radar designer, and allow use of solutions common to other radar fields, such as defense and aerospace. For example, the large range swath in constraint 1 can require an additional 60–70 dB of dynamic range[1] over that present in the scene. If the clutter scene was sparse or not of interest (constraint 2), the large range swath may not be an issue as the radar could process only exoclutter returns, or sequentially process various parts of the range swath by spatially nulling the small number of scatterers in the sparse scene to reveal other returns. Similarly, if the radar were able to operate with very short pulses (thus removing constraint 3), it could again sequentially process sections of the range swath, choosing not to

1. Path loss from a target at 320m is 60 dB greater than that from one at 10m.

process large close-in returns when trying to view smaller returns at the end of the range swath and "range-gating out" the close returns in this situation.

7.1.1 Range Swath

Radar can contribute to the AV sensor suite both as a short range, perimeter sensor and as a long range all-weather sensor. The ranges of concern for an AV extend from a meter or less in urban environments to a maximum sensing range determined by several vehicle and scenario-related parameters, including the assumed maximum closing rate, the assumed maneuverability of the vehicle (e.g., stopping distance, ability to execute a lane change), and operating conditions such as type of roadway (e.g., surface streets, freeways, terrain), and current weather conditions (e.g., raining, wet road). Speed limits in United States freeways are posted up to 85 mph (136.8 kph). For passenger vehicles, this can easily translate to a maximum sensing range requirement of more than 300 meters when reaction time and the latency in physical actuation of maneuvers are considered. For large trucks on a freeway this maximum range can extend to nearly 1,000m. While this range swath of 1,000m or less is short compared to many other types of radars, the ratio of the minimum to maximum range is extreme, and as just discussed can drive an additional 60 dB or more of dynamic range for the radar design.

7.1.2 Imaging Dense Clutter

For fully autonomous vehicles the system's reliance on radar at night or in weather may increase, and the radar may be required to detect and localize a range of road users from pedestrians to large trucks whether they are moving or stationary. Therefore, a radar capable of both being an imaging radar as well as providing moving target indication (MTI) may be desirable, as shown in Figure 7.1. Fortunately, this aids AVs in other ways, as at shorter ranges the image of the radar clutter can provide significant context for the radar information, including the location of roads, curbs, and other stationary objects for use in the perception of the scene.

This information is useful at longer ranges as well, however the cross-range resolution degrades with range, complicating the localization of objects. For example, at longer ranges a radar may have difficulty determining if a stopped car is in its lane or on the side of the road, as well as the number of objects at that range.

Another observation from the images is that they are very dense. Even in a visually sparse scene (a test track) Figure 7.2, the range-azimuth image shows that there is significant clutter at all ranges and at many angles.

7.1.3 Simultaneous Transmit and Receive

Many automotive radars use simultaneous transmit and receive (STAR) to increase the energy on targets while still observing returns from close-in objects. However, STAR can significantly increase the dynamic range needed to observe a small target far away in the presence of a larger target at a close range (called the near–far problem).

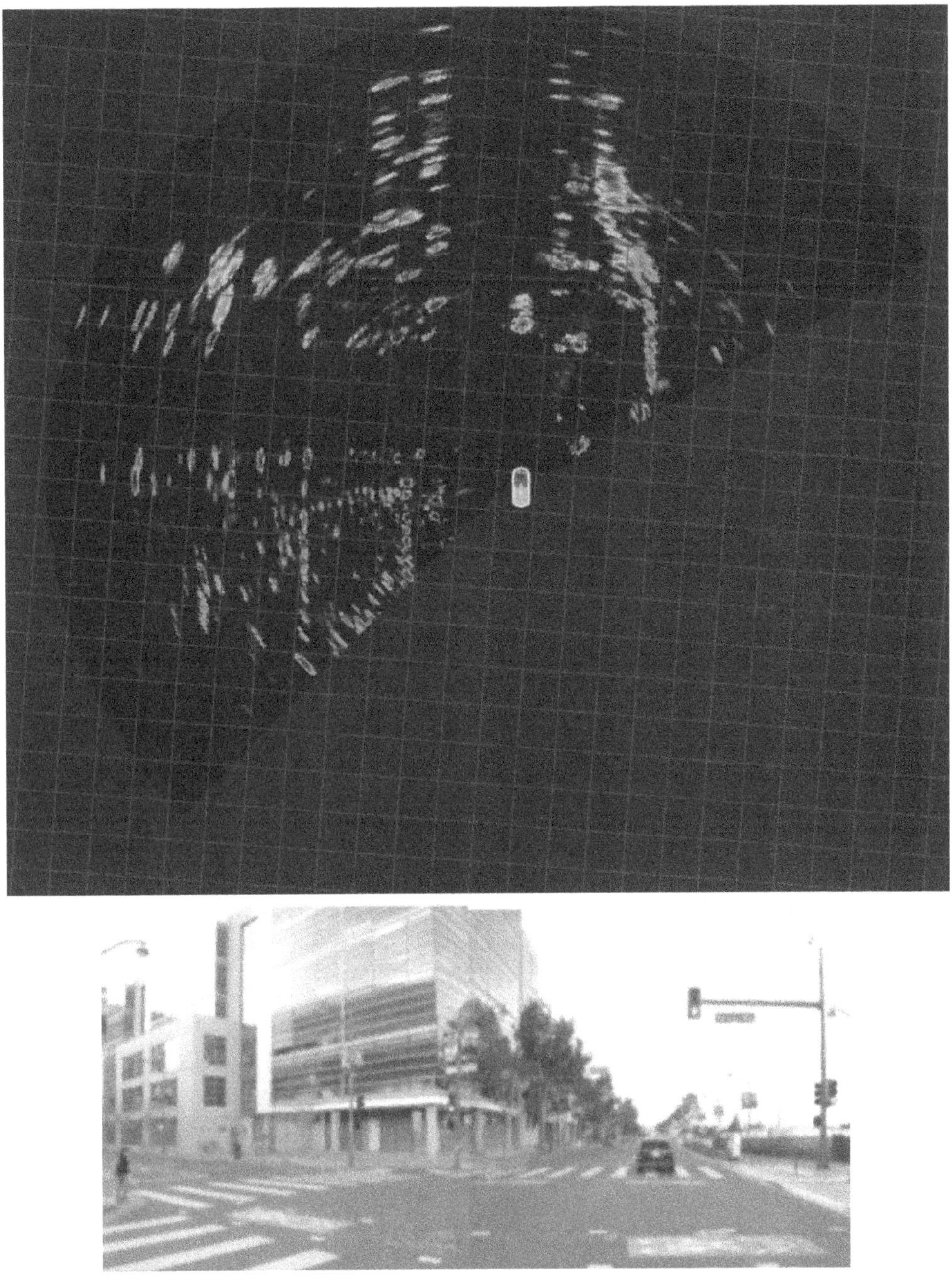

Figure 7.1 Short range scene, with the range-azimuth radar view above (gray scale is backscattered power) and camera below. Strong returns from both moving and stationary objects are present, and the radar image provides significant contextual information about the scene with pedestrians, passenger cars, and stationary objects all visible. (Source: Waymo.)

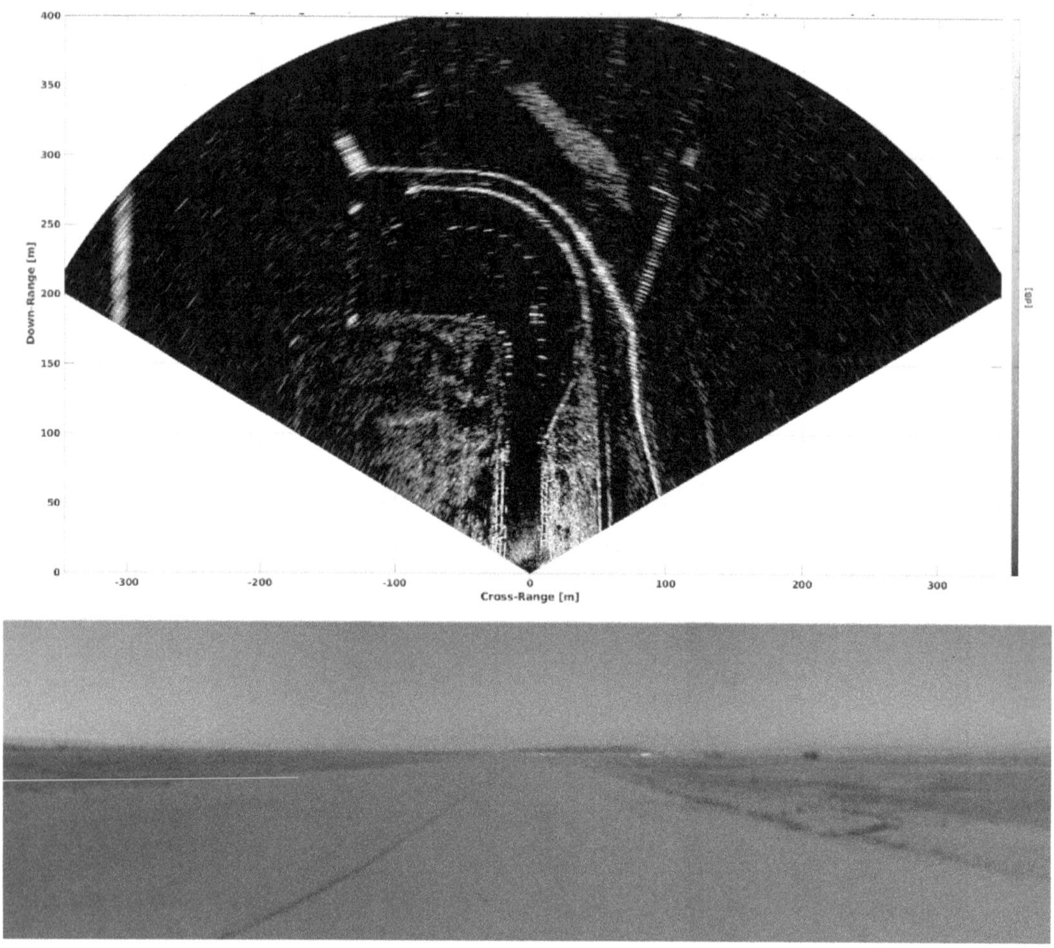

Figure 7.2 A down range, cross-range automotive scene shows the density and complexity of the environment. (Source: Waymo.)

For direct sampled waveforms—where the ADC bandwidth encompasses the entire pulse modulation—the near–far problem can be mitigated by avoiding STAR and range gating[2] the near target. Range gating in direct sampled waveforms involves opening the receiver window after the return of a near target is past the radar. For direct sampled waveforms range gating requires the transmit pulse length to be equal to or shorter than the length of the range swath, or the radar must operate as simultaneous transmit and receive and coupling for the transmit to the receive must be accounted for in the hardware design. In addition, the shorter the transmit pulse length, the less radar energy is reflected from the target, impacting the signal to noise ratio.

Alternatively, for linear frequency modulated (LFM) stretch waveforms range gating can be achieved by the addition of a high pass filter before the ADC. The transmit-to-receive leakage must be accounted for in the high end of the dynamic range of the RF components before the high pass filter of the stretch receiver but

2. Sometimes called *range gating out* the near target.

the high pass filter protects the analog-to-digital (A/D) converter from the effects of the leakage. However, depending on the range to the near target, multiple high pass filters may be necessary and even then the high pass filter may not be adequate in its suppression. Unfortunately, in the automotive environment, the number of situations when radar designers can avoid STAR and the near–far problem is small, and there are challenges to doing so.

In the remainder of this chapter, we present data on three commonly occurring challenges encountered by radars for fully autonomous vehicles. We show in detail the dynamic range encountered from everyday driving and the challenges that presents to radar, especially when coupled with the long range swath needed. Second, we highlight the special specular characteristics roads exhibit when viewed from the extremely low grazing angles used in automotive radar, and the challenges this presents for antenna and signal processing design from these unresolved reflections. Finally, we discuss complications from resolved multipath reflections.

7.2 Scene Dynamic Range

The automotive environment is filled with objects of dramatically different RCS; overhead signs, the backs of box trucks, and buildings can be very large, while motorcycles can be several orders of magnitude less and pedestrians even smaller, and they all can be present at significantly different ranges. The resulting large–small signal detection problem may be the single-most stressing requirement on the radar.

The most common source of the large–small signal problem arises when a radar must detect two similar objects, one placed near the radar and one placed very far away. The signal backscattered by the far target will suffer a relative attenuation that varies with the fourth power of the ratio of their ranges, meaning that the powers seen by the radar from an object at 160m will around 4,000 times weaker (36 dB) than that of one located at 20m.

Another source of the large–small signal problem arises when a cross-section of two objects in the scene vary by a large amount. The expected value of the RCS of mobile objects is typically −10 to 20 dBsm, corresponding to a pedestrian and a vehicle [4]. The stationary scene, or *clutter*, is observed at a very low depression angle, resulting in little to no backscatter that might yield homogenous ground clutter, rather clutter returns tend to be composed of scattering discretes like tree trunks, walls, and signs. Small clutter discretes of interest might be road debris as small as −20 dBsm [5] and large clutter discretes tend to be signs as large as 40 dBsm.

In practice it can be difficult to reason about sensor-independent RCS of large objects in the stationary scene because they may be resolved (i.e., they occupy multiple cells in the discrete range-Doppler-angle space). We have performed real-world measurement campaigns to characterize clutter RCS as observed in HH polarization with 0.5m range resolution, 700 Hz Doppler resolution, and 1 degree azimuth resolution. Data at the output of range-Doppler-azimuth matched filters are scaled to RCS using measurements from calibrated corner reflectors. This leads us to two key observations.

First, 40 dBsm scatterers are encountered often enough to be considered a canonical "large" scatterer that every automotive radar will encounter. Such scattering is typically caused by traffic control signs such as speed limit and overhead freeway signs and large structures like buildings (Figure 7.3), and bridges over freeways (Figure 7.4).

As seen in these figures, the sources of large returns are unusual but not rare; large signs, buildings, and bridges are encountered routinely. To quantify this frequency we utilized a very sensitive radar and collected all peaks exceeding the noise floor by approximately 11 dB in the range-Doppler-azimuth domain observed during unstructured freeway driving. The empirical cumulative density of such peaks is presented in Figure 7.5. From it we see that under those driving conditions we might expect to encounter one 40dBsm return for every thousand 10dBsm returns.

Second, the large dynamic range of objects in the scene presents a difficult challenge to radars used for fully autonomous driving. Since the fully autonomous vehicle must always be able to make the decision whether it is safe to continue to proceed down its lane of the road, we might require that the radar detect an object

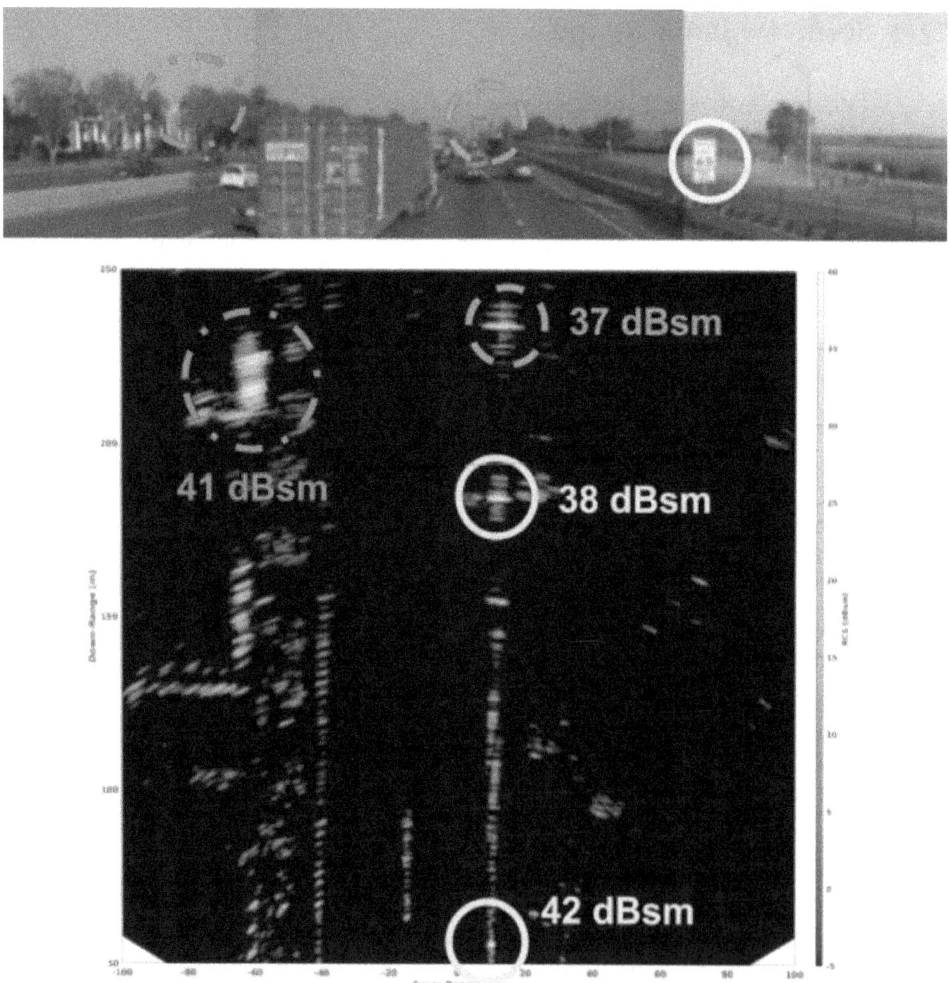

Figure 7.3 Signs and buildings commonly observed in automotive scenarios appear as very large targets, five were simultaneously observed in this single freeway scene. Camera image taken at a different time to illustrate the scene. (Source: Waymo.)

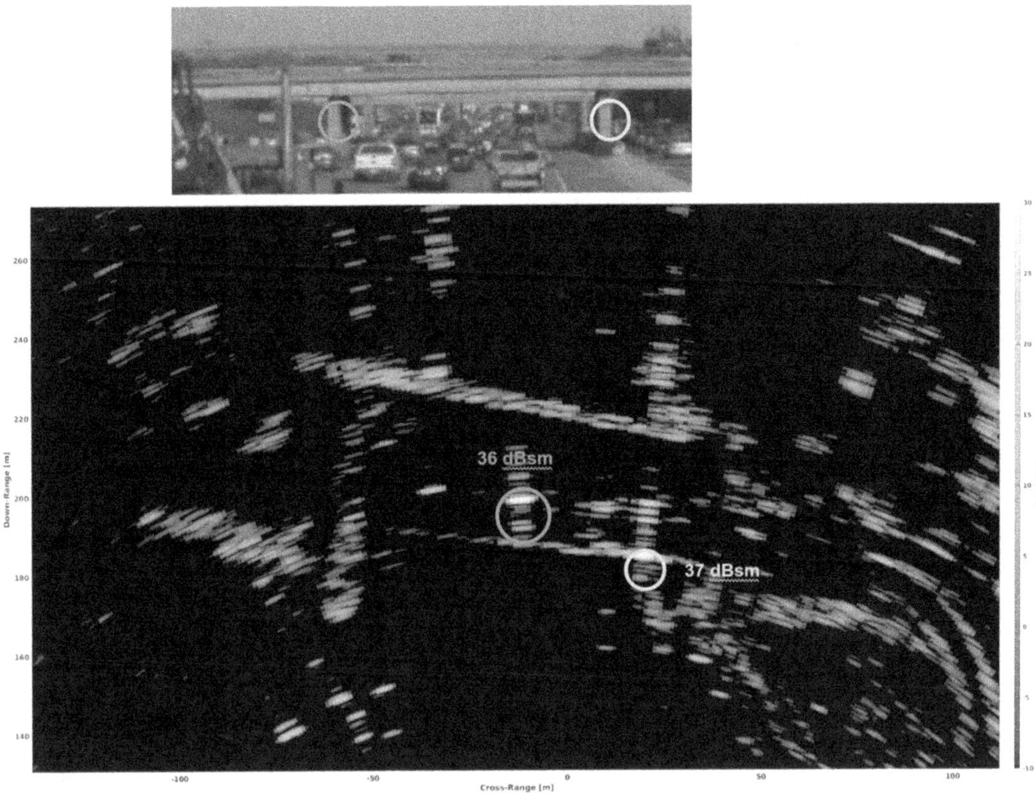

Figure 7.4 Bridges and their supporting structures contain very large RCS scatterers. Camera image taken at a different time to illustrate the scene. (Source: Waymo.)

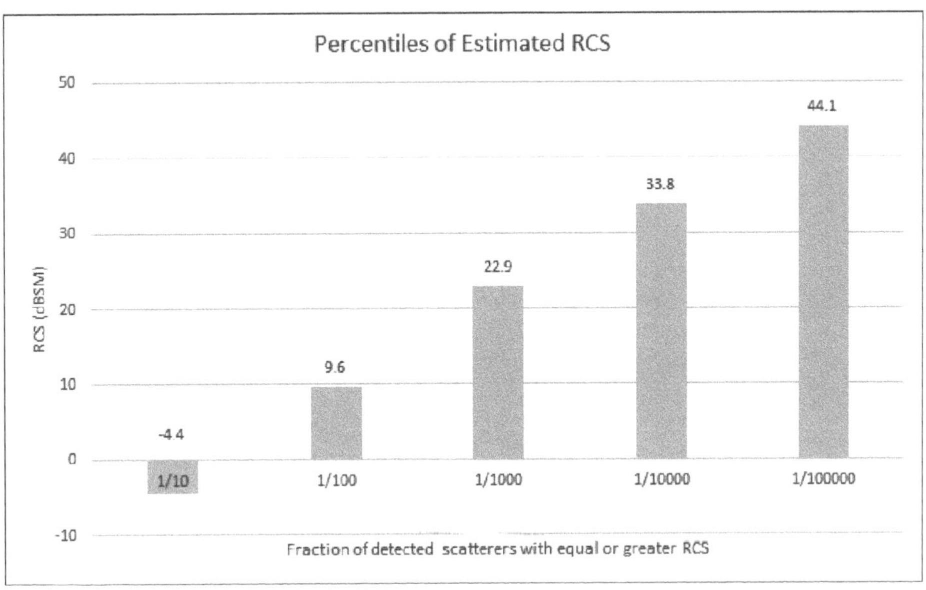

Figure 7.5 Percentiles of estimated RCS from unstructured freeway data. (Source: Waymo.)

in that lane as small as −10 to −20 dBsm (i.e., the RCS of a small pedestrian). While detecting this size object may be straightforward in the clear, it becomes more difficult in the presence of a 40 dBsm scatterer, which we have seen to occur relatively frequently. Detecting a stationary pedestrian at the same range as the 40 dBsm sign or structure next to the road requires 50–60 dB of dynamic range in the direction of arrival dimension—a challenging requirement for automotive radar to satisfy.

7.3 Ground Bounce (Unresolved Reflections)

Automotive radar operates at very low grazing angles, as shown in Figure 7.7, and road surfaces are highly specular at those angles. This can lead to strong multipath channels to objects of interest like other vehicles on the road. Under these conditions ground bounce-induced fading effects will be a large component of overall radar performance.

Even though ground roughness is variable [1], we find experimentally that typical city and freeway road surfaces are smooth enough to support specular reflection at mm-wave frequencies used by automotive radar. This means that the road surface acts in a manner similar to a mirror.

Asphalt exhibits variability in its relative permittivity owing to its mixed media construction as well as variable water content. Reported values for asphalt are 3–5 with only a small imaginary component at millimeter wave frequencies [2, 3]. Treating this problem with geometrical optics and assuming asphalt is nonmagnetic, we can predict reflectance as a function of the plane wave incidence angle to the road surface with a relative permittivity of 4 (Figure 7.6).

For many geometries of interest, the angle of incidence is very nearly 90° and so from first principles we might expect substantial reflection of the field from the road surface. These predictions are largely borne out in the field. To illustrate, Figure 7.8

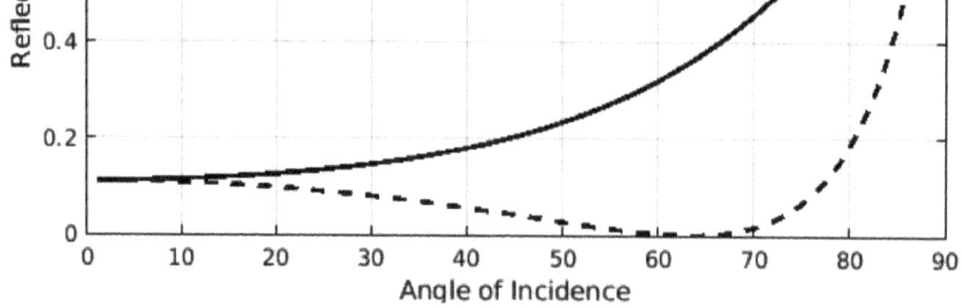

Figure 7.6 Reflectance as a function of incidence angle. As the grazing angle approaches 0 the incidence angle approaches 90°, implying high reflectance at any linear polarization. (Source: Waymo.)

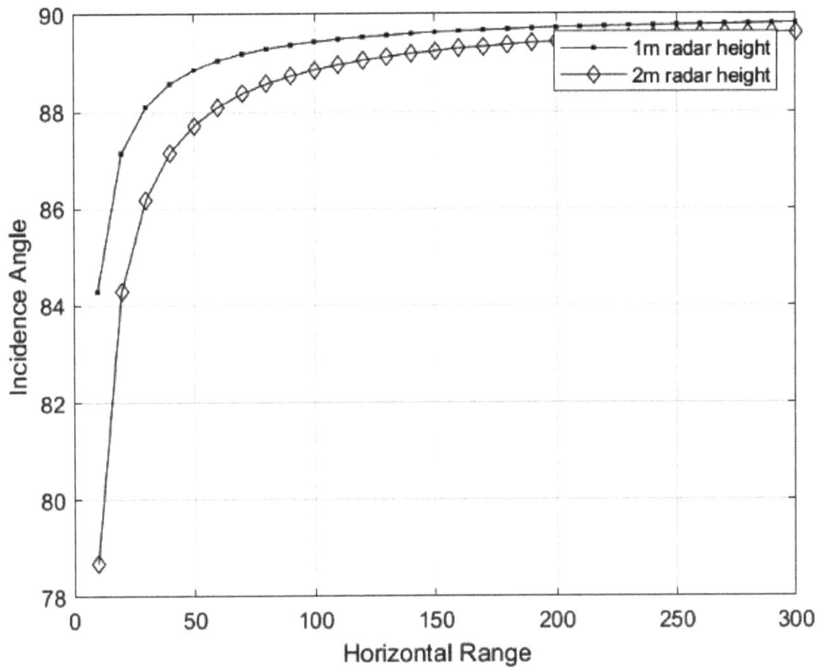

Figure 7.7 Incidence angle to the ground as a function of range and radar height illustrates the generally small grazing angles to ground clutter. (Source: Waymo.)

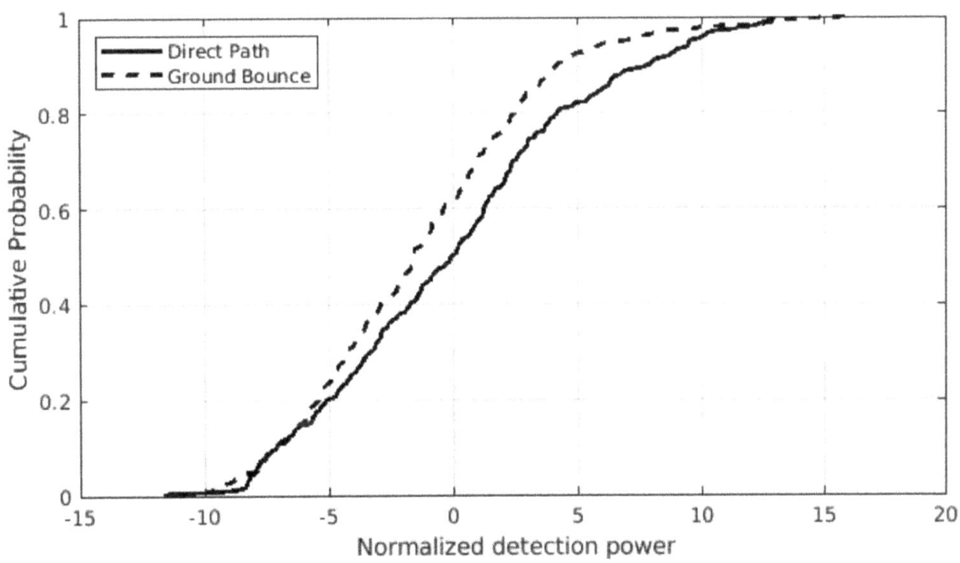

Figure 7.8 Empirical CDF of normalized detection power of an overhead bridge. Direct path and ground bounce returns are identified by virtue of the ground bounce appearing to have negative elevation. Their power distribution is similar to that of direct path returns with positive elevation angles. (Source: Waymo.)

is constructed from a segment of data taken from an overhead bridge about 150m downrange of the radar while on a freeway. The data were isolated into peaks taken from the direct path and from ground bounce using the measured elevation angle as a discriminant. Normalizing detection power by the median direct path power clearly shows that the ground bounce path suffers only about 2 dB of attenuation in the mode of the data.

A strong specular reflector like a smooth road will create a mirror image of the scene if each multipath channel is separable from the other. The resulting scene creates an image of objects "below" the ground as in the Figure 7.9 below the elevated pedestrian crosswalk.

Ground bounce returns may not always be separable from direct path. This can happen for a variety of reasons including a radar without elevation resolution and/or a waveform with low range resolution. That said, even a high range and elevation resolution radar will struggle to resolve ground bounce under all conditions. For example, consider a scatterer return from 1m from the ground observed by a radar 2m above the ground. Figure 7.10 illustrates that < 0.3deg of elevation resolution or < 0.02m of range resolution would be required to resolve a ground bounce channel. Further, their difference in Doppler would be indistinguishable. So, in general, we can expect returns from objects near the ground to exhibit a response that includes the coherent combination of the direct path, a single ground bounce channel, and a weaker double ground bounce channel that includes no direct path.

If the direct path and multipath channels are not resolved then they combine coherently in the signal processor. When the channels are in phase this will boost the signal power of the target, but when they are out of phase this will attenuate the target. The phase between channels is determined by their relative path length as well as any reactance of the road surface which might impart a phase shift during the bounce.

If a scatterer is observed while range is changing, the ground bounce channels will alternate between constructive and destructive interference with the direct path. To illustrate this phenomenon, Figure 7.11 simulates the effect for the scenario just described with a single scatterer at 1m height and an all-real road impedance.

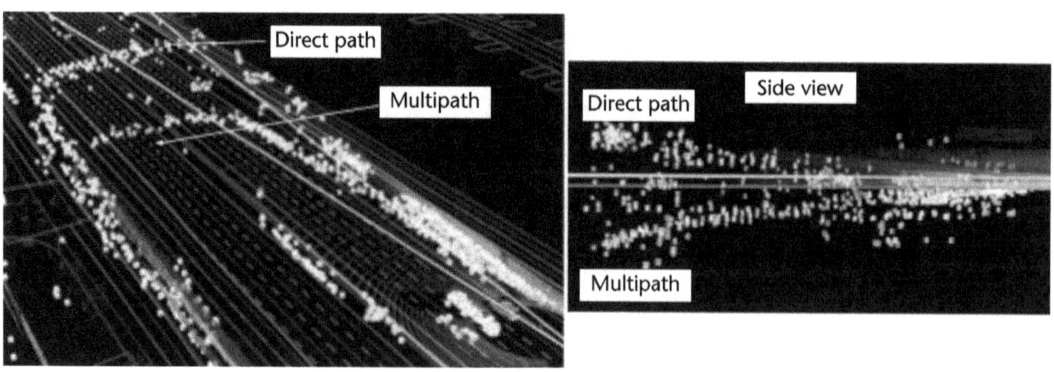

Figure 7.9 The mirrorlike nature of roads causes reflections of overhead structures to appear below the road, as well as at multiple locations between true position and the reflection due to angular glint. (Source: Waymo.)

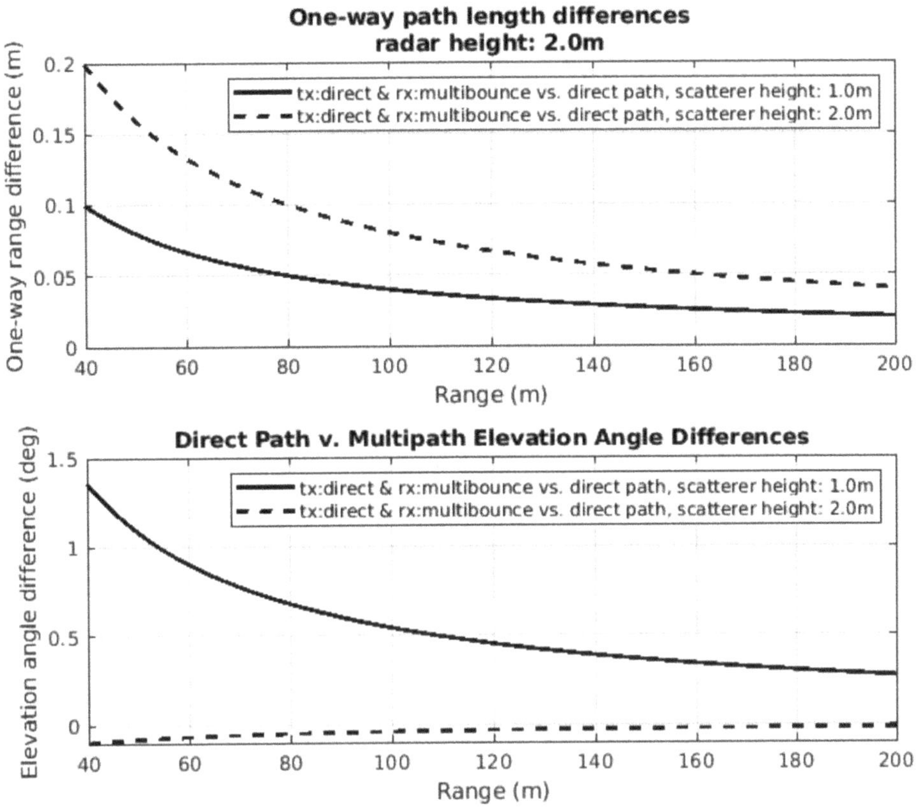

Figure 7.10 Direct and bounce path returns are closely spaced in both angle and range, making ground bounce channel unresolvable from direct path under many conditions. (Source: Waymo.)

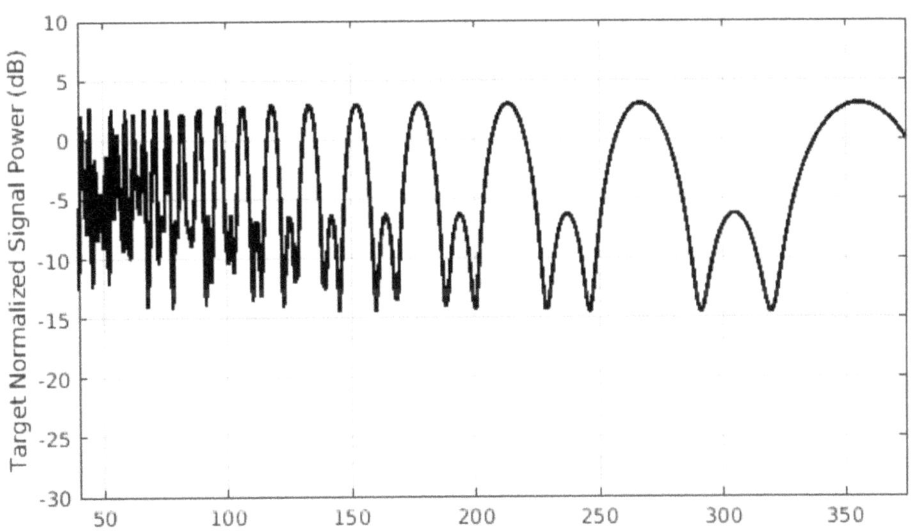

Figure 7.11 Simulated results indicate severe fading (multipath cancellation) at multiple ranges for a static return. One meter radar height, 2m scatterer height, 0.79 reflectance. (Source: Waymo.)

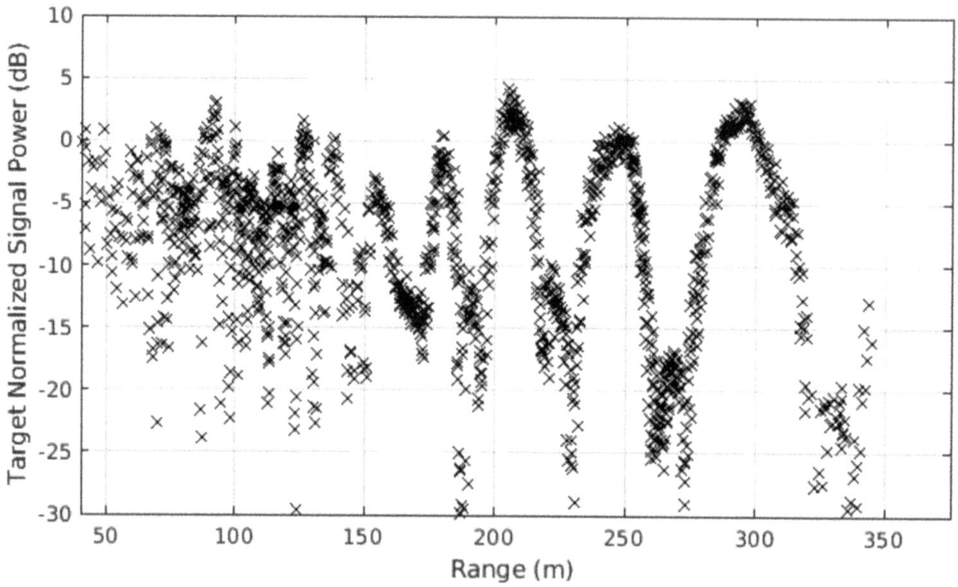

Figure 7.12 Data from demonstration radar verify severe fading (multipath cancellation) at multiple ranges for a static return. Six passes with 1m radar height and 2m scatterer height. (Source: Waymo.)

Real-world tests confirm this phenomenology, as illustrated in Figure 7.12. Here we have the same geometry as simulated in Figure 7.11, with a corner reflector as a target.

At close ranges a small change in the range causes a change between constructive and destructive addition of the multiple paths, effectively causing a single stable scatterer like a corner reflector to exhibit Swerling 1 (i.e., Rayleigh distributed) cross-section fluctuation. To illustrate this in Figure 7.13, we evaluate the statistical properties of only the near-range (40–100m) from Figure 7.12 and find that the empirical probability mass function of target power matches well with a Swerling 1 model. This is consistent with our assertion that we are actually observing two or more unresolved return paths from the corner reflector that are fluctuating

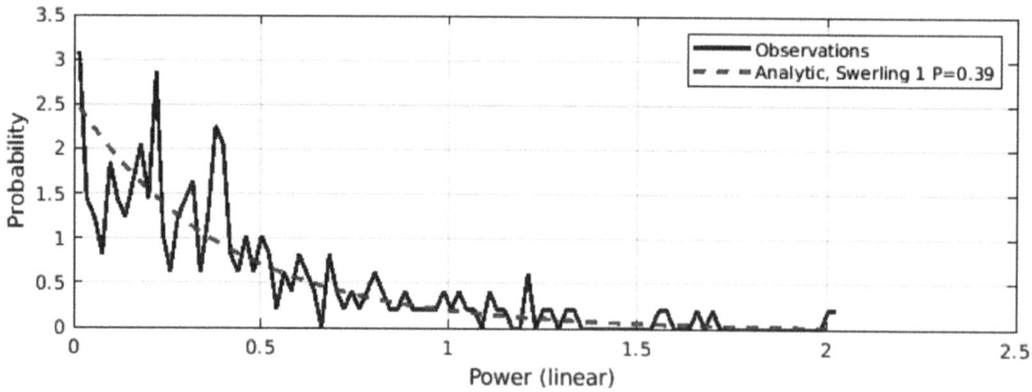

Figure 7.13 Target reflected power from Figure 7.12 extracted from 40–100m of range. These data are well matched to a Swerling 1 scattering model which posits a complex scatterer composed of several roughly equal sized scattering discretes. (Source: Waymo.)

quickly because the path length(s) of the underlying channels is changing rapidly from look-to-look.

Ground bounce is an integral element of the automotive radar environment. The radar designer must provision for multipath channels at all times, accounting for "mirrored" structures when these channels are resolved in one or more domains and for multipath fades when they aren't. Fades can change slowly at long range and very rapidly and short and medium range. Fades can have a meaningful impact on sensitivity at long range and on the relationship between RCS and the average power measured by the radar.

7.4 Multipath (Resolved Reflections)

While the ground bounce results in mostly unresolved multipath reflections, the dense automotive environment results in multipath from other vehicles and surrounding structures, which are often resolved in one or more of the radar dimensions. These resolved multipath reflections result in false detections and even false tracks, and apparent range extension of targets that the radar and perception designers must take into account. In addition, radars using MIMO angle processing (see Chapter 4, Automotive MIMO Radar) may suffer a degradation in the quality of the spatial spectrum, depending on the transmit and receive topologies and the angles between the direct and reflected paths. To illustrate, for the radar-reflecting surface-target-radar path, the angle from the transmit antennas to the reflecting surface will not equal the angle from the reflecting surface to the radar, which violates the underlying MIMO assumptions used for the array design.

In Figure 7.14 the vehicle containing the radar is stationary on the side of the road. The target vehicle circled in the figure is moving down the hill. The radar observes the target vehicle, the left circle in Figure 7.15, as a direct path signal and as a multipath signal, the right circle in Figure 7.15, that bounces off the buildings on the right.

Figure 7.16 shows the corresponding range Doppler map for the image in Figure 7.14. In this scene the target vehicle is moving and the radar is stationary. However, instead of a single return appearing at the corresponding range and Doppler frequency, both the direct path and the reflected paths are present due to the buildings providing several opportunities for multipath to the moving vehicle. The reflected paths Doppler frequencies correspond to the rate of change in their longer reflected path, which is a function of the geometry of the scene. In this scenario there is sufficient angular separation of the direct path and two multipath scatters to produce multipath targets that are distinct in Doppler.

Figure 7.17 shows a particularly challenging multipath environment where the radar is traveling on the lower deck of the San Francisco Bay Bridge. The supports at the edge of the roadway and the bottom of the upper deck form a large number of metallic corner reflectors—an ideal environment for multipath. In range-azimuth radar image in Figure 7.17, the row of bridge supports along either side of the road are clearly visible, as are the two vehicles immediately ahead of the radar. All of the radar returns outside of the bridge supports (the two bright vertical "lines" in the radar image) are multipath false targets that appear in the free space to the left and

Figure 7.14 The buildings on the right of this scene provide a reflective surface likely to produce multipath returns from the vehicle (in the circle) traveling down the hill. (Source: Waymo.)

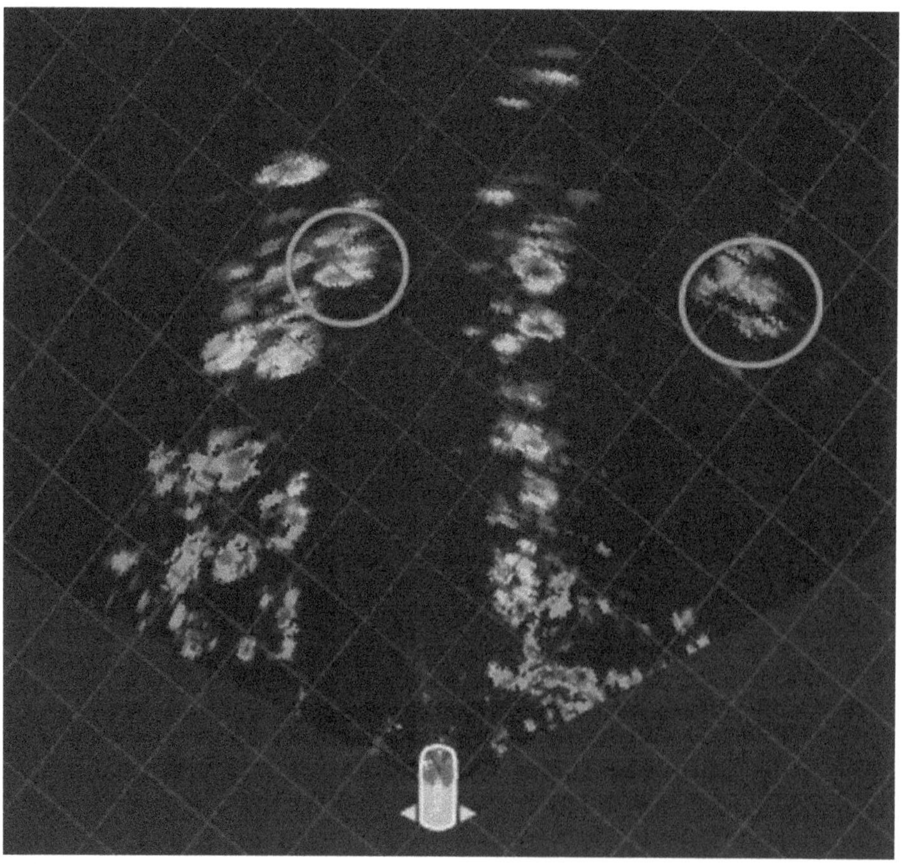

Figure 7.15 Range-azimuth map of the scene in Figure 7.14. The left circled target is the direct path vehicle image and the right is the multipath vehicle image. (Source: Waymo.)

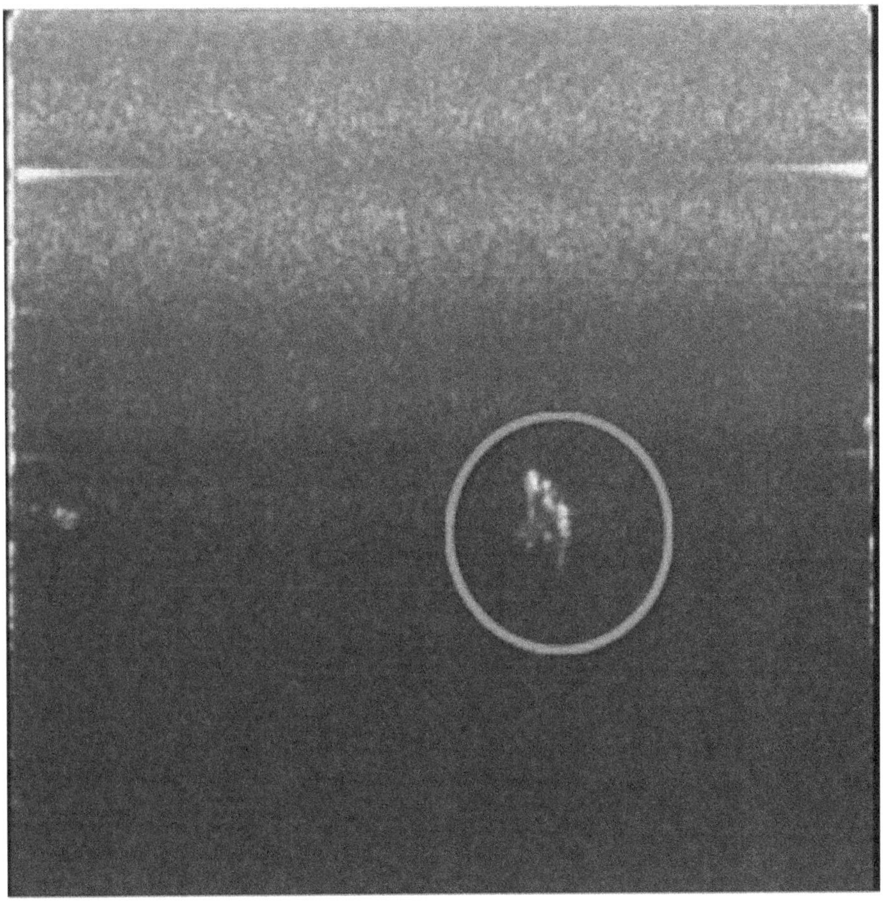

Figure 7.16 Corresponding range-Doppler map for Figures 7.14 and 7.15. Near range is at the top of the image and zero Doppler is at the left of the image. The direct path as well as two distinct multipath returns at slightly farther ranges are visible in the image. (Source: Waymo.)

right of the bridge. In this case, correlating the radar data with a priori knowledge of the roads can help interpret the radar data. In general, false multipath targets that are sufficiently offset in angle to not interfere with the planned motion of the AV may be of a lesser concern to the radar and AV designer. Correlation of the radar data to other sensors, when available, and to a map of the roads may lessen the impact of these false targets.

Multipath can also result from reflections within a vehicle structure or between a vehicle and its immediate surroundings, as shown in Figures 7.18 and 7.19. In these images a passing utility vehicle has an open bed that presents a corner reflector to the radar. For reference, Figure 7.18 shows the range-azimuth radar image with the truck when there are no significant internal reflections in the bed of the truck. The dimensions of the truck are visible in this image. In Figure 7.19, the geometry supports the radar receiving multiple bounces, likely from within the bed of the truck, leading to an apparent extension of the vehicle dimension in the range dimension of the radar.

Figure 7.17 Range-azimuth plot of radar return on the lower deck of the San Francisco Bay Bridge. Returns from the supports at the edge of the lanes can clearly be seen from the image, as well as a significant amount of multipath return from the free space surrounding the bridge. (Source: Waymo.)

Figure 7.20 shows another form or multipath that is common to automotive radar that occurred in this scene. The corner reflector formed in the bed of the utility vehicle provides a large enough return that a double bounce return at double range (circled in the radar image) is created from the radar signal bouncing off the target vehicle, off the AV carrying the radar, and a second bounce off the target vehicle. These double bounce returns can be particularly challenging because they can present an inbound false target heading toward the AV. While the detection of these second bounce returns generally requires that they originate from a very

Figure 7.18 Radar range-azimuth image of a utility vehicle showing the outline of the truck as visible to the radar. The grid size in this image is 5m. (Source: Waymo.)

large scatter, overhead signs, bridges, and large vehicles provide ample scatters in the automotive environments that can cause false inbound targets that appear in the same lane as the AV.

Multipath cannot be avoided in the automotive environment but options for mitigating the effects involve minimizing the number of false targets that make it through the tracker, exploiting target kinematics and/or fusion with map data (sometimes called *knowledge-aided*) or other sensors [6]. When the scene is very dynamic, false detection from multipath will tend to be ephemeral and may either not form tracks or result in very short lived tracks. Large permanent scatters like signs, bridges, and parked cars are more likely to form more stable false targets that are kinematically consistent and can result in false tracks. Correlation of these

Figure 7.19 The radar range-azimuth image of the same vehicle as Figure 7.18 with internal reflections in the bed of the truck creating an apparent extension of the vehicle dimensions in the range dimension. The grid size in this image is 5m. (Source: Waymo.)

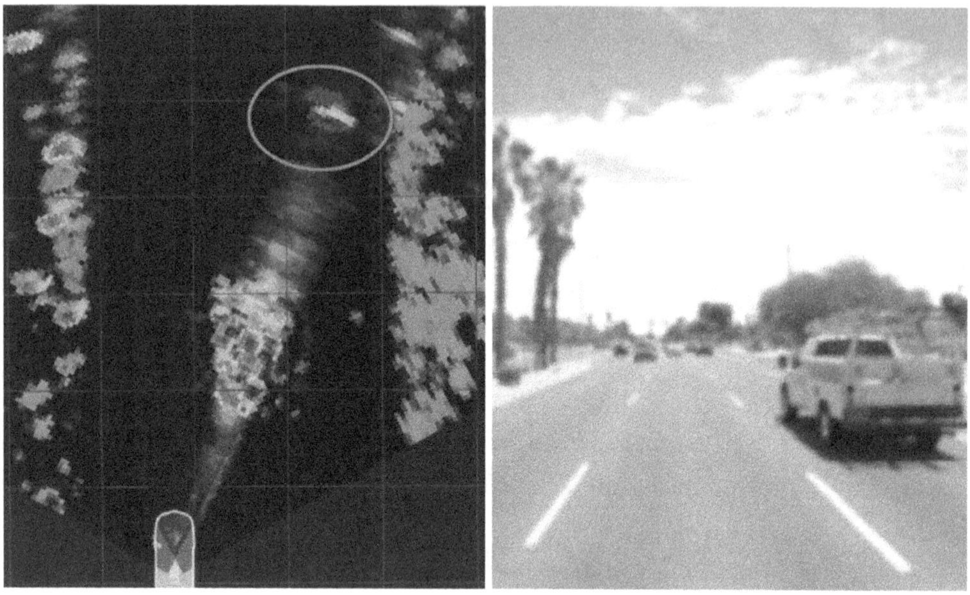

Figure 7.20 The strong return from the bed of the utility vehicle causes a false return (circled in the radar range-azimuth plot) at twice the range of the strong scatter in the bed of the truck. (Source: Waymo.)

false tracks with the true target tracks can help better interpret the false tracks. Other options include correlation of the radar tracks with a priori knowledge of the roads and correlation to other sensor outputs [7]. The existence of multipath false targets is inherent in the environment that automotive radars operate and must be considered in the system design for autonomous vehicles.

References

[1] Smith, K. D., and P. Ram. *Measuring and Specifying Pavement Smoothness:[techbrief]*, No. FHWA-HIF-16-032. U.S. Federal Highway Administration, 2016.

[2] Sarabandi, K., Li, E. S., and A. Nashashibi. "Modeling and Measurements of Scattering from Road Surfaces at Millimeter-Wave Frequencies." *IEEE Transactions on Antennas and Propagation,* Vol. 45, No. 11, 1997, pp. 1679–1688.

[3] Jaselskis, E. J., Grigas, J., and A. Brilingas. "Dielectric Properties of Asphalt Pavement," *Journal of Materials in Civil Engineering*, Vol. 15, No. 5, 2003, pp. 427–434.

[4] Matsunami, I., Nakamura, R., and A. Kajiwara, "RCS Measurements for Vehicles and Pedestrian at 26 and 79GHz." *2012 6th International Conference on Signal Processing and Communication Systems,* IEEE, 2012, pp. 1–4, doi: 10.1109/ICSPCS.2012.6508004.

[5] Shibao, M., Uchiyama, K., and A. Kajiwara, "RCS Characteristics of Road Debris at 79GHz Millimeter-Wave Radar." *2019 IEEE Radio and Wireless Symposium (RWS)*, IEEE, 2019, pp. 1–4, doi: 10.1109/RWS.2019.8714559.

[6] Smith, L. "Multibounce Target Mitigation," United States Patent Publication No. 20210055375 (priority date August 19, 2019).

[7] Smith, L. "Multibounce Target Additional Data," United States Patent Publication No. 20210055373 (priority date August 19, 2019).

CHAPTER 8

Radar Interference

Matt Markel, Ralf Reuter, and Arunesh Roy

8.1 Introduction

This chapter discusses the various ways that radio frequency interference (RFI) can manifest in a victim radar, focusing on the different domains of observability the radar has for data, such as fast-time and slow-time, range and Doppler, and so on. It then surveys some canonical approaches that radars can employ to mitigate the interference. Finally, and perhaps most importantly, it looks at what the radar community can do going forward to improve the RF environment and radar performance for fully autonomous vehicles.

8.2 Motivation and Definitions

RFI occurs when the transmissions from one radar, which we will call the interferer, negatively affect the operation of another which we call the victim.[1] To illustrate the magnitude of the radar interference problem, consider a situation like that pictured in Figure 8.1 (a crowded freeway). Typically, a lane will be around 3.6m wide, and a standard car is around 5m long. If we add in another few meters in front to represent driving in crowded freeway traffic and make the lanes a little wider—primary for easier math—then each car occupies an 8m × 4m box. It is easy to see that on a crowded freeway like the one depicted in the figure with five lanes each direction, the number of vehicles in close proximity (say, within 300m) of the vehicle with our victim radar is well over 700.

A naïve way to ensure interference-free operation for all could be to try to dynamically divide up the 5 GHz of spectrum between these 700 radars, leaving each with about 7 MHz of spectrum in which to operate. However, this provides an underwhelming 20m of range resolution, which isn't adequate for fully autonomous driving (recall that 20m would be about four car lengths). And even these numbers are optimistic, since RFI can be an issue at ranges far greater than 300m, and we didn't address the mutual interference between multiple radar sensors inside a car

1. The association of the names *victim* and *interferer* is solely based on the perspective with which you are viewing the situation—the victim may actually be the interferer to the other radar.

Figure 8.1 A typical large-city freeway shows that there are too many cars to just divide the spectrum.

or the multiple possible paths that the various other radars' signals could take to arrive at the victim (Figure 8.2).

We can draw an interesting parallel between hostile jamming (often called electronic attack, historically called electronic countermeasures)—where an adversary intentionally seeks to deceive, deny, degrade, or otherwise impact a radar's operation (see [1–3]) and RFI. Currently we are not experiencing hostile jamming in the automotive space, but as we will see in this chapter many situations exist where RFI can appear as if it were an intentional attack. Fortunately, the world of radar antijamming technology (often called electronic protection, historically called electronic counter-counter measures) has been maturing for decades and provides a rich source of tools for our RFI mitigation toolbox.

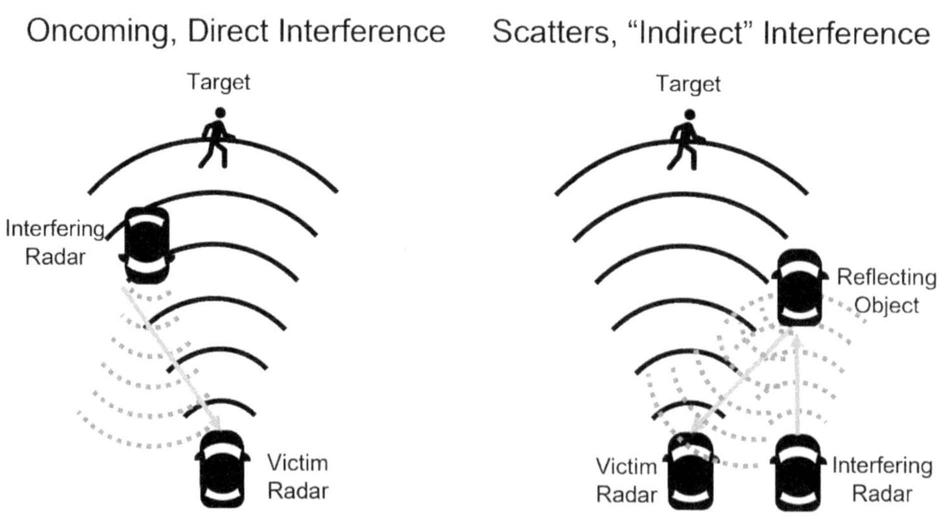

Figure 8.2 Interference can take direct and indirect paths to arrive at the victim.

8.3 Impacts and Manifestation

8.3.1 LFM/FMCW

A very useful way to *visualize* how interference manifests in a victim radar is in the range-Doppler map because many radars perform detection in the range-Doppler domain and then perform the computationally costly angle estimation only on selected range-Doppler cells. Range-Doppler processing is a two-step approximation of the CPI matched filter where the responses from each pulse are first matched filtered in range (often called *range compression*, since the entire pulse energy is compressed into a single range bin), typically with the assumption that there is zero target velocity. When LFM—also called FMCW—is used this can be accomplished with some mixing followed by a DFT or a FFT implementation. After this step each of the range responses are—range bin by range bin—matched filtered with different constant velocity assumptions. As with range, this step can be accomplished with a DFT or FFT.

We will quickly review some of the fundamental equations in this process and show representative illustrations of the processing steps. Define our radar's (which will be the victim radar, since it is to this system we are evaluating the RFI effects) transmit waveform to be of the form [4, 5]:

$$S(t) = \exp\left(j \cdot \left(2\pi f_o + \gamma t\right) \cdot t\right) \quad 0 < t < T_{\text{pulse}} \tag{8.1}$$

And receive waveform when a target is present is

$$R = \exp\left(j \cdot \left(2\pi f_o + \gamma(t - t_0)\right) \cdot \left(t - t_0\right)\right) \quad t_0 < t < T_{\text{pulse}} + t_0 \tag{8.2}$$

Where t_0 corresponds to the range r of the target, $t_0 = 2 \cdot r/c$. The range matched filter for stretch processing[2] is typically implemented in two parts. First, the receive signal is multiplied by the transmitted waveform in a process called *de-ramp*. The de-ramp equation is

$$D = S^*(t) \cdot R(t) \tag{8.3}$$

which in practice is followed by a narrow IF filter. When a target is present the post-de-ramp signal this the form:

$$D = \exp\left[A + Bt\right] \tag{8.4}$$

Where

$$A = \left(\gamma t_0 - 2\pi f_o\right) t_0 \tag{8.5}$$

2. In stretch LFM processing the return signals are mixed with the transmitted signal followed by a relatively narrow low pass filter in the analog domain. This greatly simplifies the hardware, especially the ADC, but limits the amount of hardware correction possible for nonlinearities in mixers and other analog components.

$$B = -2\gamma t_0 \tag{8.6}$$

which by inspection is a sinusoid. The second part of the matched filter, a DFT, utilizes the fact that the result after de-ramp is a collection of sinusoids, each of whose frequencies corresponds to the delay (range) from the radar, so the appropriate matched filter to this takes the form of the Fourier transform, which we use to estimate the range to the target. The range response for a single target is shown in Figure 8.3.

To determine the target's Doppler frequency—which is proportional the relative radial velocity between the radar and the target—the radar sends out multiple pulses. We can examine this stage of processing as a matrix formed by the accumulation of range-compressed data for these pulses, with range bins on the ordinate (y-axis) and pulse number along the abscissa (x-axis), as shown in Figure 8.4.

Notice that the range response from one pulse is replicated across the various pulses, with a phase progression (not shown in the amplitude plot) pulse to pulse corresponding to the target's Doppler. Also, along each column we see the peak range bin corresponding to the range of the target, and minimal energy in the sidelobes commensurate with the windowing function employed in the range DFT.

To complete the range-Doppler formation process, a second DFT is applied across pulses for each range bin (i.e., row by row), producing a single dominant focused point (Figure 8.5).

With this review of range-Doppler processing we can now look at the many ways that RFI can manifest in the victim radar. Some common misconceptions when discussing RFI are that it will *always* appear simply like additional targets, or that it will *always* appear like additive white Gaussian noise (AWGN) that just raises the noise floor. However, it turns out RFI can appear as *either* of these, and additionally it can take on a third form that can actually be more deleterious to the

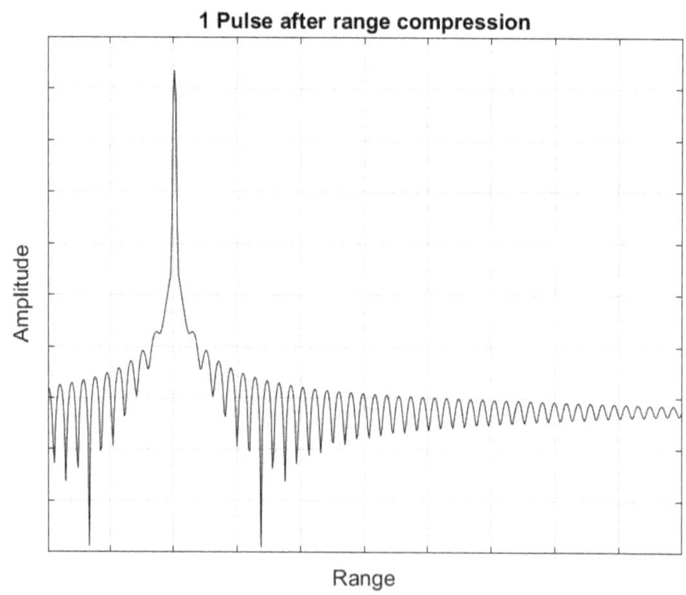

Figure 8.3 Range matched filter response to a target.

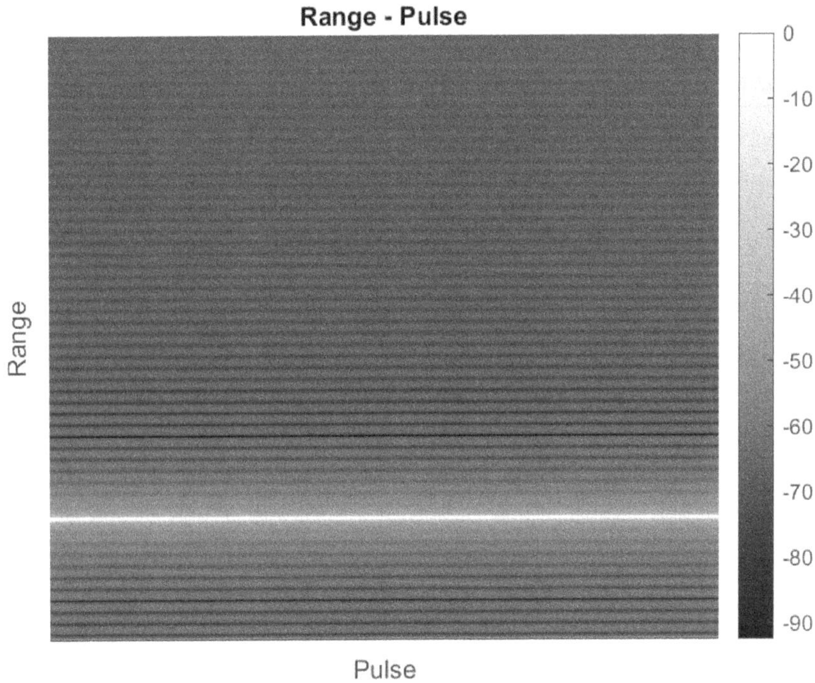

Figure 8.4 Range-pulse domain for a single target; column n represents the output of the range matched filter for pulse *n*. This and all range-Doppler images that follow are self-normalized amplitude views (in dB) to easily illustrate the appearance of the target or interference signals.

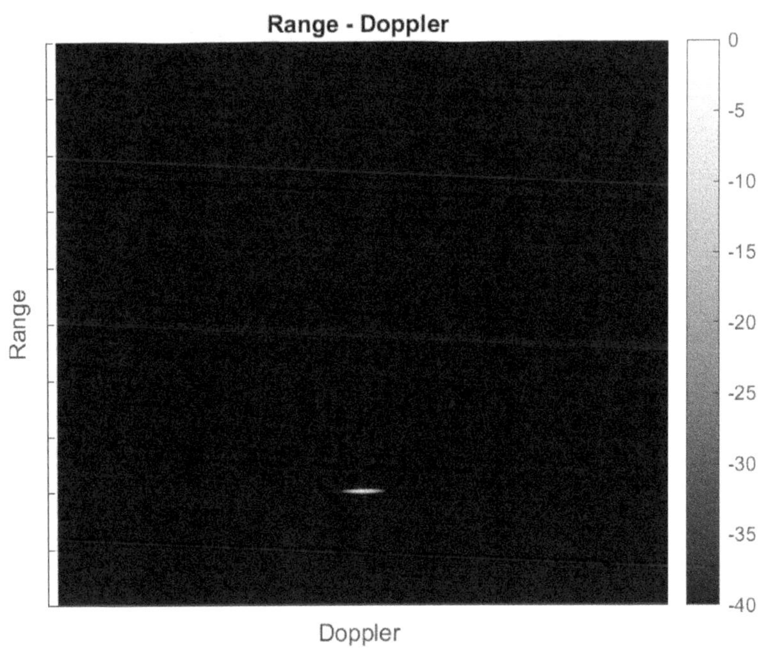

Figure 8.5 Range-Doppler view of a target formed from taking the transform, row by row, of the range-pulse view.

Table 8.1 Taxonomy of Types of Interference Based on Appearance after Range-Doppler Processing

Interference Type	*Appearance in RDM*	*Range-Doppler Processing Gain*
Coherent	Focused, target-like	Full range-Doppler processing gain
Semicoherent	Spread in range, Doppler, or both Not necessarily uniform amplitude	Receives partial range-Doppler processing gain
Noncoherent	Unfocused, noiselike	Receives little processing gain

victim radar. Table 8.1 defines a taxonomy that includes all three of these conditions, labeling the RFI as either *coherent*, *semicoherent*, or *noncoherent*. It turns out that each has a different appearance in the range-Doppler map (RDM), as well as a different *processing gain*, the improvement over noise that a signal receives through the range and Doppler processing.

To motivate the discussion, let's consider the situation where victim and interferer are both using LFM modulation, and the victim is using stretch processing. We will now work through several cases to illustrate how slight changes in the interferer waveform parameters (keeping the victim constant) can cause each of these types to appear in the range-Doppler map.

8.3.1.1 Case 1: Matched Slope, Frequency, and PRI

Here the interfering radar will appear just like a target, although potentially much brighter. This is an example of coherent interference. (See Figure 8.6.)

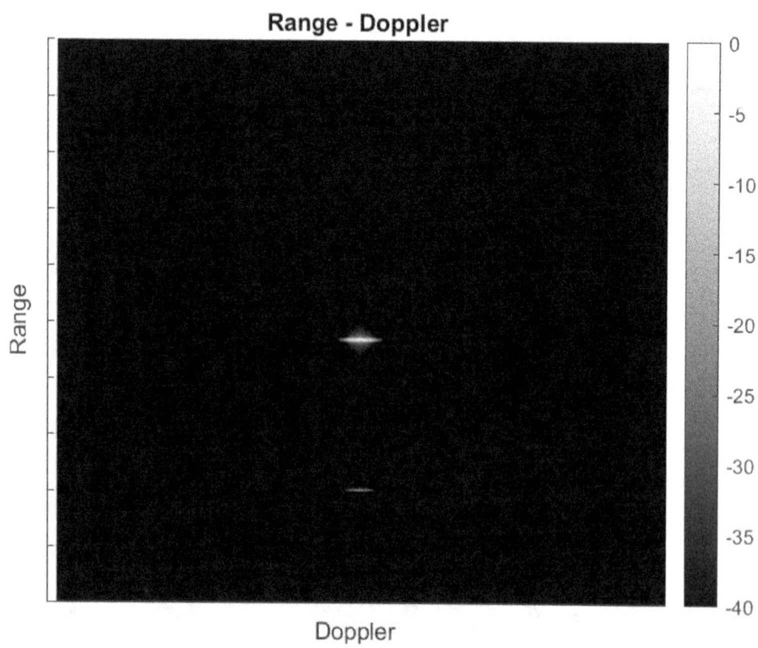

Figure 8.6 Range-Doppler view for a single target and coherent RFI. Notice that the target energy and RFI (higher, brighter dot in the figure) are focused in range and Doppler.

8.3.1.2 Case 2: Mismatched Slope, Matched Frequency, and PRI

In this case, the interferer has the exact same carrier frequency and PRI as the victim, but the frequency slope is slightly different. Recall that the victim transmit waveform is of the form:

$$S = \exp\left(j \cdot \left(2\pi f_o + \gamma t\right) \cdot t\right) \quad 0 < t < T_{\text{pulse}} \tag{8.7}$$

The return from the interferer will be identical, except that

$$\gamma_{\text{interferer}} = \gamma + \Delta \tag{8.8}$$

such that the return from the interferer is

$$R = \exp\left(j \cdot \left(2\pi f_o + (\gamma + \Delta)(t - t_0)\right) \cdot (t - t_0)\right) \quad t_0 < t < T_{\text{pulse}} + t_0 \tag{8.9}$$

After de-ramp, we notice that in addition to linear phase term there is a quadratic phase term as well.

$$D = \exp\left[A + Bt + Ct^2\right] \tag{8.10}$$

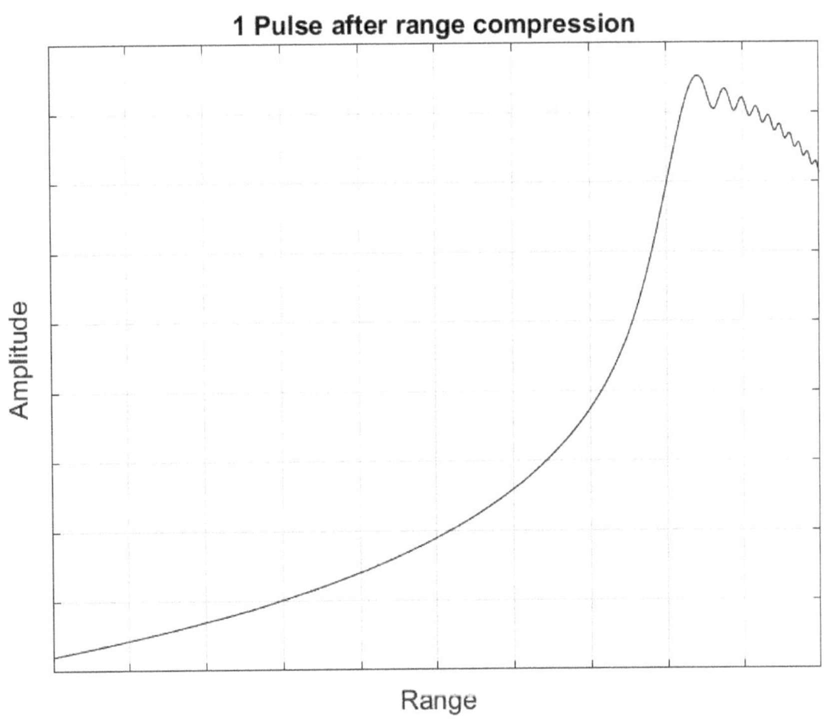

Figure 8.7 Range response where victim and interferer are matched in frequency and PRI, but slightly mismatched in chirp slope. Notice the range response has a significantly wider response than expected from a target.

Where

$$\begin{aligned} A &= \left((\gamma + \Delta)t_0 - 2\pi f_o\right)t_0 \\ B &= -2(\gamma + \Delta)t_0 \\ C &= \gamma \end{aligned} \tag{8.11}$$

This quadratic phase term will cause a smeared response in the range domain, and it will receive some processing gain from the range compression but not to the level of a fully coherent signal (Figure 8.6). It is important to notice that this response looks neither target-like nor noise-like but is what we call semicoherent. Since the PRI between the interferer and the victim are the same in this case, each pulse will have the same shape (Figure 8.7), and the result after Doppler processing in the RDM is a return that is well contained in Doppler but smeared in range (Figure 8.8 and Figure 8.9).

The radar will likely have difficulty detecting a target whose range and Doppler is near the interference for two reasons. First, the interference may be significantly stronger in power (even after losing some processing gain) than the target, completely masking it. Second, even if the target has sufficient power, the nonflat shape (i.e., not like what noise would look like) of the RFI can cause difficulty in a CFAR detector when the RFI occupies the RDM cells used for noise estimation, resulting in a missed detection.

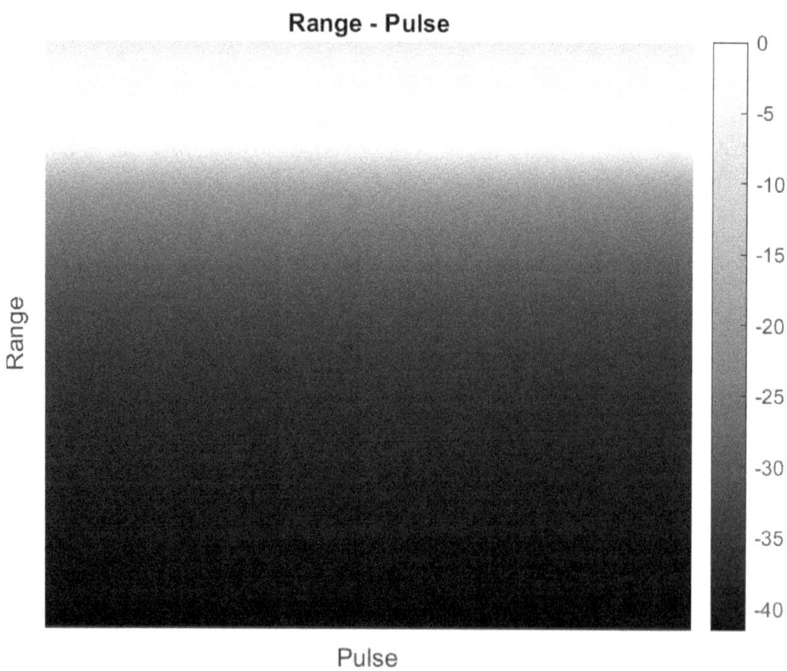

Figure 8.8 Range-pulse view where victim and interferer are matched in frequency and PRI, but slightly mismatched in chirp slope. Notice the response for each pulse is identical in amplitude, but not focused like a point-scatterer target would appear.

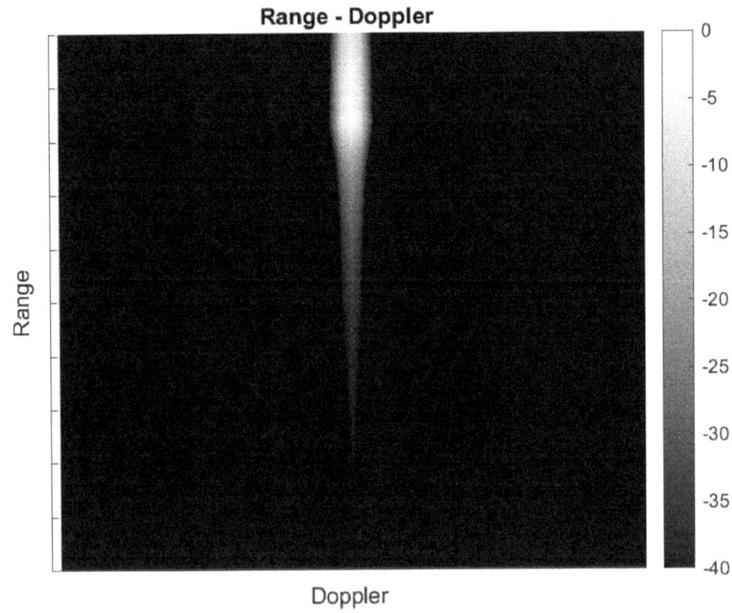

Figure 8.9 Range-Doppler view where victim and interferer are matched in frequency and PRI, but slightly mismatched in chirp slope. Notice the interferer response is focused in Doppler but not in range.

8.3.1.3 Case 3: Matched Slope and Frequency, Mismatched PRI

Next we consider the case when the interferer waveform pulse is identical to the victim, but the PRI is slightly different. The way the RFI will present in the RDM will depend on how different the PRIs between the victim and interferer are, but we can gain insight from looking at a few canonical examples. In general, for this case, the interference will be focused in the range domain, but spread in some way in Doppler.

To visualize interference in the Doppler domain, we consider how it manifests, post range compression, in a particular range bin for each pulse in the CPI. Recall that Doppler processing involves taking the transform (DFT or FFT) across slow time of each range bin. If the response in a particular bin is identically the same in amplitude with a phase that is either constant or linearly varying—that is, it is a complex sinusoid, then after Doppler processing it will compress into a single or small number of filters and receive the full processing gain—a SNR gain roughly commensurate with the number of pulses (window losses or straddle losses will decrease the SNR gain a bit). Target returns received by the radar behave like this, which is why range-Doppler processing is so useful. However, for an interfering source to cohere in Doppler it must be very well matched to the victim's PRF.

Consider the case where the interfering radar used the identical waveform but had a very slightly shorter PRI. After range processing the peak for each pulse will be moving closer, appearing in a different range bin each pulse. For a particular range bin the peak may only be present, say, on the nth pulse and the rest are lower amplitude sidelobes, which will have a Doppler response resembling that of a delayed delta function[3], which is a constant across frequency with no additional Doppler

3. Recall the delta function is only nonzero for a single sample.

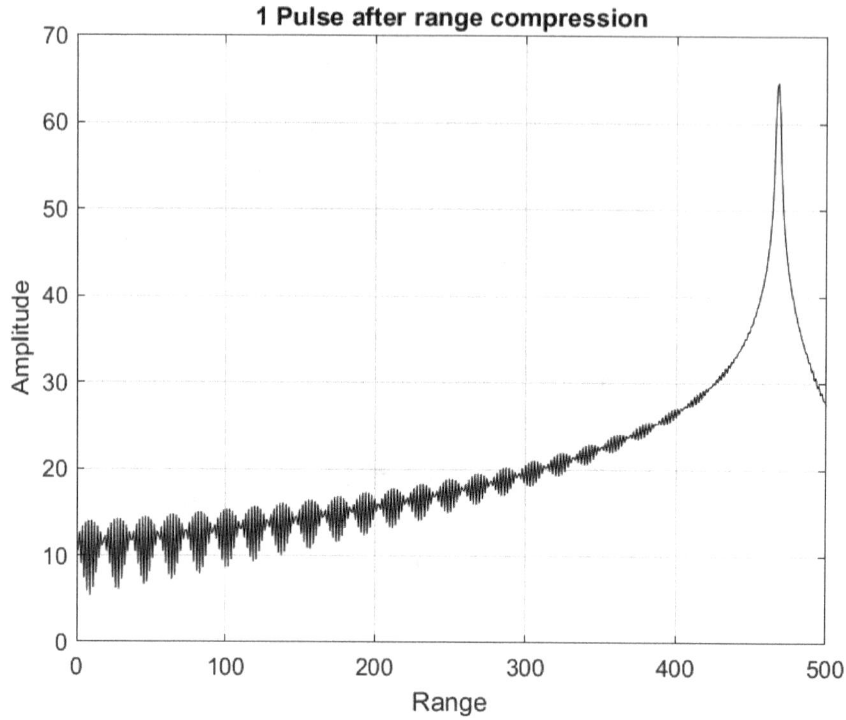

Figure 8.10 Range response from first pulse in the CPI where victim and interferer are matched in frequency and chirp slope, but slightly mismatched in PRI. Notice the interferer response appears target-like in range near the far edge of the range window.

processing gain (Figures 8.10–8.13). After range and Doppler processing, we find that the changing range response pulse to pulse creates a large interference area in range-Doppler space.

Another way interferers can appear in Doppler occurs when the victim PRF is a multiple of the interferer, and therefore the interferer only appears on certain pulses (e.g., if the PRF of the victim is three times that of the interferer, the interferer will only appear on every third pulse the victim receives, Figure 8.14). Finally, Doppler artifacts can be generated if only a fraction of the pulses have energy from the interferer (even if everything else is matched) because the CPI start times are not aligned.

8.3.1.4 Case 4: Matched Frequency, Mismatched Slope and PRI

When both the frequency slope and the PRI between the victim and interferer differ, then we see the RFI defocused in both the range and Doppler domains. Again, however, even though it is not completely focused in the RDM, it is still incorrect to view this as just a noise floor increase.

Consider first the situation where the frequency slope is slightly different, as in Case 2 (8.3.1.2), and the PRI is slightly different, as in the first example from Case 4 (8.3.1.4). Here, the wider range response caused by the mismatched chirp slope will migrate across range from pulse to pulse, resulting in a response that is defocused across a significant portion of the range-Doppler map (Figure 8.15 and Figure 8.16).

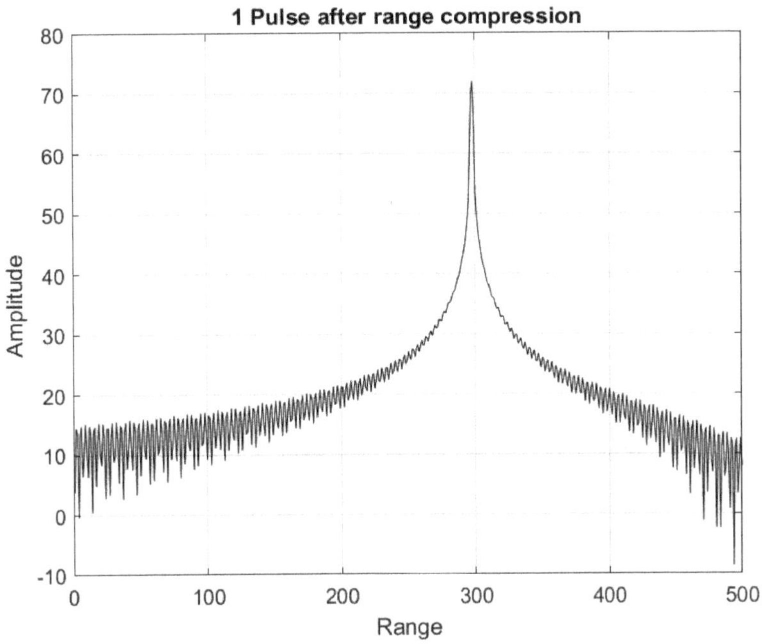

Figure 8.11 Range response from 32nd pulse in the CPI in the CPI where victim and interferer are matched in frequency and chirp slope, but slightly mismatched in PRI. Notice the interferer response appears target-like in range but has migrated to near the center of the range window, due to the mismatch in PRI.

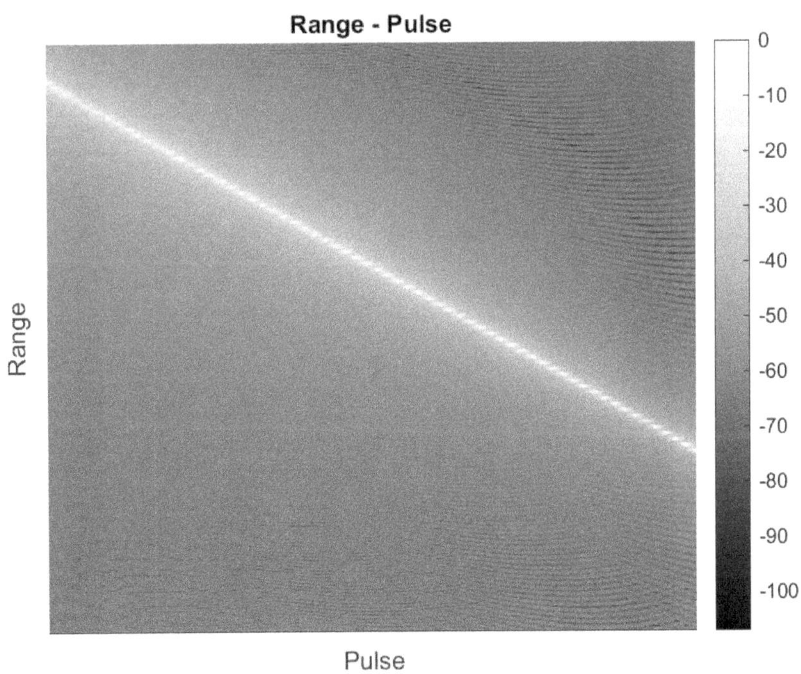

Figure 8.12 Range-pulse view where victim and interferer are matched in frequency and chirp slope, but slightly mismatched in PRI. Notice the interferer response appears target-like in range but appears to move through the range window during the CPI, due to the mismatch in PRI.

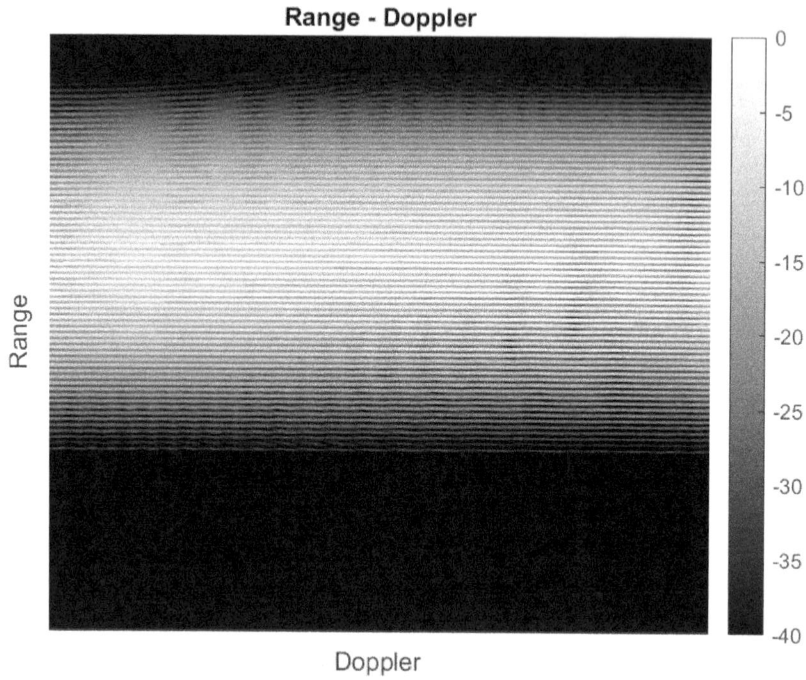

Figure 8.13 Range-Doppler view where victim and interferer are matched in frequency and chirp slope, but slightly mismatched in PRI. Notice the semicoherent interferer response appears defocused over a significant portion of the range-Doppler matrix but has more structure than thermal noise.

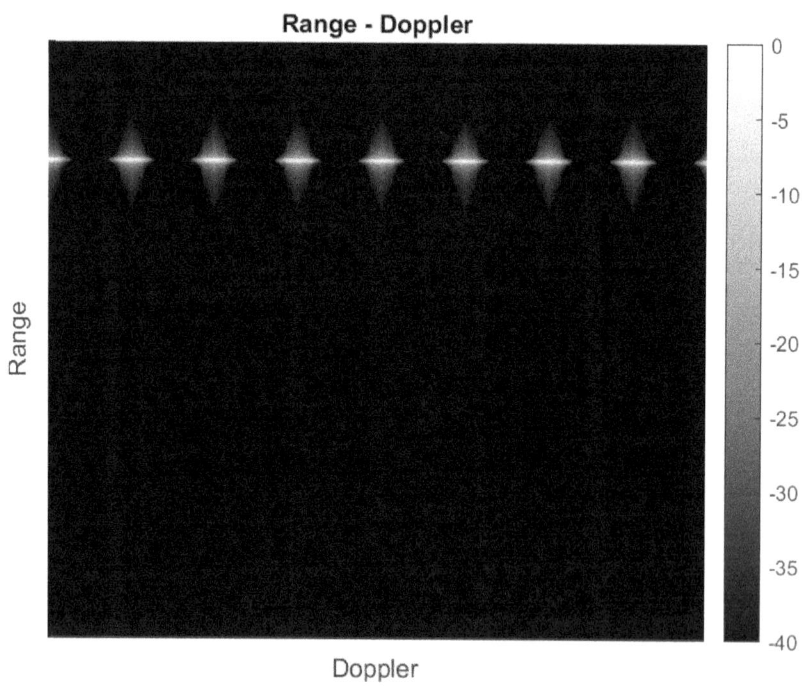

Figure 8.14 Range-Doppler view where victim and interferer are matched in frequency and chirp slope, but the interferer PRI is a multiple of the victim (here every eighth pulse of the victim contains interferer). Notice this results in the generation of multiple target-like returns.

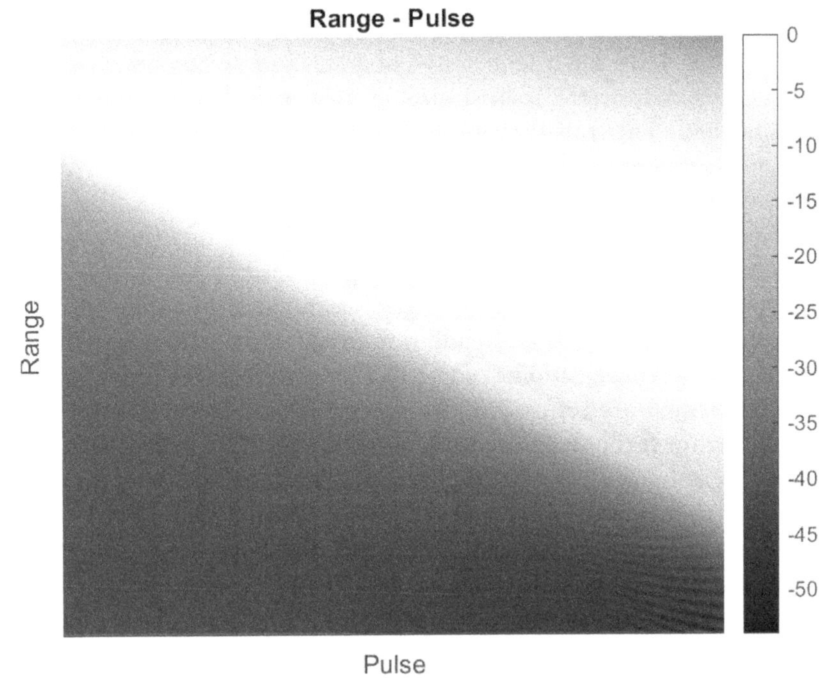

Figure 8.15 Range-pulse view where victim and interferer are matched in frequency, slightly mismatched in chirp slope and PRI. Notice the interferer response appears smeared in range and walks through the range window during the CPI, due to the mismatch in PRI.

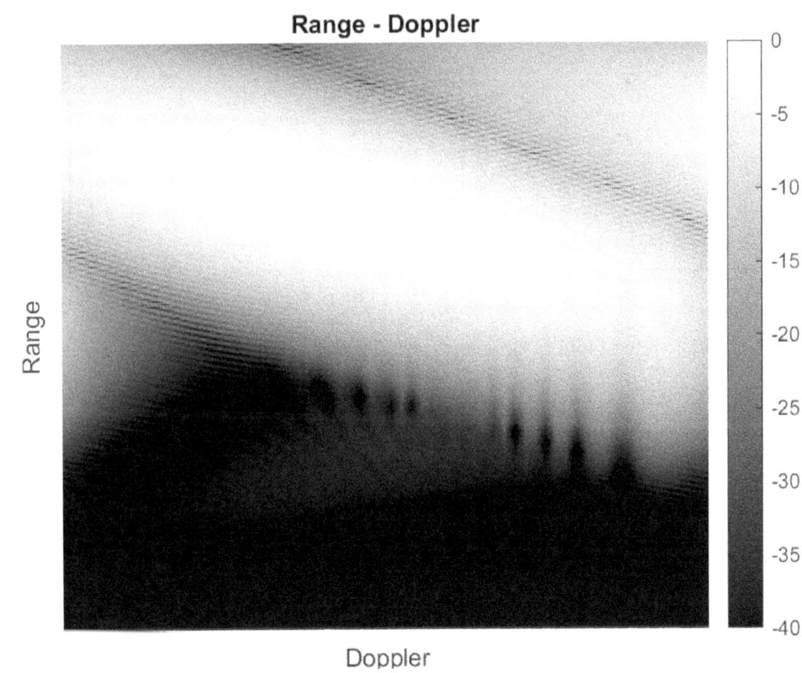

Figure 8.16 Range-Doppler view when RFI has mismatched slope and PRI. Notice that a significant portion of the range-Doppler map is obscured with RFI.

8.3.1.5 Case 5: Mismatches in Multiple Parameters, Including Frequency

In Cases 1–4 we have seen how coherent and semicoherent RFI appears when the victim and interferer both use LFM waveforms that are matched or slightly mismatched. This raises the question as to what would happen if the waveform parameters were *significantly* different, especially in center frequency. To answer this, we need to review how the stretch receiver operates. Recall that the stretch receiver enjoys the (cost, power, complexity) benefits of a lower ADC rate at the expense of a limited instantaneous range coverage. A key part of this approach is the narrow IF filter that follows the de-ramp and limits the spectral observability to (typically) 10–40 MHz. In addition to the lower ADC rate, this mechanism provides an inherent rejection of some RFI, as interferers that do not lie in that narrow spectral region are rejected. One can consider the victim receiver experiences the RF spectrum through a narrow "soda straw," and interferers that don't fall into it aren't seen and don't impact the victim (i.e. the timing and parameters could occur in a way that the interferer never is in the victim IF bandwidth (Figure 8.17)).

Now consider the case where the interferer only enters the "soda straw" for a small fraction of the pulse (Figure 8.18). This short burst of interference will appear impulse-like in the time domain and will spread over the entire range swath and Doppler spectrum, resulting in a noncoherent appearance (i.e., it results in an increase in the noise floor, but no discernable structure in the range-Doppler domain). The processing gain it receives is minimal, but it can still be a problem for the victim, especially when attempting to see objects with low SNR, such as pedestrians or motorcycles at longer ranges.

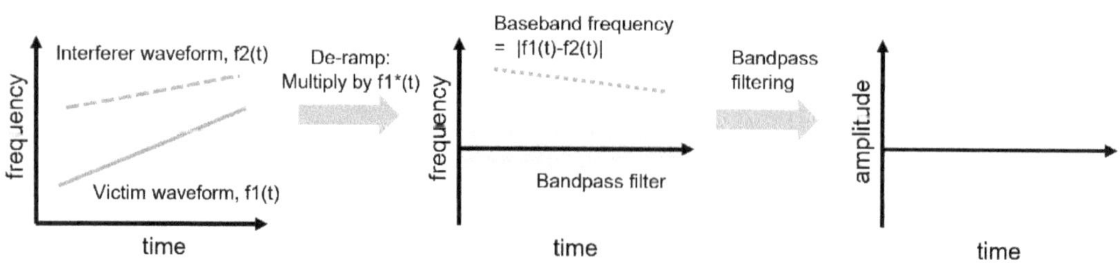

Figure 8.17 Pictorial view of victim processing chain where interferer doesn't enter victim IF bandwidth.

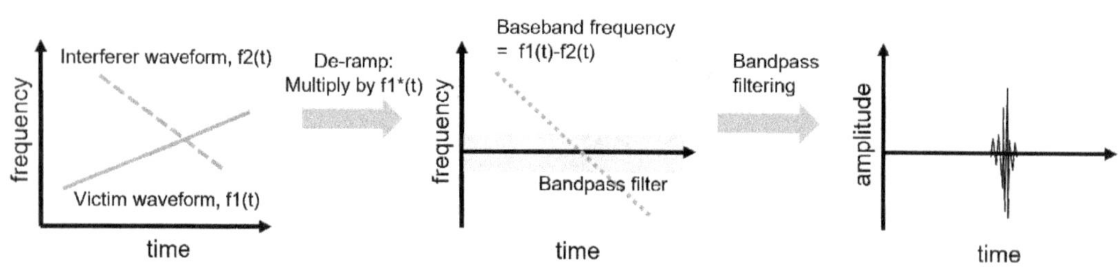

Figure 8.18 Pictorial view of victim processing chain where interferer only enters the victim's IF bandwidth for a small fraction of the pulse.

8.3.2 PMCW Radar and Mixed Waveforms

We now turn from FM waveforms to code modulated waveforms (a detailed description of these waveforms was presented in Chapter 3). For this analysis, it is instructive to turn from the time-domain analysis we used in the previous section to a frequency domain view. To review, for frequency modulated radar the transmitted signal is a sinusoidal signal in which the frequency of the signal varies with time, and for PMCW radar—also called digital code modulated (DCM) radar—the transmitted signal is a sinusoidal signal in which the phase varies with time.

It turns out that nearly all victim interferer permutations of PMCW and FMCW result in noncoherent interference, similar to what we observed in Case 5 (8.3.1.5). To understand the power levels of the interference, we will look at the ratios of processing gain between the victim's own signal (as it is returned from a target of interest) and that of the interferer, calling this term the *processing gain improvement (PGI)*. We will provide some top-level heuristics for calculating these differences, which can then scale the relative transmit powers and path loss to derive an effective signal to interference ratio (S/I). That being said, there may be a few dB of difference between these heuristics and the actual S/I ratio when considering the actual waveform, filter characteristics, and mismatches in frequency.

8.3.2.1 FMCW Interfering with PMCW

When FMCW interferes with a PMCW victim, the result in the range-Doppler domain is a noncoherent interference, as illustrated in Figure 8.19. The interference power is maximized when the FMCW frequency is completely contained within that of the PMCW victim.

The PMCW signal will receive the full processing gain of the code, while the FMCW should receive none, resulting in the PGI between the victim and interferer being product of the length of the code and the number of pulses:

$$PGI_{\text{FMCW to PMCW}} = N_{\text{Chips}} \cdot N_{\text{Pulses}} \tag{8.12}$$

8.3.2.2 PMCW to FMCW

PMCW interfering with an FMCW victim also results is also a noncoherent interference in the range-Doppler domain (see Figure 8.20).

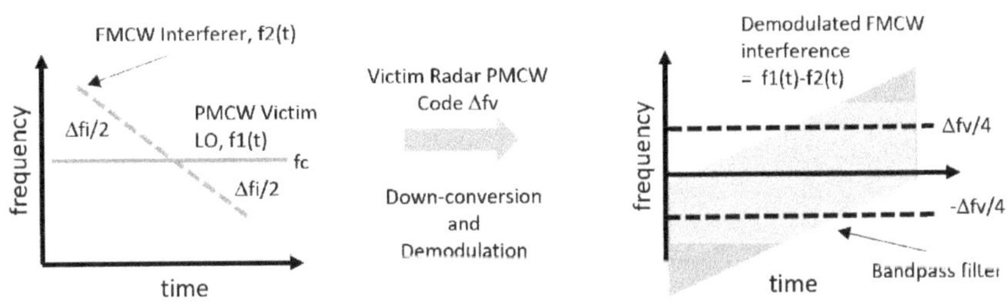

Figure 8.19 Typical LFM to DCM interference scenario and the corresponding impact on the baseband signal.

Recall that with FMCW stretch processing a narrow filter follows the de-ramping process. This filter attenuates much of the PMCW signal, resulting in a processing gain improvement related to the ratio of the bandwidth of the PMCW signal (i.e., the bandwidth of the chip) to that of the filter:

$$PGI_{\text{PMCW to FMCW}} = \frac{Bandwidth_{\text{Chip}}}{Bandwidth_{\text{FMCW Filter}}} \cdot N_{\text{Pulses}} \tag{8.13}$$

8.3.2.3 PMCW to PMCW

When PMCW interferes [6] with another PMCW victim, the result is strongly a function of the differences in the phase codes employed by the two radars. If they use the same code or codes comprised of similar segments, then the possibility for coherent or semicoherent interference exists. In practice, the likelihood of this is small, as randomization of codes is typically incorporated in their use.

If the codes between the victim and interferer are different, then the interference will appear as noncoherent (see Figure 8.21), similar to the other examples in this chapter, with processing gain related to the number of chips and pulses used by the victim:

$$PGI_{\text{PMCW to PMCW}} = N_{\text{Chips}} \cdot N_{\text{Pulses}} \tag{8.14}$$

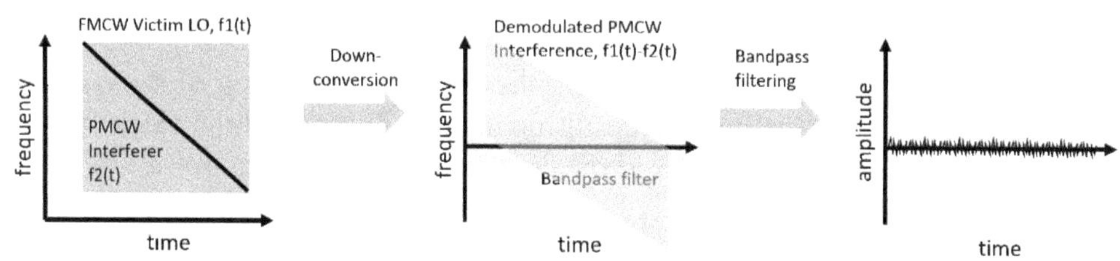

Figure 8.20 Typical PMCW to FMCW interference scenario and the corresponding impact on the baseband signal.

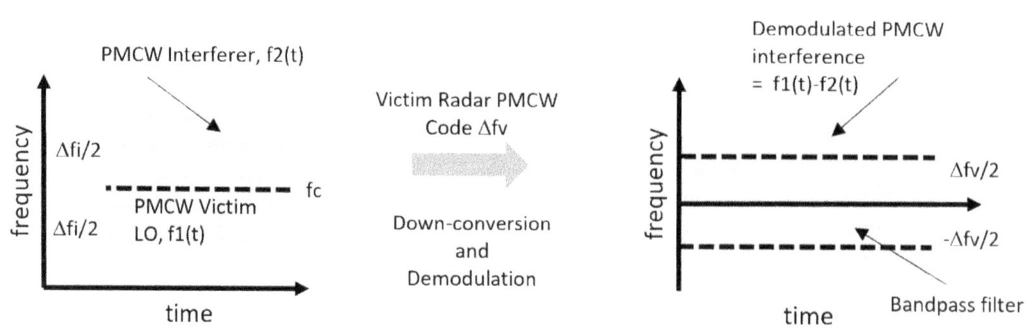

Figure 8.21 Typical PMCW to PMCW interference scenario and the corresponding impact on the baseband signal.

8.4 RFI Mitigations

In the previous sections we showed that RFI can take many forms in range-Doppler space: coherent, semicoherent, and noncoherent. Each of these can lead to three deleterious effects; the radar can experience type 1 errors (false positive detections), type 2 errors (missed detections), and an overestimation of the expected radar performance when compared to a quiescent environment. These can lead to suboptimal braking, deceleration, or acceleration of the fully autonomous vehicle.

In this section, we examine what the radar community can do to buy back all or some of this performance in ways that are not merely self-serving. For example, a selfish approach could be to design a radar that when it detects RFI simply increases its transmit power—without regard to the damage that action will cause to other radars in the scene. These types of behavior are to be avoided.

We begin with a set of tools that radar designers can employ in their radars without coordination with any other parties. These form a core set of mitigations that will seek to optimize performance no matter how bad the RFI scenario is. From these, we look at a series of options that involve participation by an increasingly larger number of organizations, concluding with a discussion of overarching government regulations on RFI, and the strengths and limitations of regulations in this space. Finally, we provide our recommendations for initiatives the radar community should take to address RFI. This is our call to action for the designers, customers, and users of automotive radar, and we believe these initiatives will preserve this much-needed sensing modality for fully autonomous vehicles of today and tomorrow.

8.4.1 Mitigations Local to the Radar

Through thorough planning and incorporation of technology, the radar designer can incorporate several layers of RFI protection into the radar [7]. These techniques require no coordination with other radars, global timing strategies, or overarching regulations, but they do come at a price: every mitigation will add complexity and potentially cost to the radar system. If applied correctly, they can substantially increase the radar's performance in an interference situation (e.g., they can increase signal to interference plus noise ratio [SINR]), although they may not repair the radar data products to the level that they were had no interference been present (i.e., they may pay a price in SNR). Table 8.2 contains a sampling of typical types of RFI mitigation; for more information, see [8].

As we have shown in Case 5 (8.3.1.5), some situations exist where the interferer only appears within the victim IF for a small fraction of the pulse. In this situation, an effective mitigation can be to *excise the RFI in fast time* by (1) identifying the area of the pulse where the overlap occurs, usually through comparing the amplitude to a threshold, (2) removing (zero-out) that section of the pulse, and (3) applying a correction or repair to the removed section if necessary. The effectiveness if this method is a function of the fraction of the pulse containing interference; the smaller the fraction the less damage to the radar impulse response, sidelobe levels, and SNR. Implementation of this mitigation can be challenging, as it requires insertion early in the processing chain to compare the fast-time samples from each

Table 8.2 Many Methods are Available to the Radar Designer to Protect from Interference from Other Radars

RFI Mitigation Method	*Description/RFI Targeted*	*Strengths*	*Limitations*
Excise RFI in fast time	Exploit the fact that RFI with large difference in waveform parameters (e.g., chirp rate) will only appear in the victim for a fraction of the pulse; identify and remove that fraction.	Works well if fraction of pulse contaminated is small and RFI easy to identify.	Potentially difficult to insert a thresholding stage early in the processing pipeline; can be difficult to estimate the appropriate RFI threshold; may impact range impulse response.
Excise RFI in angle (adaptive beamforming)	Exploit the fact that RFI may be coming from a single angle; use multiple spatial channels to remove that angle from processed data products.	Significant reduction of RFI power possible.	Processing intensive; loss of visibility in RFI direction; may require more extensive radar calibration.
Increase processing gain	Increase energy from the target through longer pulsewidth, more pulses, or more receive channels to improve SNR against RFI that appears noiselike.	Linear increase in SNR across each dimension (e.g., two times pulsewidth yields 3 dB SNR increase).	Impact to radar timeline (fewer looks); may not be sufficient to overcome noise RFI power; limits exist on use before more processing required (e.g., target acceleration compensation).
Tracker rejection	RFI with closely matched parameters can appear target-like, but may not persist as waveform parameters cycle through modes; target tracker eliminates sporadic detections.	Effective at preventing sporadic detections from contaminating downstream processing.	Effectiveness degrades if target detections persist.
Confidence estimation	Report that RFI is present, and potentially the magnitude of degradation, to downstream processing to aid in decision making.	Enables better reasoning about radar data quality.	Doesn't actually improve radar data quality.

pulse to a threshold and then keep, remove, or modify them. Finally, the choice of the threshold used to differentiate between RFI and non-RFI can be difficult in a strong cluttered environment.

When excising in fast time, we exploit the fact that the interference is concentrated in the fast-time domain, which allows us to excise it without damaging the rest of the pulse. Similarly, we can exploit that concentration in the spatial domain and *excise in angle*. In modern radars this concept is often called *adaptive beamforming*, and multiple spatial channels are used to (1) identify that a strong source is coming from a particular angle, and (2) process or reprocess the data in a manner that minimizes the contributions from that angle. For example, consider a radar that forms a range-Doppler map for each spatial channel and PDIs them together for detection, then returns to the spatial data for angle estimation on the detections (a common processing method for some automotive radars). If this radar had identified that a strong source was coming from a particular angle, θ, then it could whiten the data across channels before the PDI process. Say $v(\theta)$ is a column vector that represents the spatial channel response to the RFI, then the whitening would take the form of premultiplying the vector of data from each range-Doppler cell by a projection matrix W that removes contributions from the RFI angle, where

$$W = I - \left[\frac{v(\theta) \cdot v(\theta)^H}{v(\theta)^H \cdot v(\theta)} \right] \tag{8.15}$$

and H represents conjugate transpose. Each range-Doppler cell of the data cube x would be premultiplied by W, as

$$x(range, Doppler)_{\text{post-mitigation}} = W \cdot x(range, Doppler) \tag{8.16}$$

Leaving the postmitigation data cube void of contributions from the angle θ that contained the RFI. Note that this will remove the contributions from all sources at that angle, not just the RFI (see [9] for additional details on adaptive radar processing).

Both of these mitigation techniques will reduce the SNR of the target to some degree, as they inevitably remove some target signal with the RFI. A strategy to help overcome this loss is through increasing the received energy from the target. This can be accomplished through an increase in the pulse width, increasing the number of pulses integrated, or increasing the number of spatial channels. Each of these will increase SNR to attempt to recover the SNR lost from excising RFI as well as overcome interference that manifests as noise, such as DCM to DCM.

In some situations, the interferer waveform is well-matched to that of the victim—they could be employed on the same fleet of fully autonomous vehicles or developed by the same manufacturer. In these cases, the RFI can appear in the range-Doppler map similar to a well-focused target and be detected as such. Here the target tracker provides an important layer of protection, as it can prevent track formation or promotion on sparse detections that may occur from an occasional frequency collision. The tracker can reject tracks that are not kinematically consistent, meaning that they have nonrealistic range/Doppler/acceleration profiles. By appropriately tuning the tracker the radar can prevent occasional false targets due to RFI from entering the perception stack.

Finally, *confidence estimation* can alert the downstream processing (e.g., perception, decision planning) of the degradation in quality due to RFI, enabling that processing to make more informed decisions about how to use radar data. Typically the tracker reports track covariances based on various measurement noise and process noise models characterized in a quiescent environment. However, in an RFI setting, it may have difficulty properly estimating these parameters. In an RFI setting, even tagging measurements or track outputs with a simple "RFI-present" flag can make a substantial improvement in how the fully AV reasons about its actions. We will revisit confidence estimation later in this chapter as a recommended mitigation all radars should implement.

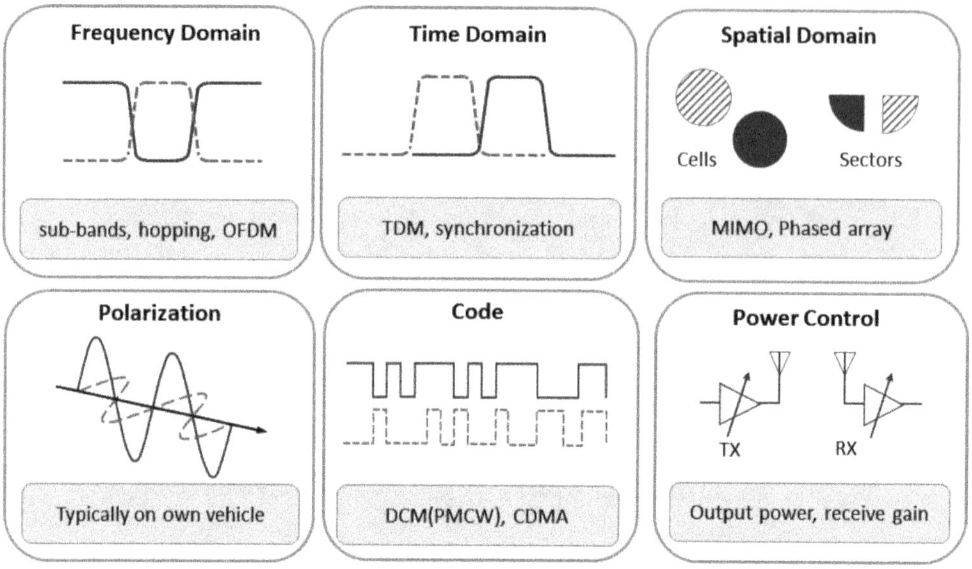

Figure 8.22 Six domains of potential sharing.

8.4.2 Global Mitigations: Noncooperative Countermeasures

The previous section looked at techniques an individual radar can employ to reduce the impact of the RFI it encounters. In this section, we turn to more global solutions that can prevent or reduce the RFI seen by a radar and lessen the burden on the individual mitigations of 8.4.1. The noncooperative countermeasures can be grouped in six domains (Figure 8.22).

Electromagnetic signals are typically discussed as existing in multiple domains, including frequency, time, space, and polarization (Figure 8.22). These domains suggest some ways to intrinsically reduce RFI through signal design. For example, one powerful combination of frequency and time used for signal isolation is coding, which is widely used and well characterized in communications and some sensing applications. If a radar employs a code on its signal this can reduce the power received by another radar after its processing. Similarly, the first radar could reduce its transmit power, or reduce it in certain sectors to minimize its impact on other radars. All of these could make the RFI problem less severe.

However, noncooperative countermeasures to reduce interference are not for free. Some methods might be beneficial only for the interferer, some might be helpful for the victim but with trade-offs on the interferer itself. To make noncooperative countermeasures beneficial they typically have to be combined with sensing of the environment to make the right decision on the countermeasure. This increases the cost of implementation, with higher power consumption, more hardware, and increased software complexity. Additionally, the resources are limited (e.g., there is only 5 GHz allocated to the 77 GHz band for automotive radar), which limits the overall utility of noncooperative countermeasures. Table 8.3 provides an overview of potential ideas and concepts with their pros and cons. Not all methods are applicable to all waveform types.

Table 8.3 Summary of Potential Noncooperative Countermeasures

Method	*Pros*	*Cons*
Fixed Polarization: Select polarization based on vehicle installation location	Highly effective of up to 30 dB in simple static scenarios. Works well on own vehicle.	Reflections will change polarization and limit the effect to 10 dB. Polarization change will impact reflections and SNR of a scenario.
Adaptive Polarization: Sense and adopt polarization of transmit and receive channels	Highly effective of up to 30 dB in simple static scenarios.	Transmit and receive antenna must be controllable in terms of polarization. Extremely costly in terms of implementation, power consumption, and sensor size. Polarization change will impact reflections and SNR of a scenario.
Use Less: Usage of radar only where is required and beneficial	Highly effective.	Need to make decision on radar usage based on other methods. Potential impact on safety of the vehicle. Will increase cost because additional intelligence is required to make the decision.
Bandwidth: Limit the occupied bandwidth of the radar sensor only to the required bandwidth to fulfill application use cases	More efficient usage of available bandwidth. Reducing probability of interference. Limit the bandwidth to meet feature requirements, to increase efficiency of other countermeasures.	
Transmit Power: Limit to application use cases and accordingly to the environment	Highly efficient. Reducing level of interference.	Potential SNR degradation (lower signal strength). Might miss objects with low RCS due to reduced transmit power. Need to sense the environment.
Beamsteering, Phased Arrays: Focus the beam in the area of interest, in the direction of main important target, scenario	Increase SNR in the direction of the most important target. Reducing probability of interference.	Potential risk to miss targets in the area the beam is not focused. Interference suppression may be limited when the interferring radar is on the target the victim is trying to see. Reduced total FOV, requires multiple beams. Complex implementation.
Frequency Hopping: Pseudorandomization of the carrier frequency from scan to scan or even from PRI to PRI	Reduces probability of persistent interference from scan to scan.	Increases the usage of frequency ranges. Limited available spectrum. Less efficient for highly dense scenario.
Time Hopping: Pseudorandomization of sensor cycle times and especially cycle time start	Reduces probability of persistent interference from scan to scan.	Less efficient with increasing numbers of interferers. Might be in conflict to perception requirements (constant frame rate).
Time Synchronization: Coordinate scan start/stop	Will reduce interference on own vehicle.	Works only on own vehicle. Requires synchronization capability across sensors from different vendors.

Table 8.3 *(continued)*

Method	*Pros*	*Cons*
DBF Adaptive Nulling: DBF main beam null in direction of the interferer to reduce width of angular sector of interference	Remove illumination (interference) of other sensors.	No improvements for targets in the same angular sector as interferer. Signal loss of target. Complex implementation.
Pseudorandomization of Chirp Timings: Jittered PRI	Reduces probability of persistent interference from scan to scan.	Increase complexity of chirp generation and baseband processing.
Pseudorandomization of Chirp Slopes: Alternating up/down or a more random scene	Reduces amount of interferer energy in victim passband, spreads interference energy across range and Doppler domains.	Increase complexity of chirp generation and baseband processing.
Clear Channel Sensing: Identify a frequency band and/or time slot with the least interference probability and move radar operation into this band	Not a standalone countermeasure but the key element to drastically increase the efficiency of mitigation techniques.	Might lead to an endless control loop in high density environments. Will increase power consumption. Not able to predict how the interferer will behave in the future.

8.4.3 Global Mitigations: Cooperative Countermeasures

Cooperative countermeasures involve a near-real-time negotiation among all radar-using parties needing to utilize the RF spectrum in a local area in order to allocate needed resources. Conceptually, this approach holds significant promise: Each situation is unique (number and types of vehicles, type of road, time of day, weather condition), so each solution can be uniquely crafted to the particular situation. The additional complexity needed to implement a communications side channel to exchange information, coordinate modulation, and coordinate frequencies and other important sensor parameters is small compared to that required for full autonomy, making it even more attractive.

In practice, however, there are several drawbacks that must be addressed to effectively and safely *rely* on a cooperative solution. And that reliance is a key distinction—fully autonomous vehicles (by definition) do not have a human driver as the last line of defense—if we adopt a particular technological solution, we must be able to trust in all situations we plan to encounter. From that mindset there are four striking challenges to relying upon cooperative approaches for RFI mitigation.

First, implementation of a cooperative approach is only useful if everyone incorporates it. Imagine how ineffective it would be if some vehicles were using—and relying on—a precise timing scheme (a potential cooperative approach where users have negotiated timeslots for operation) to prevent RF collisions, but others in the scene were operating wherever in frequency and time that they desired because they were either developed before the cooperative protocol was implemented or chose not to implement it. This would fail to achieve the goals or justify the investment in the approach, which could easily take the better part of a decade for the community to evaluate concepts, select, and ultimately work to implement the solution—far longer than society wants to wait for fully autonomous vehicles. Second, as we discussed in the introduction to this chapter, the RF spectrum is a very finite resource; there are far more users than can be supported without spectral collision or time multiplexing that impairs update rate and recall (detection probability) and ultimately vehicle safety.

Third, the latency and timing involved may be significantly greater than the AV may be able to tolerate. For example, a global synchronization approach may require greater timing precision than the radar would otherwise need—potentially microsecond timing if attempting to align pulses across platforms (see [10] for a solution utilizing this type of approach). Alternatively, if the cooperative countermeasure involves an ad hoc network the time to compute and share these allocations (e.g., time, frequency) among all the participants needs to forecast far enough into the future to allow radars to appropriately plan their operations, but that is at odds with the flexibility needed to rapidly include new users entering or users exiting the area of arbitration.

Finally, this approach has significant issues with accountability: Who has the responsibility to ensure that the cooperative approach successfully reduces or eliminates RF interference, and what happens if the approach does not? Additionally, who bears the responsibility for ensuring that all participants' communications are operating as desired? If a particular vehicle's ability to transmit the needed information is impaired, it can wreak havoc on other radars in the scene—a situation not easily addressed by the network.[4]

A fair question is if cooperative countermeasures cannot provide the entire solution, are they still a worthwhile investment? At this time, we believe the best answer is "maybe." There could be beneficial infrastructure[5] that can aid fully autonomous vehicles in very complex situations—maybe like a cooperative version of the crowed intersection traffic officer that explicitly guides vehicles through very complex situations—but it remains to be seen whether the benefits of this outweigh the costs. However, from a strictly RFI-mitigation standpoint, cooperative countermeasures have significant challenges in providing a robust solution.

8.4.4 Global Mitigations: Regulations

Reasonable and effective regulations have been and continue to be important in shaping the welfare of economies and society; their use has had numerous benefits, including safety (e.g., seat belts and air bags in automobiles) and management of resources such as the RF spectrum through band allocations and communication protocols. Given these successes, one may consider that the best way out of the automotive radar RFI conundrum is to simply turn it over to regulatory agencies. However, this thinking is flawed for multiple reasons.

4. One could envision an implementation that required a radar to receive handshake information from other radars or an infrastructure arbitrator in the scene before operating, but this seems unnecessarily complex, can fail in sparsely populated roadways, and can result in traffic gridlocks should the network to the arbitrator experience an outage.
5. These are often described as vehicle to vehicle (V2V) or vehicle to some other entity (V2X).

Similar to the complexity and slow pace discussed in incorporating cooperative countermeasures, a regulatory solution for RFI would likely approach or exceed a decade before adoption en masse.[6] This imposes an unacceptably long wait for the benefits fully autonomous vehicles can bring to society, especially since the *development* of fully autonomous technology requires the radars to provide key data to the vehicle's perception system. A further complication to regulation is the millions of radars that exist on vehicles on the roads today; we can neither mandate nor easily enforce a regulation that would require their removal from service, and some of these vehicles can easily still be operational 10 or more years, making a regulatory solution for the current 24 or 77 GHz bands partial and incomplete at best.

Another challenge for a regulatory solution in the near- to mid-term is the complexity of the problem; as we discussed in the introduction to this chapter, there are no obvious straightforward solutions. Historically, governments have struggled with regulations in complex technical areas,[7] not due to a dearth of talented and bright individuals, but because of inadequate understanding of the subtleties of regulatory influence—especially unforeseen consequences—and political pressures that can overshadow sound technical reasoning.

From a practical viewpoint, the safety of the passengers in a fully AV is ultimately the responsibility of its developer and maintainer. If after the long and expensive process of creating, socializing, and implementing regulations to eliminate RFI the fully AV developers find that either (1) the regulations do not enable sufficient radar data (they are too restrictive in the use of radar to provide a beneficial data product) or (2) do not sufficiently prevent RF interference (they are too lenient and allow radars to interfere unacceptably), then the vehicle developers will be forced to abandon radar as a useful sensor for autonomy,[8] which would obviously be an outcome counter to the intentions of the regulators.

The authors believe that for the current automotive spectra a regulatory approach to countering RFI isn't wise. However, a forward-looking incorporation of an *additional* spectrum for automotive radars can be extremely beneficial. Allocating bands in the 100+ and 200+ GHz bands for automotive applications—given sufficient bandwidth, definition of protocols and waveforms, requirements on RFI identification and mitigation, and investment in technology—can be a very effective *long-term* solution that avoids several of the pitfalls described in this section.

8.5 Recommendations for the Future

We have covered why RFI can be a substantial problem for the fully AV's radars, and some of the various ways radars can protect themselves at the local (individual radar) level and at a larger community level. We now turn to recommendations for the future. Those who are reading this work are likely the exact audience for starting

6. Consortiums such as MOSARIM and IMIKO have already been studying the RFI problem for more than a decade with little actionable results.
7. Recent examples include net neutrality, school testing, and digital privacy.
8. An alternative outcome of (2) would be that the developers would create their own mitigations similar to those discussed in this chapter, but that would seriously call into question the benefits of the regulations in the first place.

the dialogue pushing forward these recommendations, and working to make for a smarter, safer environment.

This will not be easy, as these recommendations center around a mindset shift, and that is always problematic because we become entrenched in our ways. But we need to shift away from the mentality that things will be like they were before RFI started becoming a problem. That mentality just is not supported today—there are too many cars with radars, the spectrum too scarce a commodity for all the users, and the radar's importance to the safety of fully autonomous vehicles—or even ADAS enabled vehicles—is too high. Additionally, we need to recognize that there is no silver bullet approach, and that we will have to employ multiple of levels solutions; we should think about ways to improve the RF environment and reduce the effects of interference, not remove it to pre-RFI levels. Granted, there will be some situations when the vehicle is relatively isolated and a completely clear environment is available, but these will become scarcer as time goes on, and should not be the situations we design to.

The following are three areas we believe should be universally adopted by all radar developers and users. Through these we can all work together to ensure radar continues to be a valuable and critical sensor for the operation of fully AVs.

8.5.1 Use Less Energy and Power

Although it sounds like a strange statement to be in a book about radar, but one of the easiest things we can do to help mitigate RFI is to reduce how much we use the radars. This can come in many forms. For example, a fully autonomous vehicle likely has 360° radar around the vehicle, but in some situations, this isn't necessary; radars that are focused on side coverage can perhaps be turned off when driving on the freeway and only forward and aft coverage is necessary, or when stopped at an intersection. A second area to consider is using or changing the radar frame or CPI rate. If the system can handle a lower update rate in certain situations, the radar can choose not to transmit for certain CPIs (say, every third one). In this mode it can even maintain timing with other radars on the vehicle if they are synchronized. Finally, if designed appropriately, radars can choose to employ power management to lower the transmit power if long-range capability is not needed at that time, say when in a dense urban environment. This last area is not as effective as the others; the victim receives power at r^2, while the offending radar must deal with an r^4 drop, so the amount it can drop the power is limited.

While not a panacea, these approaches can safely and significantly—both are critical—reduce the amount of RF energy each radar puts into the environment. They do come with challenges that must be addressed. First, the AV must be able to identify that it is in one of the environments where it can begin modulating the radar's behavior and may require exogenous information about the environment (e.g., detailed maps). In addition, the transition between modes needs to be implemented so that any hysteresis isn't detrimental to the vehicle. Second, since it can take time (several CPIs) to surveil the area and establish tracks on appropriate constituents of the environment, the radar needs to be switched back on *before* the data are needed, meaning that the vehicle must anticipate some amount into the future to determine if the full radar performance is needed.

Figure 8.23 Interferer signal measured (FCM type of signal) by victim sensor.

8.5.2 Report Confidence

We discussed in Section 8.1 how a radar can enable better reasoning in the higher levels of the software stack by reporting a reduced confidence in its data when in an RFI environment. We return to that concept here as one of our key recommendations, as it adds robustness and safety to the fully AV, and when employed universally it enables safe operation for everyone by ensuring the best decisions are made with the appropriate use of sensor data.

Fully AVs use multiple sensors (e.g., radars, lasers, cameras) to perceive the environment—each with their own accuracies and capabilities. It uses multiple modalities and installation locations to overcome the real-world challenges it faces, like sensor occlusion (a sensor could be blocked by, say, a large vehicle in front of it), weather effects, hardware malfunctions, or even degradation from a bug hitting the sensor cover, and it is the job of the perception system to appropriately combine these. By indicating the presence and severity of RFI, the radar provides protection against a false negative (missed detection); without the RFI indication the software stack could potentially overweight the radar data and make an unsafe maneuver, but with it the fully AV can still make progress toward its destination safely by appropriately relying on the other sensing modalities.

The radar has many observables it can use to determine the presence or absence of RFI. First, it can look at changes in the noise floor, since this will provide an indication of noncoherent RFI. Second, it can look for large localizations of energy along a particular angle but distributed in range.[9]

Third, the radar can look for returns that are extended in range, Doppler, or both away from the clutter manifold; these non-targetlike returns are an indication of semicoherent RFI. Finally, if the radar has the hardware to sense the RF spectrum (Figure 8.23), it can not only provide an indication of the RFI but use this information to inform the choice of mitigation as well by (1) measuring the interference to provide insight on interferer type, (2) measuring the bandwidth occupied by interferer, and (3) measuring the interferer timing [11, 12].

9. In the hostile jamming world this is called a *jammer line of bearing*, or JLOB.

Table 8.4 Concepts for RFI Mitigation Taxonomy

Capability	*Possible Types of Levels*	*Notes*
Detect presence of RFI.	Probability of detection as function of interference power or energy. Ability to localize interference to a spatial region.	Detecting the presence of RFI is often an important first step in determining the appropriate responses.
Estimate corruption of RFI on observables.	Qualitative (poor, good) assessment. Quantitative increase in measurement or track covariances.	Reporting the confidence in radar data is a key component of safety for fully autonomous vehicles.
Ability to avoid interference.	Key parameters of interest include number and location of interferers, their frequency extents (e.g., chirp bandwidths), frequency hopping rate, and frequency hopping extent.	This capability would include the detection of RFI, estimation of key parameters, and choice of appropriate clear frequencies and spatial regions.
Mitigation of interference.	Metrics would include ability to restore quiescent capability (that obtained in a noninterference situation) up to a certain range (e.g., no degradation up to 150m). Likely include levels for reduced operation when unable to restore full quiescent capability.	Of importance here is the combination of capability restoration and accurate estimation of residual interference effects on data quality.

8.5.3 Create a Useful Taxonomy for RFI Mitigation

Having firmly established the benefits of the radar providing confidence metrics on its performance given the RFI it is currently experiencing, we would like to extend this concept of "confidence" to the radar design itself. Currently the automotive radar community does not have well established standards on how to quantify the level or types of RFI mitigation a radar possesses, leaving the designers to use only qualitative terms like "avoids RFI" or "best RFI performance," neither of which are particularly useful.

We believe the automotive radar community should develop a taxonomy of RFI mitigation levels, as well as a series of standard, repeatable tests that can be performed to verify whether a radar has achieved this level. With these levels, a radar designer could clearly communicate the types and amounts of RFI mitigation offered a particular radar model or mode (this could be specified as either), allowing those using the radar (e.g., vehicle OEM, developer of autonomy stack) to better understand what they are receiving. In addition, the research community could use these levels to focus efforts on developing the most needed solutions and communicate their effectivity in a standardized manner.[10]

There is no doubt that developing these levels will require effort. However, since this is a means of *communication* and not *regulation*, the community can start enjoying the benefits of these immediately without having to wait for global acceptance or governmental approval. In addition, the levels do not have to necessarily be ordinal,

10. This has a parallel in the medical field, where the developer of a proposed treatment specifically clarifies the treatment's areas of effectiveness, such as increasing the five-year survival rate for women with Stage 3 breast cancer.

addressing different types and amounts of interference in whatever ways optimize communication of the radars' capabilities. Table 8.4 provides an example taxonomy to illustrate these concepts. For all of these, the community could develop a canonical set of radar waveforms and modes against which to demonstrate the capability.

References

[1] De Martino, A. (Ed.), *Introduction to Modern EW Systems*, Norwood, MA: Artech House, 2013, https://studylib.net/doc/18347338/introduction-to-electronic-warfare-scenarios.

[2] Poisel, R., *Information Warfare and Electronic Warfare Systems*, Norwood, MA: Artech House, 2013.

[3] Avionics Department, "Electronic Warfare and Radar Systems Rngineering Handbook," Naval Air Warfare Center Weapons Division, Tech. Rep. NAWCWD TP 8347, 2013, https://tinyurl.com/navy-ew-2013.

[4] Murali, S., et al., "Interference Detection in FMCW Radar Using a Complex Baseband Oversampled Receiver," *2018 IEEE Radar Conference*, April 17–23, 2018.

[5] Yang, Z., and A. Mani, "Interference Mitigation for AWR/IWR Devices," Texas Instruments Application Report SWRA662, January 2020.

[6] Skaria, S., et al., "Interference Mitigation in Automotive Radars Using Pseudo-Random Cyclic Orthogonal Sequences," October 2019, https://www.ncbi.nlm.nih.gov/pmc/articles/PMC6832580/.

[7] Electronic Communication Committee (ECC), "Studies Related to Surveillance radar Equipment Operating in the 76 to 77 GHz Range for Fixed Transport Infrastructure," ECC Report 262, January 27, 2017.

[8] Neri, F., *Introduction to Electronic Defense Systems*, Raleigh, NC: Scitech Publishing, Inc., 2006.

[9] Guerci, J. R., *Space-Time Adaptive Processing for Radar*, Norwood, MA: Artech House, 2003.

[10] Khoury, J., et al., "RadarMAC: Mitigating Radar Interference in Self-Driving Cars," *2016 13th Annual IEEE International Conference on Sensing, Communication, and Networking (SECON)*, June 20–27, 2016.

[11] The National Academies of Science-Engineering-Medicine, "Strategy for Active Remote Sensing Amid Increased Demand for Radio Spectrum," September 2015.

[12] Hakobyan, G., Armanious, K., and B. Yang, "Interference-Aware Cognitive Radar: A Remedy to the Automotive Interference Problem," *IEEE Transactions on Aerospace and Electronic Systems*, Vol. 56, No. 3, June 2020, pp. 23–26.

CHAPTER 9

The Impacts of Water (Weather) on Automotive Radar

Tim Campbell, Michael Fina, Gary Clayton, and Matt Markel

9.1 Introduction

Radar is considered by many to be an "all-weather" sensor. However we can't ignore the effects of weather on the radar, particularly at millimeter-wave frequencies. Airborne rain, snow, and sand all present challenges to the automotive radar designer. Further, water, snow, and ice accumulation on radomes, the ground, and objects in the scene create further challenges that must be considered by the radar community, especially those designing radars for fully AVs.

In this chapter, we consider the effects of water on automotive millimeter-wave radar. First, we review the effects of weather on radar from a perspective of signal loss, using a partitioning of the radar wave propagation path. Second, we examine the impacts of water on radars that use antenna array processing to image a scene. Third, we consider backscatter from airborne water. Finally, we explore options for mitigation of some of these effects.

9.2 System Losses

Object recall (familiar to the radar community as probability of detection) is a key performance parameter for a Level 4 and 5 self-driving system. As a consequence, radar designers need to be cogent of additional system losses present in the presence of precipitation. Three primary sources of loss arise in the presence of precipitation: *transmission loss* caused by absorption and scattering of airborne water droplets, *lower target radar cross-section* created by wetted surfaces, and by *reduced radome transmission efficiency* caused by water accumulation on the radar.

9.2.1 Transmission Loss

Free space transmission loss caused by rain absorption and scattering is well characterized. A number of authors have published models on the subject [1, 2, 3]. Models for transmission loss typically posit a raindrop distribution probability density function (most commonly the Laws and Parsons Distribution [4]) and apply the Mie theory of scattering and absorption. The resulting function is well approximated by a power law function. For example, Figure 9.1 is derived from the power law fit made for rain attenuation (dB/km) given rain rate r (mm/hr) at 80 GHz carrier frequency in 20°C assuming the Laws amd Parsons droplet size distribution [2],

$$L = 0.9436 \cdot r^{0.769} \text{ (dB/km)} \tag{9.1}$$

One approach to verifying this analytic model is to construct a test with both a radar and a corner reflector shielded from the rain, with the rain rate measured by a calibrated disdrometer[1] and time aligned with radar measurements of the power reflected from a 30 dBsm trihedral corner reflector. Measurements were taken at slant 45° linear polarization at fixed frequencies of 78.37 GHz and 78.41 GHz with the corner reflector 60m from the radars.

In Figure 9.2, the radar and corner reflector are placed under nonreflective enclosures to assure they remain dry during the experiment. This is done to eliminate the target and radome loss sources in order to isolate the atmospheric loss. Independent testing of reflected power with and without weather coverings of the radar and the corner reflector indicated no measurable effect on power received from the corner reflector, owing to the very high directivity of the corner reflector and the radar used in this experiment. A ground bounce channel is likely present in this collection; however, one can assume that the ground reflectance remains stable over the course of this stationary experiment.

Over the course of a rain event, observed attenuation as a function of rain rate correlates well with published models (Figure 9.3). The rain event included collecting continuous data where the rain rate changed from no rain (0 mm/hr) to heavy rain (~45mm/hr) and back to no rain (0mm/hr).

This validation is well scoped to the near-ground geometry of AV operational domains. That said, a large body of empirical evidence supports caution in the application of such models to requirements analysis because raindrop distribution models and parameters can vary dramatically with region and weather patterns [2]. Despite the expected variation, we can conclude transmission loss can have a major impact on radar performance for certain operational driving domains (e.g., detection of objects at long range in 25 mm/hr of rain). Such conditions require either substantially derating sensitivity requirements or substantially overprovisioning radar sensitivity under nominal conditions.

9.2.2 Target Loss

The radar cross-section of objects in the scene is of paramount importance to radar system design. We have found through testing that the accumulation of moisture on objects in the scene will have a material impact on their radar cross-section.

To motivate this claim, consider a trihedral corner reflector. This retroreflective scatterer provides a very stable cross-section to an illuminating radar. With our test radar we collected power measurements from a dry corner reflector and then simulated natural accumulation of rain droplets by spraying the reflector with

1. Model 5.4110.00.300, Thies Clima.

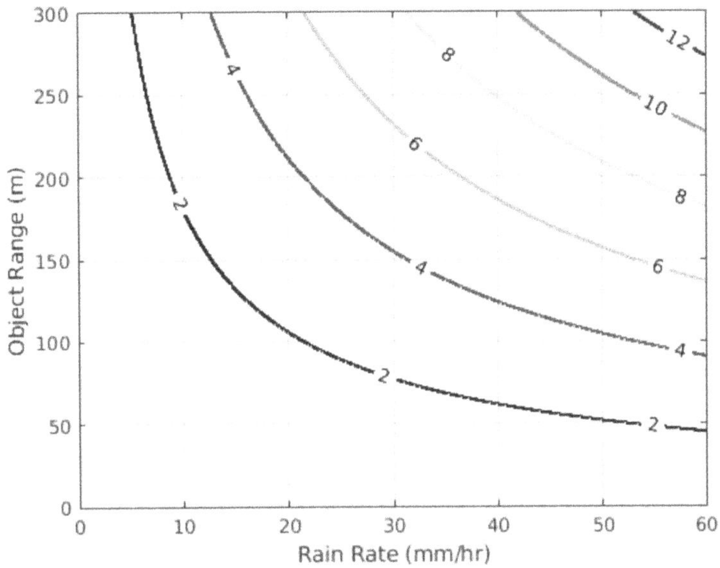

Figure 9.1 Transmission attenuation (dB) as a function of rain rate and object range: 80 GHz carrier frequency, 20°C, Laws and Parsons rainfall distribution. (Source: Waymo.)

Figure 9.2 Radars under test are mounted above the front windshield (top-left and top-right) and protected from precipitation. Rain disdrometer is located a few meters away. The retroreflective trihedral is mounted downrange and protected from precipitation. (Source: Waymo.)

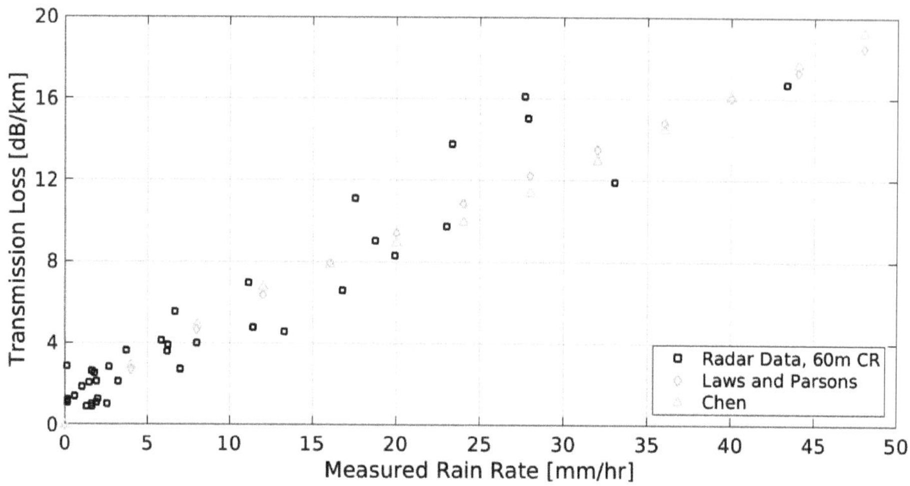

Figure 9.3 Experimental results align well with published models. (Source: Waymo.)

water. The geometry of the corner was such that droplets could only accumulate if they became naturally pinned against gravity by lateral adhesion to the painted metal surface; no accumulation or pooling of water in the corner reflector was possible (Figure 9.4).

In Figure 9.5, the corner was simultaneously illuminated by two radars at different locations and offset in frequency. We observed that naturally pinned water droplets caused up to 6 dB of loss as compared to the reflector when dry.

From this result it is clear that discrete scatterers may present drastically different cross-sections when wet. An important follow-up question is how this impacts the cross-section of objects of interest for automotive radar like cars, motorcycles, and pedestrians. Radar systems engineers often model such objects as the superposition of the response of a number of rigidly connected scattering discretes. Such models are typically dichotomized into objects composed of a number of roughly equal-sized discretes (the so-called Swerling 1 and 2 models) and objects dominated by a single scattering discrete (the Swerling 3 and 4 models) [5]. We have found empirically that objects like motorcycles and cars are well modeled as Swerling 1 objects; they present as a number of roughly equal-sized scattering discretes and are relatively stable over a coherent integration interval of 10–20ms. To motivate this conclusion, Figure 9.6 is a set of motorcycle data collected at nose aspect in the 76–81 GHz band at slant-45° polarization. In it we see that the empirical probability density function of SNR is well approximated by the Swerling 1 model whose free parameter (average SNR) is fit to the data with an RMS loss function.

Having established that objects of interest are well modeled as the superposition of many equal-sized scatterers and also that any one scatterer may have its radar cross-section dramatically altered by its particular state of wetting, we might expect that the overall cross-section of an object undergoes a less drastic change of radar cross-section under wetting by virtue of an averaging process of the state of its constituent scattering surfaces. Experimentation has borne out this intuition, with complex objects suffering up to about 4 dB radar cross-section loss when wetted (Figure 9.7).

Figure 9.4 Trihedral after wetting. Notice pinning of water droplets on the inner surface of the reflector. (Source: Waymo.)

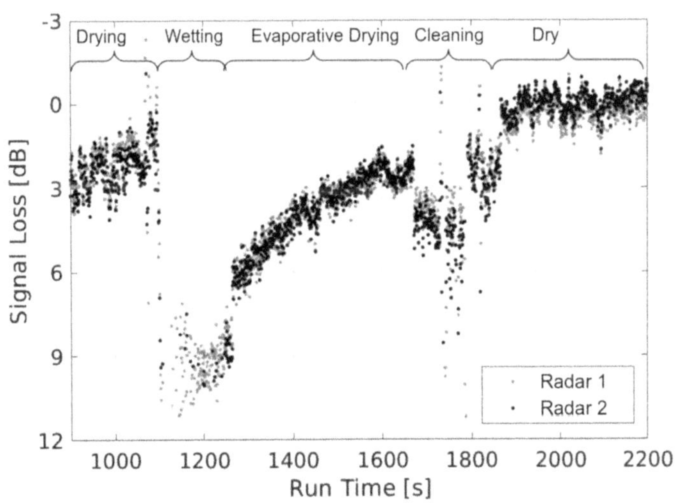

Figure 9.5 Power received from a wetted retroreflector varies substantially with water accumulation on the surface of the reflector. (Source: Waymo.)

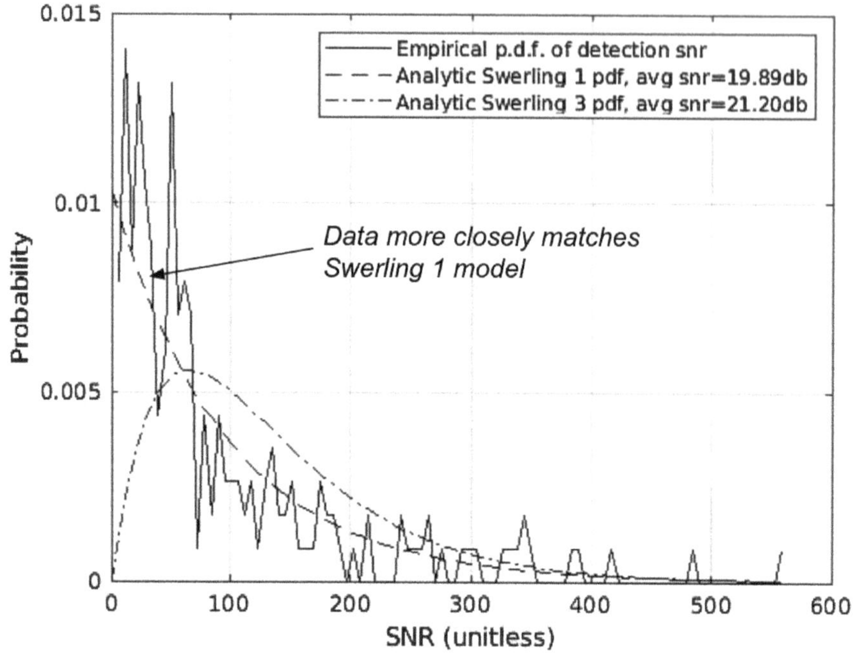

Figure 9.6 Experimental data suggests a motorcycle is well modeled as a Swerling 1 object. (Source: Waymo.)

Our experimental results, two trials of which are presented in Figure 9.7, were collected under static conditions at ranges from 50m to 150m with scattering discretes and complex metallic targets that have been wetted to saturation. That is, they have been sprayed with small droplets of water to give surfaces an opportunity to pin droplets up to the capacity of each surface to carry them. Under real-world

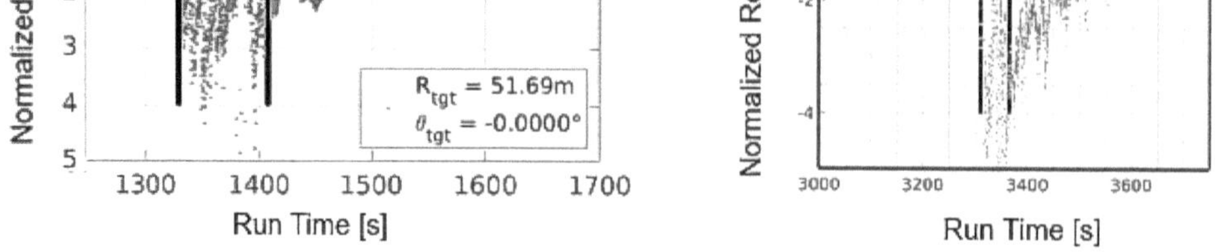

Figure 9.7 Exemplary trials collected during experiments in which test objects were wetted and their reflectivity in the 76–81 GHz band was measured. On the left a car is dry, then wetted, and then allowed to dry via evaporation. On the right a similar process is performed on a motorcycle. (Source: Waymo.)

conditions continuous precipitation will cause surfaces to exceed droplet carrying capacity causing running droplets down the surface. This mechanism causes a cascade of shedding of the surface clearing a portion of it, at which point accumulation can begin again. Also under real-world conditions the movement of an object will create additional aerodynamic forces that will change the droplet carrying capacity of surfaces around the object as the motion of the object changes. The end result is a dynamic equilibrium of accumulation and shedding of each surface of the object the superposition of which is some average loss somewhere between 0 and the worst case that we have observed to be about 6 dB for common scattering discretes and about 4 dB for complex objects like cars and motorcycles.

9.2.3 Radome Loss

Adherence of precipitation on the radar radome can be a large source of loss, exceeding heavy rain atmospheric loss in many cases.

It can be challenging to approach quantifying this loss analytically. Models may be developed for water as a sheet on the radome [6], but in practice the contact angle of water on the surface of common radome materials encourages the formation of droplets. This results in a condition where portions of the radome are clear while others are blocked by adhered water droplets (Figure 9.8).

Figure 9.8 In this image we see a radar radome after simulated rainfall. In it we see water that has adhered to the radome as droplets. (Source: Waymo.)

Droplets on radomes are in molecular contact with the underlying radome material and this adhesive force pins the droplet in place until it evaporates or it is exceeded by another force like gravity or aerodynamic drag.

Sliding of a droplet is opposed by a force of lateral adhesion [7, 8],

$$F_{la} \approx kL\gamma\left(\cos\theta_{\max} - \cos\theta_{\min}\right) \tag{9.2}$$

where:

- k is a shape dependence constant, typically values range from 1/2 and $\pi/2$;
- L is the droplet contact width, normal to the direction of motion;
- γ is surface tension at the water-air interface;
- $\theta_{\max}$ and $\theta_{\min}$ are the maximum and minimum contact angles around the droplet contact line.

Lateral adhesion is capable of supporting droplets against gravity, however this force is strongly influenced by the difference in front and rear contact angle θ developed by gravity acting on the droplet. Untreated or painted polycarbonate can support droplets with diameters approaching a wavelength while hydrophobic treatments can reduce this force to the point that only very small droplets are able to adhere against the force of gravity.

There is a maximum supportable droplet size that perfectly balances lateral adhesion with external forces acting on a droplet. A suspended droplet is not a perfect half-sphere, but we can expect its volume to be proportional to L^3, in which case the maximum droplet size balances gravity with lateral adhesion,

$$-\alpha\rho L_0^3 g = kL_0\gamma\left(\cos\theta_{\max} - \cos\theta_{\min}\right) \tag{9.3}$$

- ρ is water density;
- α is a constant encompassing tilt and droplet deformation;
- g is gravitational acceleration.

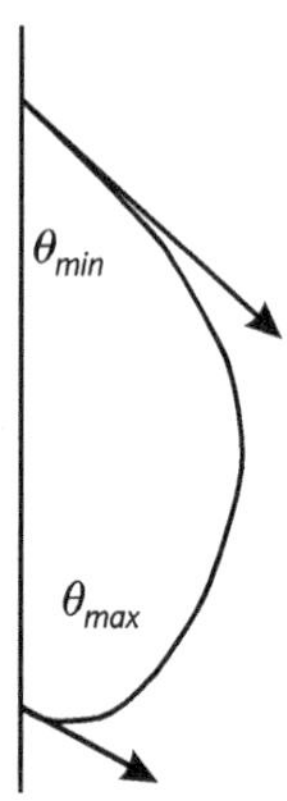

Figure 9.9 A pinned water droplet viewed in profile, with contact angles $\theta_{\max}$ and $\theta_{\min}$. (Source: Waymo.)

In the neighborhood of L_0, a small increase in the droplet vertical width L results in gravity overcoming lateral adhesion, causing the drop to begin moving.

Once a droplet is in motion a much smaller force is required to maintain motion [7]. As a moving droplet travels across the radome surface it may collide with another pinned droplet. In such instances cohesive molecular forces create a larger droplet, further extending the advantage of gravity over lateral adhesion. This droplet shedding is a cascade effect that clears pinned droplets along a line from the start of motion down to the bottom of the radome. In Figure 9.10 we can see several large droplets in the process of moving down the radome, creating a vertical path of droplet-free surface.

Under constant precipitation a radome will reach steady state loss once the rate of droplet accumulation equals the rate of droplet shedding. To illustrate the point we conducted constant precipitation tests with rain rate controlled at 1 mm/hr with two radomes: painted polycarbonate and polycarbonate treated with hydrophobic coating (on the left and right foreground respectively in Figure 9.11).

In Figure 9.12, over several trials we see that both radomes undergo an accumulation phase during which loss increases. This is followed by a dynamic equilibrium in which droplet accumulation is matched by droplet shedding. In this state loss is roughly constant in spite of continuing precipitation. Further, it is clear that hydrophobic treatment reduces loss by virtue of preventing water accumulation on the surface of the radome.

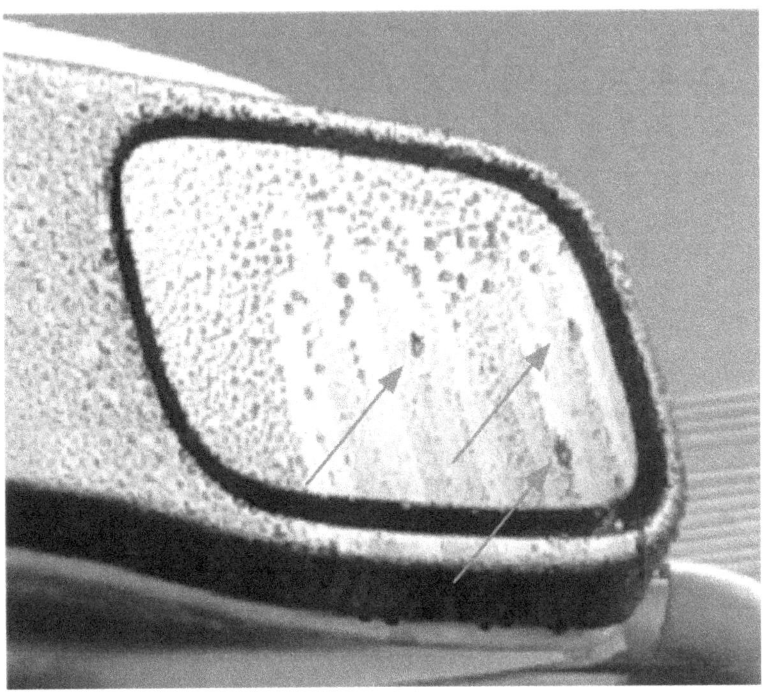

Figure 9.10 A radome with many pinned water droplets, with some droplets in the process of shedding by running down the surface of the radome. In the process of shedding smaller pinned droplets are cleared. (Source: Waymo.)

From Figure 9.12, it is apparent that surface adhesion plays a critical role in the performance of the radar radome in the presence of precipitation.

Figure 9.11 Start of rain at left and saturated radomes at right. The background radome is untreated polycarbonate, the foreground radome is treated with hydrophobic material. (Source: Waymo.)

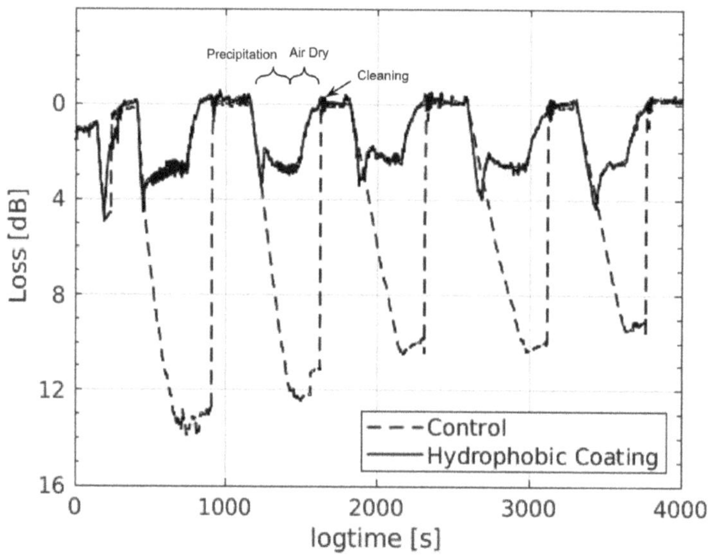

Figure 9.12 Two-way radome transmission loss, with and without hydrophobic treatment. Five trials, each trial includes 5 min 1 mm/hr simulated precipitation and 2 min air dry. (Source: Waymo.)

9.3 Array Performance

Fully autonomous vehicles are likely to require radars with high angular resolution and low angular sidelobes to facilitate the detection of closely spaced objects of disparate radar cross-section. This in turn implies that we can expect such radars to include electrically large apertures. One key assumption in antenna and antenna array design and operation is that the free space transmission medium is homogeneous; the far field radiation pattern is designed with that assumption. Of particular concern is the effect of perturbations of radiation very near the aperture.

To characterize these effects, we designed and conducted tests where a radar underwent intermittent wetting and clearing cycles at an effective rain rate of approximately 50 mm/hr. An isolated corner reflector was placed in the scene. The radar under test included a MIMO array topology that permitted us to collect many samples across the aperture of both the combined outgoing and incoming wavefront. Each array element was linearly polarized.

The range-Doppler matched filter corresponding to the test target was isolated and loss relative to nominal received power from the target was calculated (Figure 9.13). The experiment was structured in three phases, starting with dry collection for approximately 2 minutes followed by active precipitation for 2 minutes followed by cleaning of the radar under test. Based on radome loss results in the previous section it is reasonable to attribute most of the observed loss to water droplets adhering to the polycarbonate radome.

We have sampled the radar aperture with MIMO channels formed of tuples of transmit and receive elements colocated in the aperture. Denoting the i-th channel response to the test target at time t $s_{i,t}$ then we begin by calibrating the array response to the test target using the first snapshot of the experiment, $\hat{s}_{i,t} = s_{i,t}/s_{i,0}$. To correct for small changes in the absolute range of the test target caused by wind and vibration we further correct the data by removing the average response of the N elements of the array, $\tilde{s}_{i,t} = N\hat{s}_{i,t}/\sum_{k=0}^{N}\hat{s}_{k,t}$. The resulting phase and magnitude of all elements of the array as a function of time are presented in Figure 9.14. From this figure, it appears that precipitation has the effect of increasing amplitude and phase noise across the radar aperture.

In Figure 9.15 each snapshot of the array is processed to calculate the unbiased sample variance of array phase and amplitude. From this figure it is evident that

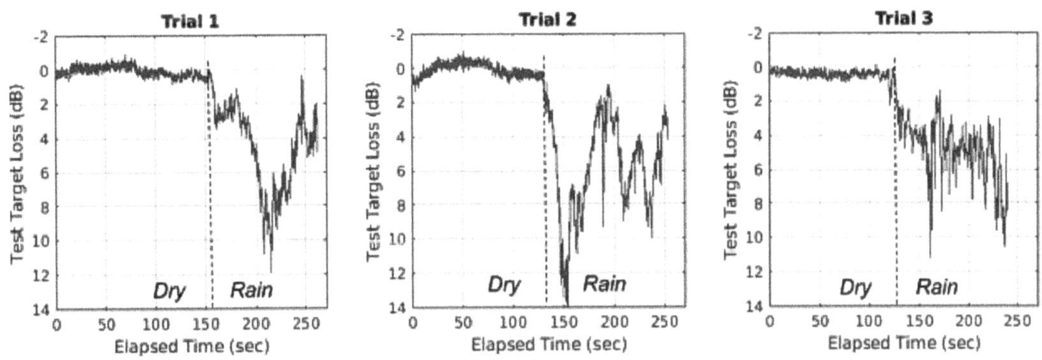

Figure 9.13 Test target loss during the experiment, with experiment phases highlighted. (Source: Waymo.)

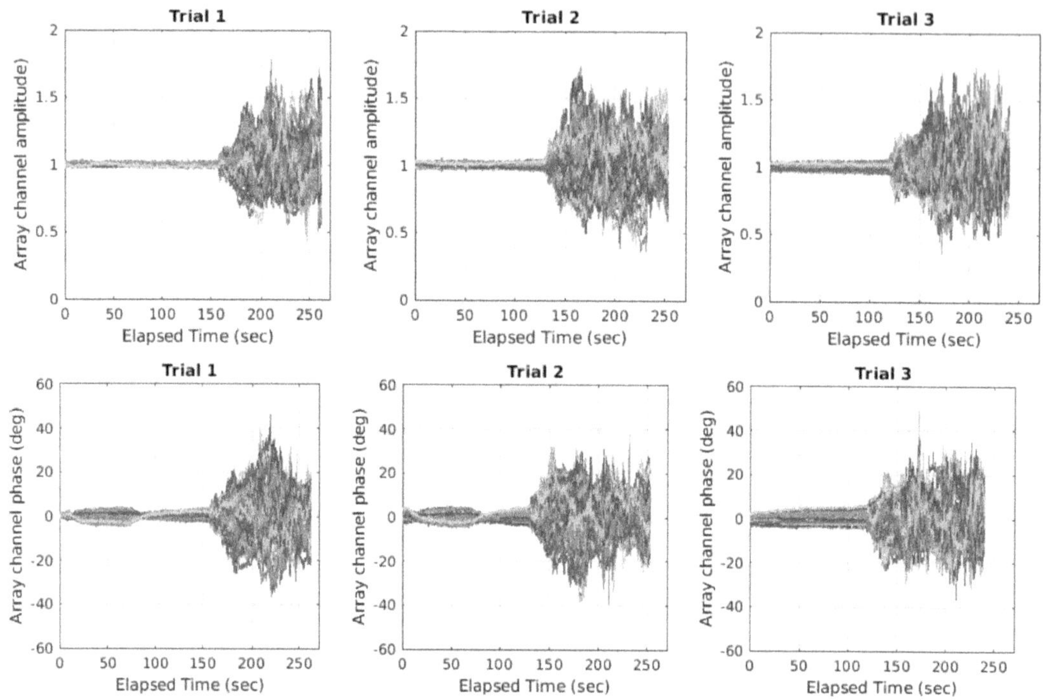

Figure 9.14 Sampled radar aperture while undergoing alternating periods of dry operation and precipitation events. Multiple transmit/receive tuples plotted; refer to previous figure for dry/rain boundary times. (Source: Waymo.)

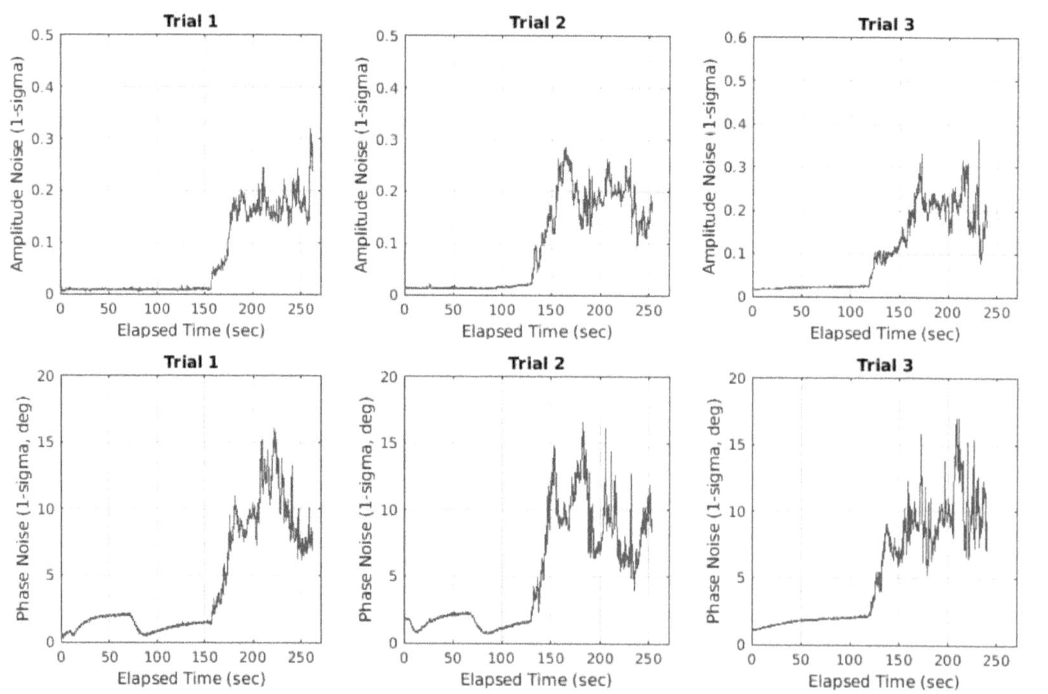

Figure 9.15 Array amplitude and phase standard deviation as a function of time. (Source: Waymo.)

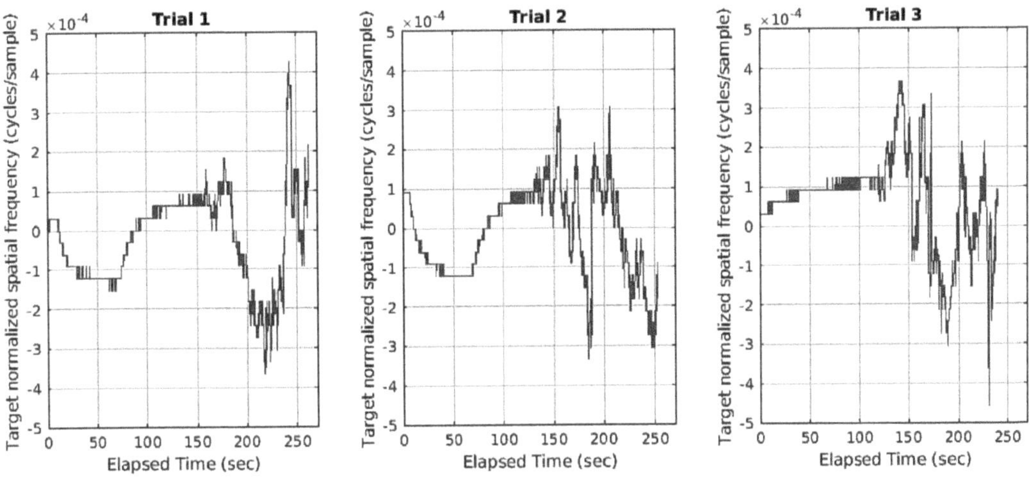

Figure 9.16 Test target normalized spatial frequency as a function of time. (Source: Waymo.)

water droplets adhering to the radar radome and in the air near the radar induce a substantial channel-to-channel phase and amplitude noise source. Ultimately this will degrade sidelobe control of both electrically large antennas and arrays.

Finally we consider whether it is possible for the array response to become biased. On the surface this seems unlikely, and the data bears out this intuition. In Figure 9.16 we have calculated the observed normalized target spatial frequency (with Nyquist interval $[-0.5,0.5]$) for each array snapshot with an overresolved angle-of-arrival matched filter bank. During the precipitation period of each trial we see more noise in the measurement of target frequency, but no definitive bias induced by rain.

9.4 Backscattering Phenomenology

9.4.1 Rainfall Backscatter

Rainfall is capable of generating backscatter whose power rises above the receive noise floor of a high performance automotive radar. Nonhomogeneity of rainfall returns can cause image artifacts that the radar designer must be cognizant of. For example, in Figure 9.17 rain backscatter presents over limited Doppler bandwidth at a power level well above receiver noise.

As with rainfall transmission loss, the scattering of droplets is well approximated with Mie scattering theory, owing to their large size relative to wavelength and their nearly spherical shape. To bring Mie scattering to bear we need only the index of refraction of water [9, 10], droplet diameter, and carrier frequency. For reference the backscattering cross-section of a water droplet at a temperature of 20°C is provided below in Figure 9.18 [11].

Droplet diameter is not uniform and is typically modeled as a random variable. The seminal characterization of this was performed by Laws and Parsons [4] (with tabular data reproduced in [12]), who determined the differential number of

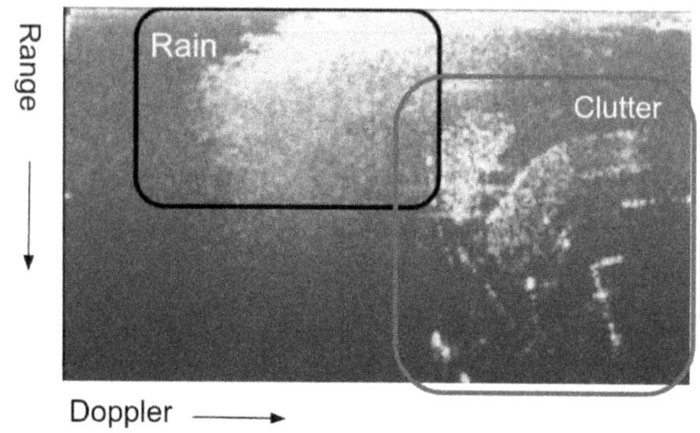

Figure 9.17 A range-Doppler image illustrating both normal clutter backscatter as well as a large region of rain backscatter. (Source: Waymo.)

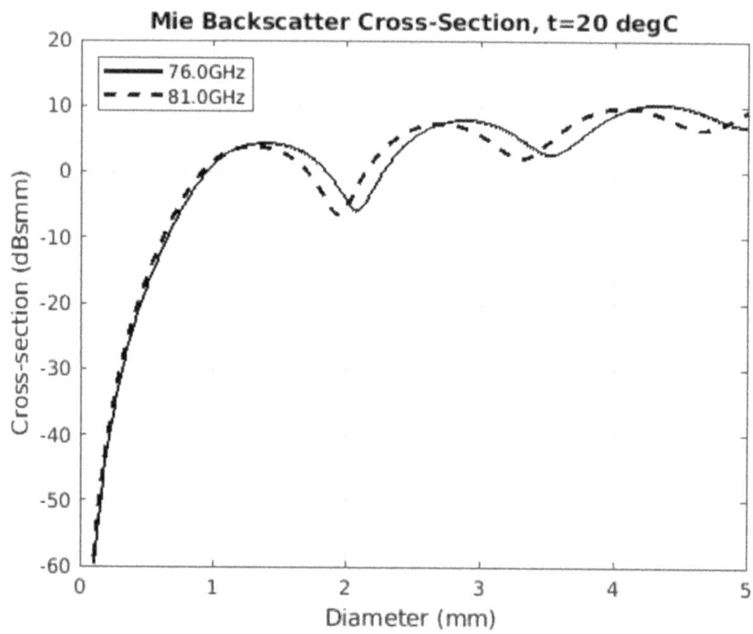

Figure 9.18 Mie backscatter cross-section of water droplets at 20°C. Note units are in dB square millimeters. (Source: Waymo.)

droplets per unit volume $N(D)$ (m^{-4}), a function of droplet diameter random variable D. Integrating over some differential diameter dD we can obtain the number of droplets per unit volume $N(D)dD$ (m^{-3}). Analytic approximations of the Laws and Parsons tabular model have been developed that are accurate at MMW frequencies. The analytic approximation developed by de Wolf is reproduced here for reference [13, 14]. Olsen performed a comprehensive comparison of analytic models [2].

We can combine the backscatter of a single droplet with a droplet distribution to reason about the reflectivity of rain per unit volume of space. For some droplet diameter D and integral diameter dD, the number of droplets per unit volume of

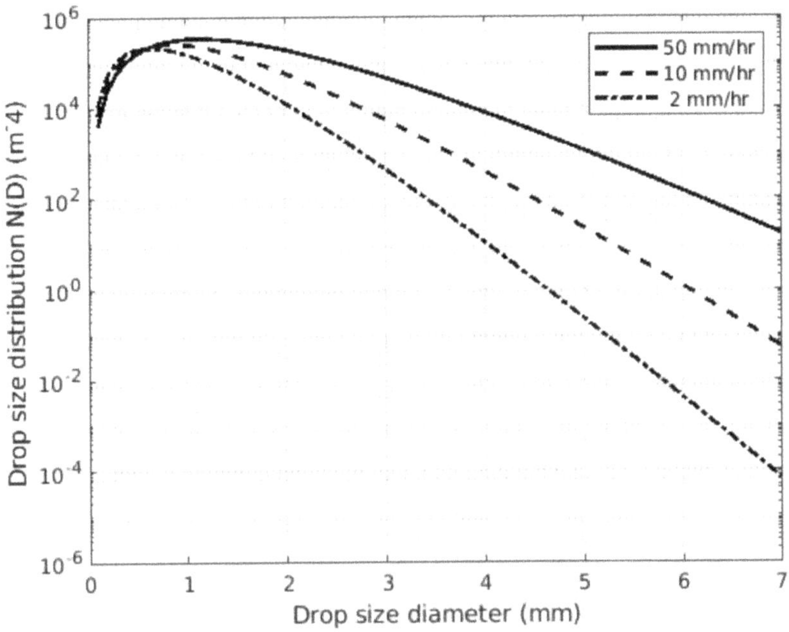

Figure 9.19 The de Wolf approximation of the Laws and Parsons drop size distribution. (Source: Waymo.)

that diameter is $N(D)dD$ and their backscattering cross-section is $\sigma_{bs}(D)$, so total backscatter from droplets of diameter D per unit volume is $\sigma_{bs}(D)N(D)dD$. Total rain reflectivity, termed the backscatter coefficient η (m^2/m^3 or m^{-1}) is the integral of this expression over droplet diameter, $\eta = \int_0^{inf} \sigma_{bs}(D)N(D)dD$ [9, 14]. The integrand of the backscatter coefficient (Figure 9.20) can give some insight into the sensitivity of rain reflectivity as a function of droplet diameter. The combination of a strong mode at about 1 mm in both Mie scattering cross-section and the rain droplet distribution causes droplets of this size to be the largest contributor, with another mode at about 2.5 mm at higher rain rates. As one might expect, backscatter increases with rain rate, with only small variation across the automotive band (Figure 9.21). Empirical models of η fitted to peak scattering power have been developed at millimeter-wave frequencies [15].

The backscatter coefficient provides information on precipitation backscatter cross-section per unit volume. This must be combined with information about a particular radar system to be useful. In particular for a matched filter volume defined by a modulation bandwidth b, vertical two-way half-power beamwidth ϕ, and horizontal two-way half-power beamwidth θ, we can express the volume contained by a matched filter as,

$$V \approx \frac{\pi}{8ln(2)}(\phi R)(\theta R)\left(\frac{c}{2b}\right) \tag{9.4}$$

In this approximation [16] it should be noted that a small angle approximation has been made, a correction for the elliptical shape of the beam ($\pi/4$), and a

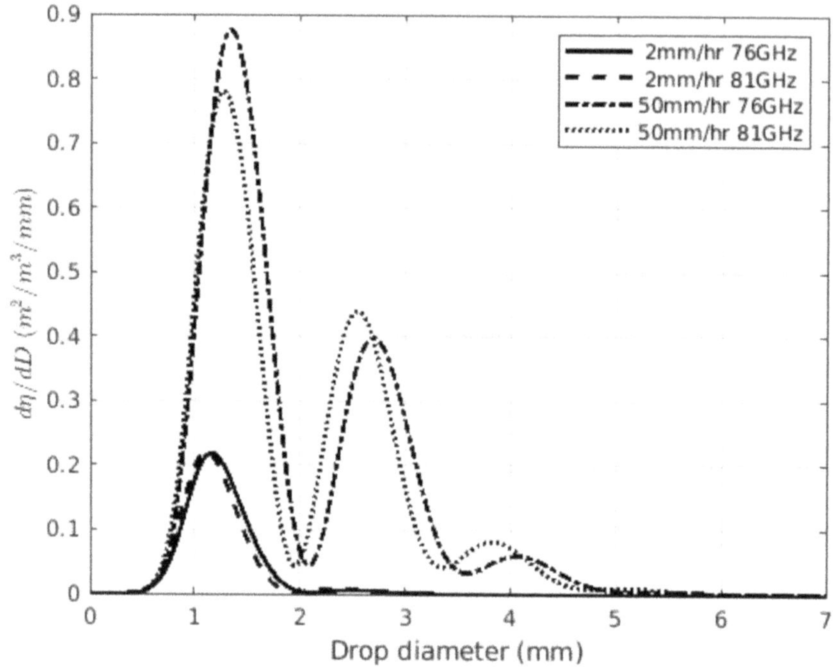

Figure 9.20 Sensitivity of radar reflectivity to raindrop diameter at 20°C. Strong modes occur at approximately 1.3 and 2.5 mm diameter. (Source: Waymo.)

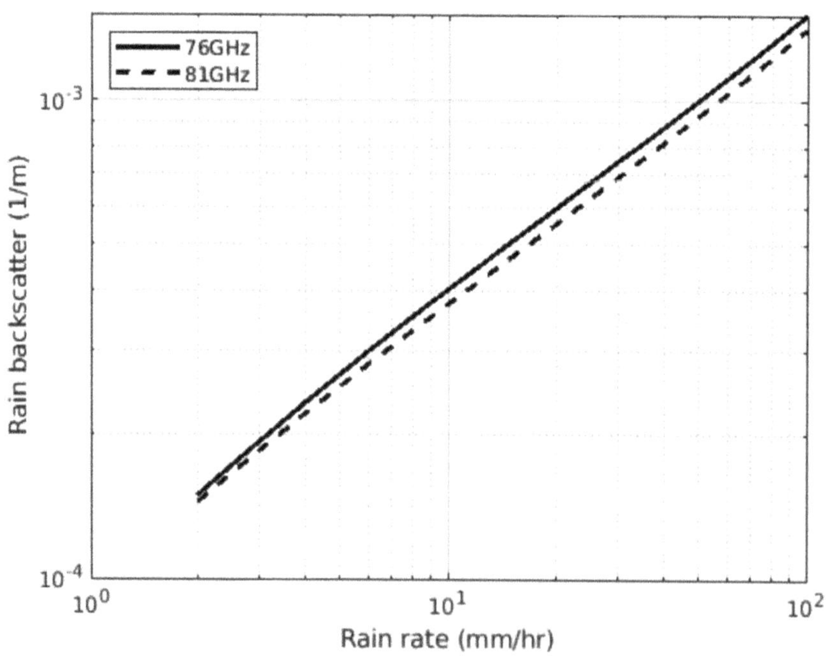

Figure 9.21 Backscattering coefficient constructed with Laws and Parsons raindrop distribution at 20°C. (Source: Waymo.)

further correction to account for the nonuniform illumination of the volume by a Gaussian beam ($1/(2ln2)$).

Illumination volume and backscatter coefficient can be integrated into a basic radar equation. Starting with a standard received power equation where we account for average transmit power P_t, two-way directivity gain G^2, carrier frequency wavelength λ, scatterer radar cross-section σ, and range R we have,

$$P_r = \frac{P_t G^2 \lambda^2 \sigma}{(4\pi)^3 R^4} \tag{9.5}$$

This standard equation may now be amended by setting $\sigma = V\eta$,

$$P_r = \frac{P_t G^2 \lambda^2 \phi \theta c}{1024 ln(2) \pi^2 R^2 b} \eta \tag{9.6}$$

In this expression we have a tool for both reasoning about the import of radar backscatter at the radar design stage as well as for linking rain rate to measured signals during operation.

In summary, rainfall backscatter is likely to rise above the noise floor of a sensitive automotive radar. Fortunately, the backscattering cross-section of rain is well understood. The combination of Mie scattering theory and well validated rain distribution models provide the radar designer a tool for predicting the magnitude of rain backscatter during radar design as well as for developing onboard algorithms for estimating rain rate from backscatter or vice versa, depending on the needs on the system.

9.4.2 Road Spray

Road spray, which occurs when water and contaminants that have collected on the road are dispersed into the air by vehicles in the scene, can have a material impact on the performance of an automotive radar. Such events take place when water that is collected on the road is sprayed into the air by other vehicles in the scene. This is common during and after rainfall but can also be caused by water spilled in clear weather. Once airborne and in the beam of the radar, water droplets induce a transmission loss that reduces sensitivity and create backscatter that can generate interference above the noise floor of the system.

To characterize these effects, we designed and conducted tests to investigate spray and compare nonspray conditions. A mobile water truck was used to create a large puddle on the road. The same vehicle was used to collect passes of data on clear road and then under road spray conditions. In Figure 9.22 we see this water truck as observed by two forward-looking radars as well as by camera on a dry road while approaching a puddle of water.

A few seconds later the water truck enters the puddle but is not actually actively spraying water, so all water droplets in the air are transferred there by the wheels of the truck.

Two observations may be made from the range-Doppler and range-angle imagery data (Figure 9.23): first, there are large amounts of interference above the receiver

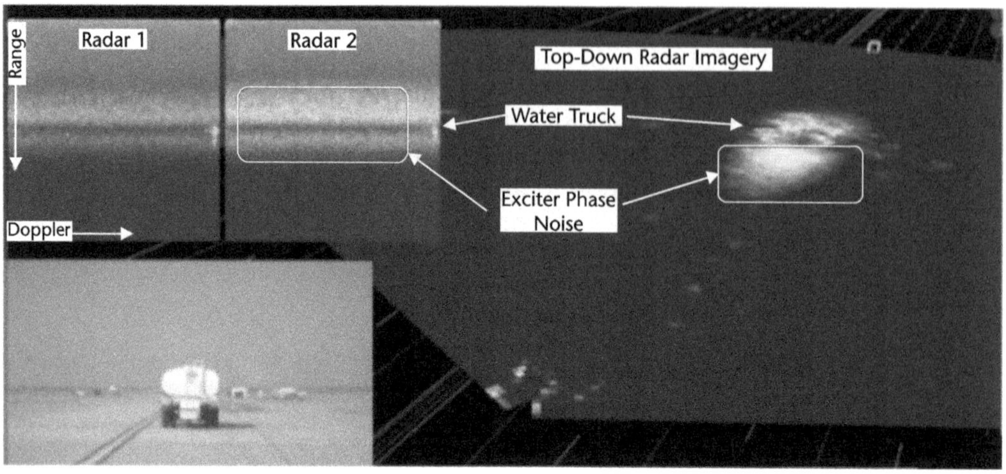

Figure 9.22 Two forward-looking radars trailing a large truck on a dry road. Range-Doppler imagery at top left and fused range-azimuth imagery on the right. Camera image of the scene as viewed by the radars at the bottom left. (Source: Waymo.)

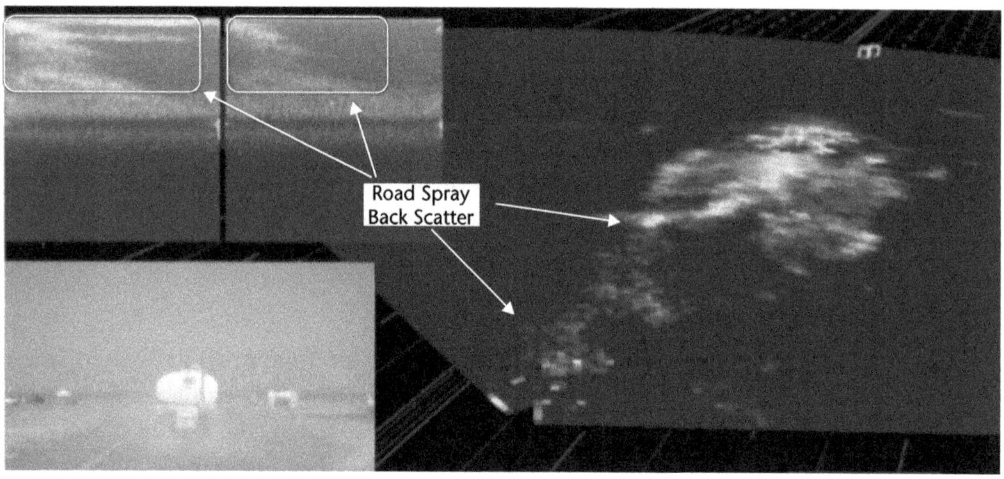

Figure 9.23 The scene in Figure 9.22 viewed a few seconds after, the truck is now driving over a wet road surface. (Source: Waymo.)

noise floor in the radar data, and second this interference is not homogenous; it can be highly localized in range-Doppler-angle matched filters.

Airborne droplets from road spray may generate material transmission loss when attempting to detect objects whose line of sight passes through the spray. During the course of driving through a puddle the target truck's peak return power was decreased by 10 dB (an example is illustrated in Figure 9.24). After passing through the road spray peak return power is observed to be –4 dBc, which is attributed to the wetting of both the truck and the radar radome.

This experiment serves to highlight the potentially large impact road spray may have on radar performance. We believe it does establish the importance of considering road spray in radar design and in considering radar in the context of autonomous systems.

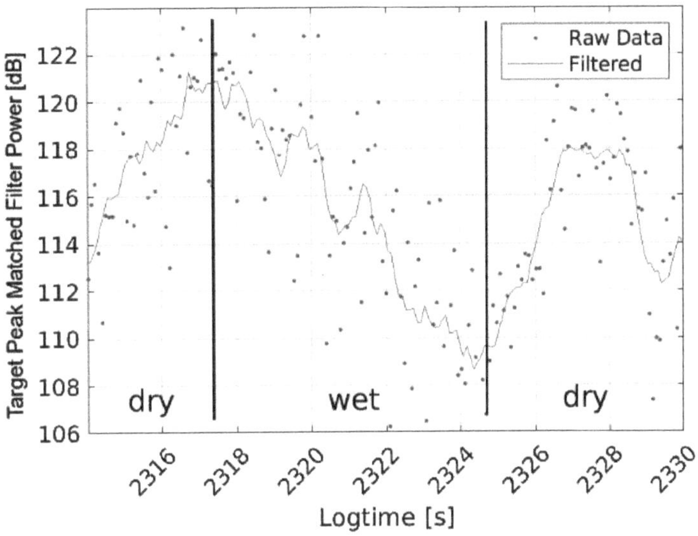

Figure 9.24 Truck peak scattered power before, during, and after driving through a puddle. (Source: Waymo.)

9.5 Potential Mitigations

To wrap up this treatment on the effects of water and weather on automotive radar, an obvious question would be "so what can we do about it?" As with nearly everything in engineering, the answer is "it depends," and in this case it depends on what "it" (the weather effect) is, and whether we can determine through some method that we are experiencing it.

Earlier we discussed many of the ways that water can desensitize radar. One approach would be to determine the maximum degradation expected and incorporate this loss into the sensitivity requirements of the radar. This is generally not preferred, as it leads to a radar that is significantly overspecified (and likely more expensive, power-hungry, and complex) in most of the operational scenarios.

The alternative is to design the radar with variable sensitivity and include a way to estimate the current weather conditions to adjust the sensitivity appropriately. Many radars have the capability to increase sensitivity through integration time, update rates, and beamsteering usually at the expense of performance in a less-import angular sector or mode. Another area to consider to mitigate sensitivity loss due to water on the radome is to incorporate some method to reduce the amount of water that can collect on it. We presented data earlier on the effectiveness of a hydrophobic treatment, but there are other active solutions such as wipers, blowing compressed air on the radome to clear it of water, and even designing structures such as overhangs that keep rain from falling on the radar surface or for shaping airflow to clear the radome surface.

Perhaps a greater difficulty is in accurately estimating the current weather experienced by the AV. While this is still an active area of research for radar autonomy, there are generally two classes of solutions: estimating the weather via the radar itself or through some exogenous means, and of course these can be combined. Estimating the weather via the radar involves exploiting the features in the various radar data domains, such as range-Doppler (Figure 9.17) or the range-angle point clouds (Figure 9.23), perhaps using presurveyed fiducials in the environment to estimate the total losses present. Exterior to the radar sensor are even more options, including use of cameras or other sensors on the fully AV and using "vehicle to everything" (V2X) communications to pull data from local weather collection sites.

Finally, there are multiple ad hoc software solutions that can be utilized to mitigate localized weather effects, such as road spray or snow blowing off of a car in winter. One could consider developing an approach to identify these phenomena (machine learned or via other heuristics) and reject them.

References

[1] Chen, C. C., *Attenuation of Electromagnetic Radiation by Haze, Fog, Clouds, and Rain*, RAND Corp. Rep., 1975.

[2] Olsen, R., Rogers, D. V., and D. Hodge. "The aR *b* relation in the Calculation of Rain Attenuation," *IEEE Transactions on Antennas and Propagation*, Vol. 26, No. 2, 1978, pp. 318–329.

[3] Dudzinsky, S. J., "Atmospheric Effects on Terrestrial Millimeter-Wave Communications," *1974 4th European Microwave Conference, IEEE*, 1974.

[4] Laws, J. O., and D. A. Parsons, "The Relationship of Raindrop Size to Intensity," *Trans. AGU*, Vol. 24, 1943, pp. 452–460.

[5] Swerling, P. "Probability of Detection for Fluctuating Targets," *IRE Transactions on Information Theory*, Vol. 6, No. 2, 1960, pp. 269–308.

[6] Chen, N., et al. "The Influence of the Water-Covered Dielectric Radome on 77GHz Automotive Radar Signals," *2017 European Radar Conference (EURAD), IEEE*, 2017, pp. 139–142.

[7] Gao, Nan, et al., "How Drops Start Sliding over Solid Surfaces," *Nature Physics*, Vol. 14, No. 2, 2018, pp. 191–196.

[8] Antonini, C., et al., "General Methodology for Evaluating the Adhesion Force of Drops and Bubbles on Solid Surfaces," *Langmuir*, Vol. 25, No. 11, 2009, pp. 6143–6154.

[9] Lhermitte, R. "Attenuation and Scattering of Millimeter Wavelength Radiation by Clouds and Precipitation," *Journal of Atmospheric and Oceanic Technology*, Vol. 7, No. 3, 1990, pp. 464–479.

[10] Liebe, H. J., Hufford, G. A., and T. Manabe. "A Model for the Complex Permittivity of Water at Frequencies Below 1 THz." *International Journal of Infrared and Millimeter Waves*, Vol. 12, No. 7, 1991, pp. 659–675.

[11] Mätzler, C., "Matlab Functions for Mie Scattering and Absorption," *IAP Res. Rep*, No. 02-08, Bern, Switzerland: Institute of Applied Physics, University of Bern, June 2002, pp. 1–18.

[12] Medhurst, R. G., "Rainfall Attenuation of Centimeter Waves: Comparison of Theory and Measurement," *IEEE Trans. Antennas Propagat.*, Vol. AP-13, July 1965, pp. 550–564.

[13] de Wolf, D. A. "On the Laws-Parsons Distribution of Raindrop Sizes," *Radio Science*, Vol. 36, No. 4, 2001, pp. 639–642.

[14] Mätzler, C., "Drop-Size Distributions and Mie Computations for Rain," Bern, Switzerland: Institute of Applied Physics, University of Bern, 2002.

[15] Richard, V. W., Kammerer, J. E., and H. B. Wallace, "Rain Backscatter Measurements at Millimeter Wavelengths," *IEEE Transactions on Geoscience and Remote Sensing*, Vol. 26, No. 3, 1988, pp. 244–252.

[16] Skolnik, M. I., *Introduction to Radar Systems*, London and New York: McGraw-Hill Book Co., 1962.

PART 3

Integration and System Considerations

CHAPTER 10

Safety Considerations for Radar in Fully Autonomous Vehicles

James McGinley and Xiaoli Lu

10.1 Introduction

Any consideration of fully autonomous[1] driving (i.e., autonomy at SAE Level 4 [1]) must encompass safety. Indeed, many would consider safety the key property. This chapter explains what safety means in an autonomous driving context and the implications for the radar system. It discusses relevant safety standards and their relationship with existing automotive development processes. It asks what lessons AV developments can take from the application of safety standards in the industry to date. It concludes with an exploration of safety as applied to an autonomous radar system, discussing challenges and trade-offs.

10.2 What Is Safety?

Safety is defined as the absence of unreasonable risk, where risk is the combination of the probability of occurrence of harm and its severity. Unreasonable risk is a level judged to be unacceptable in a certain context according to societal moral concepts. A hazard is defined as a potential source of harm [2]. While these definitions could apply to many automotive contexts, for example, crash safety or eye safety for light-emitting components, this discussion is limited in scope to hazards resulting from the functioning of a vehicle's hardware and software.

An automated driving system (ADS) is defined as hardware and software that are collectively capable of performing the entire dynamic driving task (DDT) on a sustained basis, regardless of whether it is limited to a specific operational design domain (ODD). The DDT encompasses all the real-time operational and tactical functions required to operate the vehicle in on-road traffic. The ODD defines the conditions under which a function is specified to operate, including, but not limited to, environmental, geographical, and time-of-day restrictions, and/or the requisite presence or absence of certain traffic or roadway characteristics [1]. Different

1. This chapter, in keeping with the autonomy community, considers SAE L4 high driving automation as *fully autonomous* within its prescribed ODD.

automation levels are defined in terms of the degree of human interaction; Level 4 (L4) is where the user does not need to supervise the ADS or be receptive to a request to intervene while the ADS is engaged and is considered the minimum required level for a fully autonomous vehicle.

Safety is implicit in the definition of an ADS in that it assumes that the ADS does not introduce unreasonable risk to vehicle occupants and other road users. However, as we shall discuss, achieving a safe ADS introduces many challenges, including:

- Quantifying unreasonable risk targets and translating these to measurable system performance indicators for various ADS subsystems, including the radar system.
- ODD limits must be determined in line with ADS functional and performance capabilities in a manner that avoids unreasonable risk.
- The DDT must be executed in a manner that avoids unreasonable risk, accounting for all driving scenarios that may arise within the ODD.
- Monitors must be implemented to detect when the vehicle exits its ODD and this scenario handled in a manner that avoids unreasonable risk.
- System malfunctions and performance limitations must be detected and handled in a manner that avoids unreasonable risk.

These challenges must be addressed in parallel to commercial deployment considerations, for example, time-to-destination, service availability, customer comfort, and power efficiency. The need to leverage emerging technologies adds challenges too, as the industry often requires time to establish best practices for the safe deployment of such technologies. Examples include machine learning (ML) algorithms, advanced sensor modalities, highly integrated computing systems, and complex pre-existing software offerings. Finally, the regulatory and standardization landscape, while critical to deployment may also introduce challenges.

Achieving acceptable risk levels for the radar system requires ensuring radar performance indicators are correctly specified in the context of the overall ADS. Ensuring that the implemented performance of the radar system is sufficiently characterized across the ODD is critical but often complicated, requiring an understanding of real-world effects such as object RCS variation, propagation loss, radome/fascia extra loss, and potential particle/droplet backscatter false positives. Robust design practices to ensure desired radar characteristics such as high angular and range resolution, high dynamic range, and low sidelobes are achieved within constraints including operational frequency bandwidth, MIMO channel separability, technology availability (e.g., RF chip), SWAP-C (size, weight, power and, cost) are necessary. As we will show, safety best practices require proactive structured analysis and mitigation of technical risks that could result in hazards, coupled with extensive verification and validation to support an argument that acceptable risk levels are achieved.

10.3 Safety Standards

A standard is a document, established by consensus that provides, for common and repeated use, rules, guidelines, or characteristics for activities or their results. Presently, several standards apply to autonomous vehicle radar safety, with several others in development. This chapter focuses on the international standards ISO 26262 [2] and ISO 21448 [3], given their global mandate and acceptance. This is not to suggest that other standards are not important. Indeed, it is recommended that those involved in autonomous vehicle radar safety ensure familiarity with all applicable standards, including industry guidelines and best practices, examples include [4–8]. While standards are often used in support of legislation and policy at international and national levels, such topics are considered out of scope here, given they may change quickly and can vary considerably depending on geography.

10.3.1 ISO 26262 and ISO 21448

ISO 26262 (aka Functional Safety, aka FuSa) addresses hazards resulting from malfunctioning behavior of systems comprising electrical, electronic, and software components in series production road vehicles. It considers malfunctioning behavior as originating from one of two sources, systematic faults, or random faults. Systematic faults, more commonly known as bugs, are those which arise due to issues in the development or manufacturing process, for example, an incorrect requirement, a coding error, or a manufacturing flaw. Random faults affect only hardware, occurring unpredictably during the lifetime of the hardware, driven by physical events, for example, wear-out due to operational stresses or radiation-induced effects. ISO 26262 presents a structured reference safety lifecycle spanning product inception to decommissioning, specifying requirements on technical, process, and assurance activities. It defines four automotive safety integrity levels (ASIL), A, B, C, and D, with D representing the highest level of integrity. Each level specifies a minimum set of requirements and targets to control random faults and avoid systematic faults. Methods to determine what ASIL to assign to a function are provided. The multiple levels accommodate the reality that not all safety risks are equal, for example, a blind spot detection function may require an ASIL A or B level of integrity against hazards posed by malfunction, while ADS functions would often incur an ASIL D level of integrity. In effect, an ASIL sets the industry agreed unreasonable risk targets for hazards resulting from malfunction.

ISO 26262 assumes nonmalfunctioning behavior (i.e., the intended functionality of the system) is safe to begin with. However, this may not be the case. Indeed, at the vehicle level, the behavior leading to a hazard is the same whether it is caused by a malfunction or an unsafe intended function. ISO 21448 (aka Safety of the Intended Functionality, aka SOTIF) addresses hazards resulting from such functional insufficiencies. While functional insufficiencies could be argued to exist in any automotive system, the scope of ISO 21448 is targeted at systems where vehicle situational awareness based on advanced sensor processing is necessary (i.e., L4 ADS is directly in scope). It considers functional insufficiencies as arising from an insufficient specification at the vehicle level, or performance limitations at the implementation level. Misuse (i.e., usage of the system by a human driver in a way not intended) is also considered. It introduces the concept of known and unknown hazards. *Known hazards* are those that may be uncovered via structured analysis of the system, leveraging for example, historic data and expert judgment. *Unknown hazards* are an acknowledgment that for such advanced technologies and complex use cases it is not realistic that everything can be uncovered via analysis (i.e., there will be scenarios you won't think of). A comprehensive verification and validation strategy is required to defend an argument that unknown hazards are below an acceptable level and that known hazards are handled accordingly, for example via a sufficient level of simulation and open road testing. Once a hazard is discovered, it becomes known and is mitigated if deemed necessary, for example, by adapting the system or modifying the ODD. ISO 21448 does not define unreasonable risk targets for functional insufficiencies, nor does it leverage the ISO 26262 ASIL to do this. Rather it presents approaches that may be applied to determine acceptable risk targets, for example by leveraging accident statistics and concepts including positive risk balance but defers to those responsible for the ADS to decide on defensible targets.

The application of both ISO 21448 and ISO 26262 is necessary to ensure a robust safety argument that unreasonable risk is avoided for an L4 ADS. Ideally, how to comprehensively and efficiently apply both standards would be evident, however for several reasons this is not the case. The standards:

- Are not mutually exclusive. ISO 21448 acknowledges what while it complements ISO 26262 there is significant potential for overlap. Specifically, functional insufficiencies of elements (e.g., radar systems and radar sensors) arise due to systematic faults in the design of the element and/or gaps in the understanding and/or exchange of requirements between the supplier and customer.
- Create conceptual models to simplify real-world complexities, however, how to align real-world observations with a model, or how to align different models across standards, may be open to interpretation. For example, ISO 26262 assumes faults manifest independently of the driving scenario. This allows it to decouple fault occurrence in its risk analysis, simplifying the frequency of occurrence of harm down to an exposure factor and governing the occurrence rate for faults via the ASIL. While ISO 21448 hinges on the premise that a functional insufficiency is activated by a triggering condition in the driving scenario, it seems plausible that an ISO 26262 systematic fault may also be activated by a specific condition in the driving scenario. Indeed, if systematic faults were independent of conditions how could they ever be debugged? Time will tell if current models are sufficient to enable safe L4 ADS deployments.
- Take contrasting approaches to normative (i.e., mandatory) and informative (i.e., guidance) text, with ISO 26262 having many hundreds of normative requirements, while ISO 21448 has less than twenty. While there may be good reasons for this, without explanation, it may result in confusion for practitioners.
- Present process aspects in a manner that suggests the interpretation is unambiguous, however as we will see this is not the case.

It's worth highlighting the challenge of developing standards sufficient to support L4 ADS applications. Presently the maturity and consensus on state-of-the-art remain low, with many open questions and differing interpretations from subject matter experts on the finer details. It is anticipated that both standards will evolve over time to better address the sector. Likewise, other industry guidelines may emerge as the industry matures, for example, to normalize targets for functional insufficiencies.

10.3.2 Relationship to Existing Standards and Processes

Even a casual reading of either standard identifies relationships with other standards and processes. ISO 26262 identifies itself as an adaptation of IEC 61508 [9] to address automotive-specific sector needs. It notes that safety is intertwined with common function-oriented and quality-oriented activities and work products, later requiring automotive quality management to be in place while referencing IATF 16949 [10], ISO 9001 [11], ISO 33000 [12], CMMI [13], and ASPICE [14]. ISO 21448 notes that to develop a safe product, rigorous engineering and quality management processes are essential while referencing IATF 16949 and ISO 15288 [15], in addition to ISO 26262. So, what are these other standards and processes and how do they relate to automotive safety and the safety of L4 radar?

IEC 61508 considers safety relating to the correct functioning of electrical, electronic, and software systems. It provides a generic framework applicable to all sectors, intending to facilitate the development of sector-specific standards. ISO 26262 derives its ASILs by applying the generic methods described in IEC 61508 to automotive scenarios. ISO 26262 is more prescriptive in terms of the requirements and methods to be applied (i.e., how to do it). However, IEC 61508 was not the origin of safety for electronic and software systems. It leveraged decades of experience gained across industries including process automation, nuclear, and oil and gas. Bell notes that many fundamental safety principles, including the need for measures to combat dangerous random and systematic faults stem from the 1980s, while quantitative safety methods date back to the 1970s [16].

Many automotive electronic and software developments don't incur safety requirements. Such developments are commonly referred to as quality managed (QM). Automotive QM developments are governed by the standards ISO 9001 in conjunction with IATF 16949 or VDA 6 [17], the latter two providing augmentations to better support the needs of the automotive supply chain. These standards are organization and quality-centric, hence provide only high-level guidance for electronic and software developments. They defer to other guidelines for these details, for example, advanced product quality planning (APQP) [18] and maturity level assurance (RGA) [19]. IATF 16949 requires that a software quality assessment methodology be used to assess an organization's software development process for embedded software. Several software quality assessment options exist; all have evolved over the past decades and to some extent have influenced each other, however, they remain distinct. The automotive community has created a domain-specific software assessment method, automotive software process improvement and capability determination (ASPICE). ASPICE includes a process reference model, which describes the processes together with their relationships. While its name suggests it is limited to software, its scope includes the system. It supports a plug-in model that allows an extension to other processes, with a compatible hardware model also available [20]. Due to the inclusion of the process reference model, ASPICE provides a holistic lifecycle view versus merely a quality-centric view. The lifecycle perspective makes it easier to map to hardware and software developments. A capabilities dimension enables an organization's application of the processes to be evaluated.

The international standards community developed the ISO/IEC 33000 series as a framework for the assessment of any process in any domain. Here a process is defined as a set of interrelated activities that transform inputs into outputs. A framework for software and system lifecycle processes are defined in ISO/IEC/IEEE 12207 [21] and 15288 respectively. How processes are stitched together to support product development is covered in the ISO/IEC/IEEE 24748 series [22]. These standards are valuable in that they provide a more in-depth consideration of many topics, for example, the iterative and recursive nature of processes that exist in practical deployments. These standards form the basis of systems engineering as defined by INCOSE [23]; systems engineering being the interdisciplinary approach governing the total technical and managerial effort required to transform a set of stakeholder needs, expectations, and constraints into a solution and to support that solution through its life.

There is no question that a relationship exists between ISO 26262, ISO 21448, and these existing standards, but what exactly is it and how is a comprehensive and efficient integrated application of all necessary standards achieved? After all, commercial deployment of a safe L4 ADS is a challenging enough task, without having to struggle with processes and standards. Unfortunately, the answer is not so obvious. The standards are not harmonized, hence differ in scope, format, terminology, consistency of language, and concepts. As an example, ISO 26262 encompasses process, assurance, and technical requirements and mixes them throughout its clauses. While it uses the term *process*, it does not define it and applies the term somewhat inconsistently throughout. It uses language that may lead a reader to believe that all its requirements are unique and in addition to those already applied by a quality management system, despite many sources noting the high level of commonality that exist between the two [20, 24–25]. In the absence of this clarity, the ISO 21448 claim that rigorous engineering and quality management processes are already addressed by IATF 16949, ISO 26262, and ISO 15288 may be challenging for practitioners to interpret.

However, it is critical to look past any ambiguity and recognize the fundamental synergies. Systems engineering is about ensuring the right product is built and that it is built right. In essence, it makes a conscious decision about what is acceptable in terms of systematic faults for the specification and implementation of a product. Quality management is about ensuring the right product is built, on time and on budget, by proactively managing program risks and applying an appropriate level of assurance to minimize systematic faults. Safety standards care only about those faults that may result in unreasonable risk and so demand higher levels of assurance on fault control, providing guidance on best practices, targets, and methods. They also widen the scope of what constitutes a fault (i.e., by including random sources). While it is critical to understand that safety engineering is not business as usual and that it does introduce unique requirements, it is as important to understand that many aspects are common with existing systems engineering and quality management practices. An integrated approach to safety and quality will result in an optimal safety solution.

10.4 Lessons from Industry

While ISO 21448 is a new standard, ISO 26262 has been actively applied since 2011. This section presents observations on its application to date and guidance on avoiding pitfalls when applying safety standards to autonomous radar systems.

10.4.1 Emphasize Understanding over Following Checklists

Work products capture the outcome of safety activities, thereby serving as evidence of their application. Any judgment on whether safety activities were sufficiently applied comes from a review of the content of those work products. A common pitfall for safety engineers and managers is placing too much emphasis on the mere existence of a work product, with the adequacy of its content, or whether an activity was applied in an appropriate sequence being a secondary consideration. This mentality is not unique to automotive safety; Douglass notes that while it may be common for engineering data to be organized in documents, that focusing on the documents often leads to poor management practices [26].

The existence of such a checklist mentality in automotive safety has been recognized. The second revision of ISO 26262 placed a greater emphasis on the objectives of the safety activities. While ISO 21448 makes the objectives and the work products the only normative text.

Guidance: Ensure the objective of a safety activity is understood. Ensure the methods employed, and results obtained truly satisfy the objective. If gaps exist, consider whether additional risk reduction is warranted.

10.4.2 Embrace Systems Engineering

Many involved in automotive safety seem unaware of its intrinsic relationship with systems engineering. Indeed, a lack of understanding of systems engineering principles can prevail. Some even consider safety as something unique, invented by ISO 26262 or IEC 61508. Such views risk at best, inefficiencies, at worst, poor safety practices. Inefficiencies arise when opportunities to benefit from the commonality of systems engineering, quality management, and safety processes are not leveraged. Poor safety practices may arise if safety is considered decoupled from other organizational processes and structures. This risks insufficient knowledge sharing. It may also result in a mindset that safety can be applied retrospectively, which in turn places unrealistic management expectations on the activities, potentially creating a confirmation bias. It may compound a belief that safety is responsible for significant program costs and overruns. However, it is the authors' experience that a significant portion of the overhead attributed to safety relates to gaps in underlying systems engineering and quality management processes; it is simply that safety is the context in which they were exposed.

Guidance: Embrace systems engineering and quality management and invest in understanding best practices. Consider whether the application of these processes is sufficient. Consider how safety activities align with these, the potential for synergies, and what deltas are truly important in safety engineering.

10.4.3 Address Safety in the Most Appropriate Place

The past decade has seen safety claims migrate further down the supply chain to Tier 2 and 3 suppliers. It is now the norm for suppliers at all levels to claim an ASIL capability. Indeed, most safety components today are developed as safety elements out of context (SEooC), a generic element that may be applied to different safety applications by different customers. In many respects this is a positive trend, rather than needing to mitigate faults at their level, a customer can rely on a supplier providing the necessary mitigation to ensure a component meets the safety requirements assigned to it. However, when a component is considered in a system context, it may transpire that the best place to mitigate faults is not on that component. For example, while it may be possible to implement a mechanism in a radar integrated circuit to detect faults on receive circuitry, a better solution may be to compare the outputs of the multiple receive paths at a level of granularity above the radar integrated circuit. Unfortunately, in an industry swamped in element level safety claims and a desire to push safety further down the supply chain, the best place to address safety risks getting overlooked.

Guidance: As a customer, ensure your safety analysis keeps focus on the most appropriate place to address safety concerns. As a supplier, consider if a safety mechanism on your element is truly the best solution. In both cases, while there may be commercial and competitive pressures to address a problem at a particular element, ensure both parties understand the trade-offs involved.

10.4.4 Improve Supplier/Customer Engagement

The automotive supply chain is complex. Indeed, many automotive extensions to quality management have evolved to manage these complexities. ISO 26262 and ISO 21448 provide further guidance to mitigate risks associated with distributed safety developments. However, the nature of the business environment often results in engagements that are nonoptimal for highly complex safety designs. For example, it is commonplace for an insufficient exchange of information on use cases, safety requirements, and potential system trade-offs to occur, particularly in the early stages of an engagement. There may be several reasons, for example, a customer may not have the information or see the value in such supplier engagements. Or the wrong parties (from a safety perspective) at the supplier and customer may be interacting. This can result in a supplier having a naive understanding of how their product is used, resulting in them developing a nonoptimal solution. Often customers are only able to discuss trade-offs when they have integrated a supplier's solution, however, that is often too late for a supplier, who may have locked-in a design at that point.

Guidance: As a supplier, seek out customer engagement. Don't be shy about highlighting the implications if insufficient input is received. If implementing a safety mechanism for functional safety that impacts the intended function, consider options to make this configurable to allow a customer to decide on the optimal safety trade-off for the system.

10.4.5 Recognize the Criticality of a High Quality Safety Manual

An overhead exists on the consumer of a SEooC. Temple [27] describes how an integrator must ensure the SEooC safety claims are sufficient for the system, that assumptions made by the supplier are valid for the system, and that assumptions of use are addressed with sufficient integrity. Furthermore, they need to ensure that the SEooC safety claim has survived integration, that any safety claims that depend on assumptions of use have emerged as desired, and finally that no new critical failure modes have emerged. Hence, it is critical that a supplier sufficiently convey information on safety claims and assumptions to a customer. Such information is provided in what is generally called a *safety manual.*

Unfortunately, ISO 26262 does not explicitly identify the safety manual as a work product. While it provides guidance on it, it does so in informative sections and in a somewhat loose manner. It places emphasis on assumptions that govern the context of the development (i.e., inward-facing) and not on assumptions that may arise during the development that may impact the context (i.e., outward-facing). This lack of guidance has resulted in several poor industry practices in relation to safety manuals. For example, there seems to be a lack of understanding that assumptions of use (AoUs) are often safety requirements that suppliers place on customers. Hence AoUs should adhere to requirement engineering best practices. Often this is not the case, with AoUs poorly identified and captured. It is critical for the SEooC supplier to be aware that AoUs may implicitly emerge during various activities of the safety lifecycle. As an example, if during the safety analysis, the analyst assumed a certain SEooC configuration then it would be critical to convey this to a customer. Such implicit AoUs can be easily missed.

Guidance: Recognize the criticality of having a high quality safety manual. Consider AoUs as safety requirements and ensure they meet appropriate criteria. Place yourself in your customer's shoes; could you implement the AoU as written? Be aware that implicit AoUs may emerge in the safety lifecycle.

10.4.6 Beware the Many Pitfalls of Safety Analysis

ISO 26262 safety analysis encompasses methods to evaluate various safety properties including the ASIL, robustness against systematic and random faults, and dependent faults. It is common for organizations to invest significant effort in safety analysis, yet it is often the case that issues remain. There are several pitfalls to avoid:

- *Timeliness:* Safety analysis is key to developing a robust safety concept and ensuring sufficient checks and balances in the development process. It needs to be applied at a point where it can influence these aspects. Applying retrospectively is of limited use. Carlson notes that the late application of the failure mode and effects analysis (FMEA) is a very common problem in all industries, resulting in a greatly reduced effectiveness [28].
- *Insufficient staffing:* While a safety analysis is owned and facilitated by a safety engineer, it requires significant support from various domain experts. Often this need is not understood and hence not staffed accordingly. The German Association of the Automotive Industry (VDA) notes that in the automotive industry the incorrect, incomplete, or inefficient application of an FMEA can still result in high numbers of field failures, which in turn results in dissatisfaction with the FMEA method [29].
- *Low confidence assumptions:* Many entries in a safety analysis are merely assumptions. Consider the failure mode and effects diagnostics analysis (FMEDA) as an example, it may have assumptions on failure modes, failure mode distributions, fault safeness, fault effects, and diagnostic coverage capabilities. Some assumptions may have high confidence based on industry experience, while others may not, especially when dealing with new or

emerging technologies. It is important to recognize the aspects of the analysis that require further verification to improve confidence. This is particularly important given that safety analysis is cited by ISO 26262 as a verification method for the design. There is a risk of creating a circular dependency, where analysis with low confidence assumptions is presented as a verification of a design which then justifies reduced verification testing.

- *Overemphasis on random faults:* Despite malfunctions stemming from both systematic and random faults, it is common to see safety analysis place an overemphasis on random faults; this is particularly true of the semiconductor industry. This is due in part to engineers gravitating toward the quantitative techniques and metrics associated with random faults and the potential to include more widgets in a design to address them, while in parallel often shying away from rigorous process overheads. Often an argument is presented that the application of the safety process by default achieves the necessary risk levels for systematic faults. However, this displays a lack of understanding of how safety analysis is used to manage technical risks by identifying additional fault preventions and detections for systematic faults for the specific product under consideration. Ironically, despite the overemphasis on random faults, significant inconsistencies in their estimations exist. There are several reasons, including, the source and application of reliability data, assumptions on safeness and diagnostic coverage, and whether transient faults have been considered. Despite the desire to utilize random fault targets to compare solutions, an apples-to-apples comparison is only possible if all assumptions, many of which may be obscured by the analysis are understood.
- *Application to software:* Several industry guidelines and publications exist that provide an in-depth discussion on the application of safety analysis methods to systems and hardware [2, 30]. However, equivalent guidance on how to apply techniques to software does not exist despite a requirement that they are applied. This has resulted in a significant spectrum of interpretations existing on whether and how to apply safety analysis to software.

Given that ISO 21448 introduces additional safety analysis, for example, to evaluate hazards and causes due to functional insufficiencies and that such insufficiencies stem from systematic faults, avoiding these pitfalls is critical to safe ADS deployments.

Guidance: Ensure the safety analysis is sufficiently planned and resourced. Ensure sufficient emphasis on systematic faults. Ensure all analysis assumptions are documented and communicated to customers. Don't shy away from software safety analysis due to lack of industry examples, consider the objectives of the analysis and how they might be achieved.

10.4.7 Applying Safety to Emerging or Complex Technologies

Safety standards can struggle to provide practical guidance on emerging or complex technologies, as often state-of-the-art on how to deploy them safety hasn't been established yet. Often guidance can appear unhelpful, for example, limited to recommending a technology not be used unless a robust safety argument is presented.

While not practically useful, at a minimum, this guidance serves to remind a practitioner that a lack of established best practice doesn't excuse the need to develop an assurance argument. Neither should the lack of established best practices be used to dismiss emerging technology in safety-critical application; such a mindset would not allow safety practices to advance.

Recall Section 10.2 identified several technologies that fall into this category for the L4 ADS. Taking the specific example of ML; ISO 26262 provides no direct guidance, while IEC 61508 recommends against its use unless rationales to support verification, freedom from interference, correctness, and predictability are developed. Salay and Czarnecki concluded that many ISO 26262 requirements were not directly applicable to ML, while also identifying requirement gaps [31]. Burton noted that until quantitative acceptance criteria for an ML function and ways to measure its failure rate and the impact of its insufficiencies are agreed upon, that the effectiveness of design-time or operational-time measures cannot be evaluated [32].

Guidance: Where established best practices don't exist, revert to considering the safety objectives and applying systems engineering first principles to determine what is necessary to achieve a defensible safety assurance argument. Recognize that irrespective of the desire to utilize a technology a defensible assurance argument is necessary.

10.5 Safety Concepts for Level 4 ADS and Implications for Radar

Automated driving systems are generally architected around a *Sense–Plan–Act* paradigm. In this model, *sense* constitutes all sensor modalities together with the localization and perception functions necessary to enable the vehicle to determine its location and perceive relevant objects within its proximity. *Plan* includes all functions necessary to predict the future behavior of relevant objects and create a safe driving plan to the desired destination. *Act* covers the control and actuation functions necessary to safely execute the driving plan [33]. These functions work together to support the DDT. An L4 ADS must also provide a DDT fallback function (i.e., the ability to bring the vehicle to a minimal risk condition) should the functions supporting the DDT suffer a performance-relevant system failure, or should the vehicle exit the ODD. The DDT and DDT fallback are distinct functions and may be performed within the same or distinct data pipelines. Figure 10.1 presents a potential logical architecture for an L4 ADS. Key aspects of the architecture include:

- Primary DDT channel;
- Dedicated DDT fallback channel;
- Various monitors to detect malfunctions, performance limitations, and ODD exits;
- A fault aggregation and fallback trigger function that consumes monitor outputs and determines the optimal fallback transition.

While it may not be apparent to those less familar with safety at the time, safety begins with the initial specification of the ADS (i.e., when the first iteration of use cases and requirements are documented). At this point, initial expectations on the

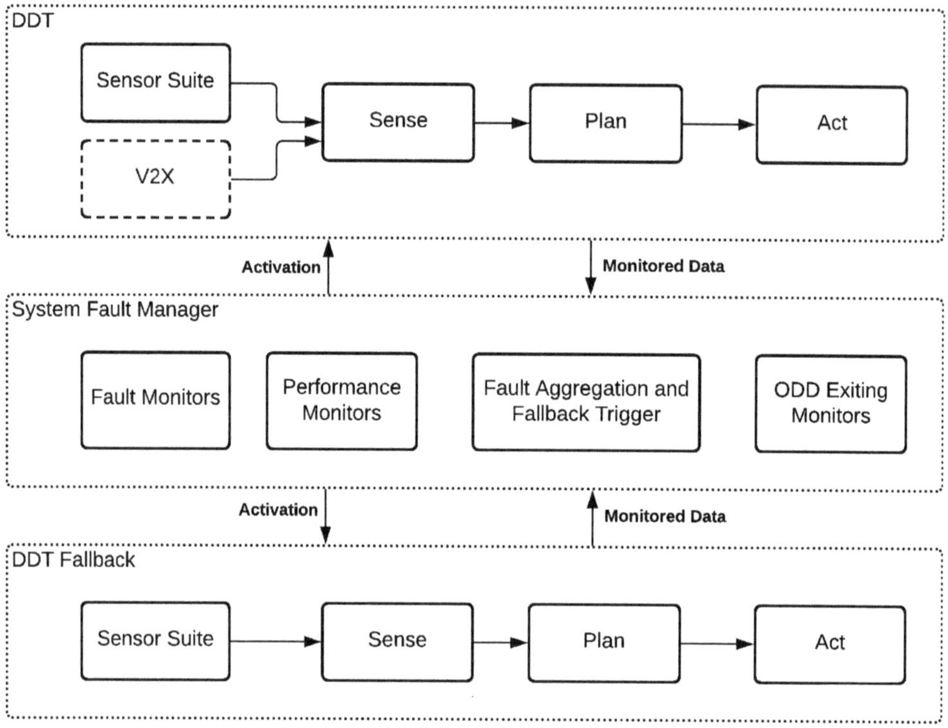

Figure 10.1 An L4 ADS architecture contains primary and fallback functions, with monitors to determine the appropriate state.

ODD are set and *Sense* targets documented based on anticipated driving scenarios. It doesn't take a background in safety to recognize that hazards may arise if the capabilities of the *Sense* function are not sufficiently specified and consistent with the ODD specification. However, this is a complex task for an L4 ADS. The ODD must account for all operational maneuvers, physical infrastructure, operational constraints, objects, and environmental conditions within which the vehicle can perform the DDT [1]. Correspondingly, the AV designer must determine the *Sense* requirements (e.g., detection and classification related precision/recall, detection range and vertical/horizontal field of view, object position/motion/shape accuracy, etc.) and the appropriate metrics so they may be appropriately allocated to sensors.

Architectural decisions, potentially implicitly driven by safety, may occur at this early point. For example, the use of multiple sensor modalities such as radar, camera, or lidar may already be proposed to support the *Sense* function. In such scenarios, appropriate performance metrics need to be allocated to each sensor modality.

For many, a more recognizable initiation of safety occurs with the application of a comprehensive structured safety analysis to the ADS specification. This analysis considers hazards resulting from both functional insufficiencies and malfunctions at the vehicle level. The analysis drives the identification of safety requirements that mitigate as appropriate the root causes of both sources. For functional insufficiencies, these mitigations may include improving individual sensor performance or adding new sensor modalities to improve detection confidence across the ODD, or restricting the ODD to align with Sense capabilities. For malfunctions, the mitigations may include additional mechanisms and measures to control random and systematic faults. Recall from Section 10.3.1 that unreasonable risk targets for malfunctions are implicitly defined by the assigned ASIL, while those for functional insufficiencies must be proposed and defended by those responsible for the deployment of the L4 ADS. The importance and impact of this distinction should not be underestimated. For functional insufficiencies, targets are not provided, rather acceptable targets must be derived and used to guide verification and validation criteria. Unfortunately, when dealing with statistics, even accident statistics, the potential for complacency arises. Robust checks and balances in an organization are necessary to ensure the functional insufficiency targets are sufficiently stringent for the life-critical applications being considered. Several approaches based on accident statistics have been proposed [34–37].

The same safety process is applied independently of the complexity of the radar; it doesn't matter whether its modes and capabilities are simple and straightforward or "cognitive" [38]. However, a more complex implementation may drive higher levels of effort in terms of analysis and verification. The remainder of this chapter presents several considerations for the application of safety to the radar system.

10.5.1 Safety Considerations on Multiple Sensor Modalities

During the specification of the Sense function, it is critical to identify whether safety requires sensor modalities as redundant or merely complementary. This holds true whether considering functional insufficiencies or malfunctions of an ADS.

Redundancy is defined as the existence of means in addition to the means that would be sufficient to perform a required function. It is a familiar design pattern in safety systems and may be implemented as either homogeneous or heterogeneous. It has many uses, for example comparing redundant outputs can increase confidence in a result or indicate the presence of a malfunction. It may also be used to increase availability; however, this may require more than two redundant channels, otherwise, it may not be possible to identify which channel is compromised. Redundancy requires that any correlation of functional insufficiencies or malfunctions be minimized, ideally to zero (i.e., the redundant channels are independent). A special case of redundancy known as *ASIL decomposition* allows the ASIL of an ISO 26262 safety requirement to be reduced if redundant requirements can be found and allocated to sufficiently independent elements [2]. ASIL decomposition presents a way to deal with the practical limitation that many sensor components presently on the market do not support an ASIL D capability. Presently Mobileye cites the use of redundant sensor modalities for their L4 ADS concept [39].

Let's consider the example where a Sense function is allocated a requirement to detect vehicles with a certain minimum accuracy when within a certain range. It is worth highlighting that both safety and nonsafety variants of this requirement may exist with differing accuracy and range limits; the limits in the nonsafety version may be driven by passenger ride experience, while the limits in the safety version are driven by the avoidance of vehicle hazards. From a malfunction perspective, the safety requirement is determined as ASIL D (recall there's no ASIL defined for functional insufficiencies). Let's assume there is a desire to apply an ASIL decomposition, where instead of a single sensor modality inheriting the most stringent

ASIL D targets for systematic fault control, a radar system and a camera (or lidar) system respectively inherit less stringent ASIL B targets for systematic fault control for the same vehicle detection requirement. Let's also assume that the SOTIF safety analysis identifies specific instances where the radar is claimed as redundant (called a *countermeasure* in SOTIF terms) for functional insufficiencies in the camera and vice versa. However, it is known that camera and lidar sensors both suffer degradation in detection range when operating in adverse weather [40] such as fog, hence cannot be considered to provide the same capability as a radar sensor under this ODD condition and therefore they are not truly redundant across the ODD. This means that in fog radar alone provides the longest-range vehicle detection capability and hence the ASIL decomposition may be invalid if this range is necessary to support the safety requirement or if the ODD includes driving in fog. A comprehensive analysis is necessary to identify the limits of redundancy of the sensor modalities. This may include quantifying the degradation the sensors experience, perhaps via the development of a weather degradation model. Where gaps are found, a first step is evaluating whether the safety requirement is violated. If it is, it may be worth considering whether the requirement is unnecessarily restrictive. If not and if a gap still exists, options include implementing measures that reduce the vehicle's speed in the presence of dense fog, to a point where the hazard would inherit an ASIL B and hence could be supported by the radar alone. However, more advanced safety arguments may also be possible. Points to consider when developing more advanced arguments:

- Does the reduction in ASIL systematic rigor achieved via the ASIL decomposition increase SOTIF risks? While this may seem like a surprising question, recall at the element level functional insufficiencies stem from systematic faults. Now consider that one of the ways ISO 26262 differentiates between ASIL B and D in terms of systematic fault control rigor, is that ASIL D requires expanded functional tests and stress tests be applied to the development. If these were omitted for the ASIL B development, might this result in an increased risk of functional insufficiencies at the vehicle level?
- Arguments that propose driving with only an ASIL B radar system in fog without reducing speed must address why any data presented is sufficient to overrule the ISO 26262 ASIL D risk conclusion, given that ISO 26262 would have already considered environmental considerations.
- Consider the possibility of confirmation bias. For example, an argument that slowing down may increase risk if the vehicle speed inhibits the flow of traffic and may even violate minimum speed laws (e.g., Vehicle Code 22400 VC—Minimum Speed Laws in California) may appear as a good argument not to slow down. However, in a safety context, the question should be viewed as whether it was the right decision to include driving in fog in the ODD given it results in both the potential for an ASIL decomposition violation and a scenario for increased risk due to slowing down. Irrespective, the need for braking studies would be necessary as there will be scenarios where you need to slow down.

Ultimately, it is critical that the safety specification is clear on redundancy expectations and that the final argument on why the solution is sufficient must be comprehensive, consistent, and backed by verification evidence. Similar consideration needs to be given to other safety functions inherited by the radar system, whether driven by perception, localization [41], mapping [42], or ego-motion [43] use cases.

10.5.2 Safety Considerations on Radar Data

Today's imaging radars provide multiple layers of data outputs including various spectra and images, as well as the traditional detections, peaks, and tracks to feed perception algorithms (see Figure 10.2). These images and lower-level data can provide the richness and context necessary for advanced ML algorithms, but this highlights two issues. The first, which is not radar specific, is the intrinsic issue of using ML in safety-critical applications [32]. Recall Section 10.4.7 provides direction on the application of safety to ML advising to revert to considering the safety objectives and applying system engineering first principles. The second issue arises from the complexity of the images formed from radar data. It is more difficult to analyze and gain statistical confidence in radar images than in traditional radar detections and tracks that follow more mature and tractable performance expectations. It is also more difficult for radar than for cameras (a traditional imaging sensor) due in part to the aspect dependence of the targets in the scene; a small change in viewing angle can result in a large change in the power reflected back to the radar and the perceived physical size of the object. How to ensure these data are of sufficient quality for use in perception algorithms is still an open area for investigation. Again, the guidance in Section 10.4.7 is applicable. While established best practices may not exist, this does not excuse the need to develop a defensible safety assurance argument by relying on systems engineering first principles.

10.5.3 Radar FuSa and SOTIF Roots Causes and Mitigations

How do malfunctions and functional insufficiencies manifest at the radar system level and how are they mitigated? Table 10.1 presents several undesired radar system behaviors with nonexhaustive examples of potential FuSa (i.e., malfunction) and SOTIF (i.e., functional insufficiency) root causes. While this table contains information derived from a safety analysis, it should not be interpreted as a complete safety analysis; for details on safety analysis please refer to [2–3].

Observe that each undesired radar behavior may arise due to either FuSa or SOTIF root causes, and that the distinction between causes can at times be negligible, driven more by safety standard conceptual models than how it manifests in the real-world. Recall Section 10.3.1 notes that at the element level (i.e., radar system or radar sensor), functional insufficiencies arise due to systematic faults; either a gap existed during the transfer of requirements between supplier and customer, or the product implementation did not fulfill its specification. Hence the need to ensure equivalent levels of rigor are applied when developing safety assurance arguments for FuSa and SOTIF causes for the radar system exists. It is a misconception that FuSa requires more rigor than SOTIF, due in part to the style used in the retrospective standards. Given that the same vehicle level hazardous behavior may arise and that radar undesired behavior may stem from both sources, this is not the case. Indeed, if different levels of rigor are applied, this is an indicator that something may be amiss and needs to be addressed.

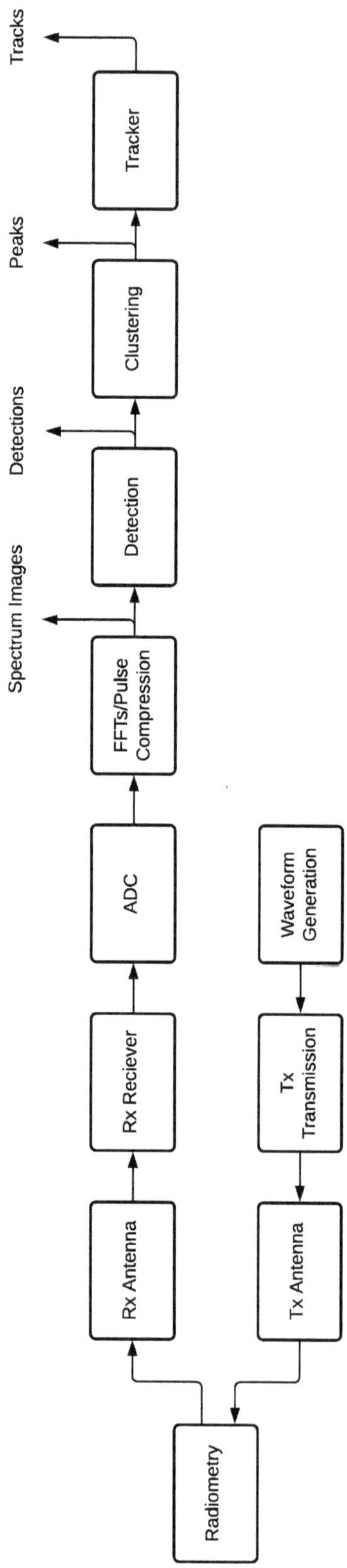

Figure 10.2 A simplified radar functional diagram.

Table 10.1 Example of Simplified Radar System Level Safety Analysis

Radar System Undesired Behavior	*FuSa Causes*	*SOTIF Causes*
Vehicle not detected (false negative).	Thermal management system malfunction causes radar to overheat and processor to stop providing detections.	Heavy rain attenuates the radar return, causing SNR to drop below detection threshold as system wasn't designed to handle this condition.
Vehicle detected when none exists (false positive).	At certain frequencies phase lock loop (PLL) spurs appear and are detected as targets.	Multipath reflection mitigation design insufficient, resulting in some reflections detected and tracked as true objects.
Vehicle detected but takes longer than allowed.	Software bug fails to limit the maximum number of targets processed in a radar dwell, causing processing to take longer than budgeted for a complex scene.	Scene more complex than planned for in specification, resulting in more detections to process than expected or budgeted in the processing timeline.
Vehicle detection reported in unexpected update rate.	For implementations involving synchronization of sensors' fields of view, errors on the radar triggering signal can cause misalignment of data and incorrect reporting of detections.	Design limitation on data bandwidth causes delay in receiving radar detections when all sensors providing maximum outputs.
Vehicle detected but with incorrect attributes.	For implementations with multiple transmit beam possibilities (with different gains), software bug results in appropriate beam not being accounted for, resulting in incorrect RCS estimate.	For implementations involving sparse arrays, the angle estimate may be incorrect (i.e., in the incorrect ambiguity) when the scene is more complex than expected.
Vehicle detection over- or under-segmentation (incorrect bounding box/incorrect clustering of energy).	Transmitter malfunctions, lowering output power and resulting in insufficient SNR on smaller (lower radar cross-section) portions of a target and underestimation of target physical and electrical size.	Segmentation design didn't include the possibility of closely spaced targets, leading to poor segmentation and incorrect physical size estimates.
Vehicle detected but with incorrect trajectory.	Radar clock has a slight timing malfunction that leads to timestamp errors, corrupting the track estimates.	In complex environments that exceed the design intention, multiple closely spaced detections may incorrectly associate to tracks, corrupting the tracker outputs.

While both ISO 21448 and ISO 26262 require a specification as an input to the safety analysis, the analysis may result in modifications to existing requirements or uncovering new radar requirements. Examples of key areas often impacted are listed here, with out-of-boundary detection criteria also in scope where appropriate.

- Detection range and horizontal/vertical FOV coverage (include dynamic range) in all weather conditions of the ODD;
- False positive and false negative thresholds for detection of moving and non-moving vehicles;
- Detection position (range and angle) and Doppler accuracy and resolution;

- Maximum sidelobe level in all dimensions;
- Process latency (including maximum capacity) and output update rate;
- Time synchronization accuracy;
- Intrinsic and extrinsic calibration accuracy;
- Operating environment (e.g., temperature, humidity, salinity, vibration, etc.);
- Mutual interference detection and mitigation;
- Multipath detection and mitigation;
- Clutter detection and mitigation.

Table 10.2 presents examples of potential mitigations.

Table 10.2 Safety Mechanisms Can Be Implemented at Multiple Layers

Root Cause	*Radar Component Level*	*Radar System Level*	*Perception System Level*	*ADS Level*
TX/RX channel degradation.	TX and RX channel power monitors.	Radar-to-radar crosscheck (use one radar to determine health of another). Check sensitivity against known RCS targets at known locations.	Utilize redundant or diverse (nonradar) sensors.	N/A
Rain accumulation on radome.	Radome/fascia design to rapidly shed water.	Overspecify radar to account for degradation from rain, estimation of current level of rain on radome. Specification enhancements to support operating in rain.	Modify ML algorithms to deal with rain accumulation.	Employ cleaning/clearing solution for the radome (e.g., wipers); reduce speed to account for degradation of sensitivity.
Excessive process latency (processing takes too long).	Increase allocation of compute resources; increase algorithm efficiency.	Employ a latency monitor.	Check timeliness of inputs to determine optimal action for latent data (skip frame if too late).	N/A
Multipath causes false returns.	N/A	Flag detections as possible multipath.	Employ "knowledge-aided" algorithms for known high multipath environments; utilize redundant or diverse (nonradar) sensors.	Descope operation (limit ODD) by rerouting around high multipath areas.

Notice that safety mitigation may be applied at different levels of system granularity. However, this doesn't mean they are equivalent. They may differ in terms of robustness, implementation and verification overheads, system resource consumption (e.g., power, processing cycles, RF bandwidth), or other criteria. While all these criteria may be important to a program, safety will emphasize the robustness of the solution. Take TX/RX channel degradation as an example. At the component level, channel power monitors could be implemented to detect issues. However, channel degradation may encompass more than a power reduction, for example, it may include phase errors, SNR degradation, unwanted tones, on many other undesired characteristics. It is critical that a structured safety analysis uncover sufficient details on root causes and considers the most appropriate place in the design to mitigate. Consideration should also be given to whether mitigation approaches risk negatively impacting system safety. Considering TX/RX channel degradation again, a component vendor may propose a self-test capability to detect more undesired characteristics. However, the self-test may consume resources that impact the nominal performance of the component. When considered at the radar system or ADS level, this may result in a poorer solution from a safety perspective. Recall Section 10.4.3 emphasized the need to address safety in the most appropriate place in a design.

10.5.4 Safety Considerations Due to Available Radar Technology

Presently most monolithic microwave integrated circuits (MMIC) available on the market are optimized for widely proliferated Level 2 (L2) driving use cases such as AEB, given the multiyear, multimillion-dollar investments required for their development. However, the hazards and radar performance expectations differ greatly between AEB and L4 ADS. For example, a less expensive MMIC with a high noise figure targeting L2 AEB may only consider mitigations for hazards related to false positives, as false negatives may be assumed as mitigated by the constant presence of a vigilant driver. Here the false positives, which result in undesirable unnecessary braking on the roads, can be mitigated by raising the detection threshold such that small SNR returns are not detected by the appropriate choice of detection thresholds. However, this means it can miss small RCS targets especially at longer ranges, which is important to the L4 use case.

Such practical realities are one of the reasons ISO 21448 was developed and provides the guidance it does. The practical constraint to use a product intended for an L2 use case in an L4 use case demands a sufficient specification for the L4 use case at both the vehicle and radar levels. Likewise, sufficient characterization of the implemented performance of the MMIC is necessary. The analysis should include a structured consideration of L4 hazards, the contribution of the radar, and whether countermeasures are necessary (e.g., restricting the ODD or including a redundant or complementary sensor modality). Finally, verification and validation are critical to identifying unknown risks that stem from the use of a component in a different context with different scenarios. This example highlights the need for close collaboration across all parties in a distributed development both for FuSa and SOTIF, as emphasized in both standards. However, recall from Section 10.4.4 that this doesn't always happen, so requires special attention for L4 ADS developments.

10.6 Safety Considerations for Verification and Validation

Verification and validation (V&V) have been at the core of automotive for decades, being a key element of quality management before the formalization of any safety standards. Automotive developments have included the notion of verification at different levels of hierarchy, for example, component level verification by the component supplier, module level verification by the module supplier, and vehicle level verification by the vehicle supplier. The latter historically included some expectations of test track and open road testing before vehicle release.

For L4 ADS safety, the potential for unknown hazardous scenarios (i.e., the ones not considered in the design) changes the safety posture. The real world presents an open-context problem, hence an ADS has the potential to encounter a driving scenario that wasn't considered during traditional systems and safety engineering practices (i.e., requirements-driven testing). This doesn't mean that traditional practices should be abandoned. On the contrary, they remain critical to a safety argument but need to be supplemented to ensure a robust safety argument that unknown hazardous scenarios do not constitute an unreasonable risk.

ISO 21448 requires the implementation of a V&V strategy to evaluate both known and unknown hazardous scenarios associated with functional insufficiencies. It promotes the value of incremental V&V for perception systems, pointing to the need to consider different approaches to ensure robust coverage, for example, bench, simulation, vehicle integration, test track, and open road testing [3].

Known hazardous scenarios and their associated functional insufficiency are identified early in the development process via a structured analysis. A common method for evaluating known hazardous scenarios is via *structured tests*. These are performed in a controlled environment, such as a test track or replicated urban environment. While a structured process should be applied to defining test parameters, test gaps may result because the process relies on human expertise, which may be insufficient for a system as complex as an ADS. This needs to be accounted for during unknown hazardous scenario evaluation planning. Simulation may also be useful when evaluating known hazardous scenarios, such as when a test may be too impractical to implement as a structured test, or when a structured test is deemed to present an unreasonable risk to the testers. Simulations may include sensor physics-based modeling [44], perception virtual reality modeling [45], and ADS end-to-end modeling [46]. How to demonstrate that a simulation environment provided sufficient fidelity or is sufficiently comprehensive and realistic to support safety verification is a challenging question. However, this does not absolve those responsible for ADS V&V from having to provide such a justification for the simulation environment. ISO 21448 proposes that techniques introduced in ISO 26262 to evaluate the confidence in the use of software tools be adapted to support such arguments [3].

Unknown hazardous scenarios are evaluated via a combination of open road *unstructured tests* and simulation. A critical aspect of unstructured testing is the need to execute it in a manner that does not expose other road users to unreasonable risk. If necessary, additional safety measures need to be in place to avoid hazards arising during such testing. A challenging question to answer concerning unknown hazardous scenarios evaluation is when is the testing sufficient (i.e., when is the remaining risk deemed acceptable?). Unlike ISO 26262 which effectively provides targets for unacceptable risk from malfunctions, ISO 21448 does not provide targets for functional insufficiencies. Instead, it provides guidance on approaches to determine acceptance criteria but leaves the targets and their defense to the vehicle supplier. A final safety argument must explain why the combined miles driven on open road and in simulation are deemed sufficient to avoid unreasonable risk.

Fault-injection simulation also offers a platform to evaluate scenarios where one or more sensors malfunction or degrade. This may be useful for both SOTIF and FuSa verification, including redundant and fail-operational evaluation. It is worth noting that simulating faults in an ADS system can be a complex task unto itself, often involving the synthesis of faulty data for replay into an ADS.

Turning to radar specifically, there are some key concepts that should be considered regarding verification. First, the performance of radar is a function of the environment (see Chapter 7 on the challenges of the radar environment). In the automotive environment, the ability to detect a particular target and the quality of the measurements regarding it are influenced (often strongly) by the other constituents of the scene. Second, radar performance is a function of the weather (see Chapter 9 on the impacts of weather). Despite radar generally being the AV sensor least impacted by weather, it still can be significantly impacted by it. Finally, it is difficult to simulate radar behavior in operational environments. While there are some efforts underway to develop simulation capabilities for radar in operational situations, none to date have shown the fidelity required to be used for in situ real-world validation. Some radar performance attributes may be characterized in a lab or controlled testing, which can validate many theoretical aspects of the radar and highlight functional errors in the system. However, this will not replace the significant amount of real-world testing on public roads required to verify the safety requirements are achieved. As a microcosm to the ADS, there is no well-established guideline for the approach or amount of public road testing needed for radar verification.

10.7 Conclusion

At its core, safety is about identifying and managing risk by applying evidence-based systems and safety engineering best practices. While the processes may seem arduous, the industry safety standards provide established guidance on these best practices. Safety should never become a paperwork exercise or about finding a path of least resistance to show compliance to a standard, but instead it should aid in the continuous improvement of an organization and its products.

Autonomous driving is currently one of the most challenging safety domains. From the radar viewpoint safety is about ensuring the ADS sufficiently understands and conveys its expectation on the radar system and that the radar system demonstrates it can fulfill those expectations. If expectations cannot be fulfilled, they must be adjusted accordingly by the ADS to align with radar capabilities. This requires thorough and frequent communication within organizations and across the supply chain. Many challenges remain for fully AVs, and undoubtedly safety methods and best practices will advance as these challenges are addressed.

As a final counsel to the reader, practitioners should never lose sight that they are working on life-critical applications and be cognizant of the potential for confirmation bias, especially when dealing with statistics. If ever in doubt, revert to safety objectives and systems engineering first principles.

References

[1] SAE J3016, *Taxonomy and Definitions for Terms Related to Driving Automation Systems for On-Road Motor Vehicles*, 2021.

[2] ISO 26262 (all parts), *Road Vehicles—Functional Safety*, 2018.

[3] ISO/FDIS 21448, *Road Vehicles—Safety of the Intended Functionality*, 2021.

[4] *Best Practice for Describing an Operational Design Domain: Conceptual Framework and Lexicon*, Automated Vehicle Safety Consortium, 2020, https://avsc.sae-itc.org/.

[5] *Best Practice for Metrics and Methods for Assessing Safety Performance of Automated Driving Systems*, Automated Vehicle Safety Consortium, 2021.

[6] UL4600, *Standard for Safety for the Evaluation of Autonomous Products*, 2020.

[7] IEEE P2846, *Assumptions for Models in Safety-Related Automated Vehicle Behavior*, 2021.

[8] ISO PAS 8800, *Road Vehicles Safety and Artificial Intelligence*, 2021.

[9] IEC 61508 (all parts), *Functional Safety of Electrical/Electronic/Programmable Electronic Safety-Related Systems*, 2010.

[10] IATF 16949, *Automotive Quality Management System Standard*, 2016.

[11] ISO 9001, *Quality Management Systems—Requirements*, 2015.

[12] ISO/IEC 33000 (all parts), *Information Technology—Process Assessment*, 2015.

[13] *Capability Maturity Model Integration*, ISACA, 2021.

[14] *Automotive SPICE* 3.1, VDA, 2017.

[15] ISO/IEC/IEEE 15288, *Systems and Software Engineering—System Lifecycle Processes*, 2015.

[16] Bell, R., "A Brief History of the Development of Guidelines and Standards," *25th Safety Critical Systems Symposium*, 2017.

[17] *Quality Management in the Automotive Industry—QM System Audit—Serial Production*, VDA, 2016.

[18] *Advanced Product Quality Control and Control Planning*, AIAG, 2008.

[19] *Joint Quality Management in the Supply Chain—VDA Maturity Level Assurance for New Parts*, VDA, 2010.

[20] Hardware SPICE—PRM/PAM 2.0, intacs.info, 2020.

[21] ISO/IEC/IEEE 12207, *Systems and Software Engineering—Software Lifecycle Processes*, 2017.

[22] ISO/IEC/IEEE 24748, *Systems and Software Engineering—Life Cycle Management*, 2018.

[23] *Systems Engineering Handbook—A Guide for System Life Cycle Processes and Activities*, v4, INCOSE, 2015.

[24] *Automotive SPICEv3.1 Pocket Guide*, Kugler Magg CIE, 2018, https://www.kuglermaag.de/fileadmin/05_CONTENT_PDF/literature_automotive-spice_pocketguide.pdf.

[25] SS 7740, *Road Vehicles—Functional Safety Process Assessment Model*, SVENSK STANDARD, 2018.

[26] Douglass, B. P., *Agile Systems Engineering*, Waltham, MA: Morgan Kaufmann Publishers, 2016.

[27] Temple, C., "Developing Complex Safety Critical Systems in Complex Supply Chains," *16th European Dependable Computing Conference*, Munich, Germany: 2020.

[28] Carlson, C., *Effective FMEAs: Achieving Safe, Reliable, and Economical Products and Processes Using Failure Mode and Effects Analysis*, Hoboken, New Jersey: Wiley, 2012.

[29] *Quality Management in the Automotive Industry: Assessment of Quality Management Methods*, VDA, 2017.

[30] *Failure Mode and Effects Analysis Handbook*, AIGA & VDA, 2019.

[31] Salay, R., and K. Czarnecki, *Using Machine Learning Safely in Automotive Software: An Assessment and Adaption of Software Process Requirements in ISO 26262*, 2018, https://arxiv.org/ftp/arxiv/papers/1808/1808.01614.pdf, CoRR abs/1808.01614.

[32] Burton, S., et al., "Safety, Complexity, and Automated Driving: Holistic Perspectives on Safety Assurance," *IEEE Computer*, Vol. 54, Issue 8, August 2021, pp. 22–32.

[33] Aptiv, Audi, Baidu, BMW, Continental, Daimler, FCS, Here, Infineon, Intel, Volkswagen, *Safety First for Automated Driving*, White Paper, 2019.

[34] Blumenthal, M. J., et al., *Safe Enough—Approaches to Assessing Acceptable Safety for Automated Vehicles*, Research Report, RAND, 2020.

[35] Kalra, N., and S. Paddock, *Driving to Safety: How Many Miles of Driving Would It Take to Demonstrate Autonomous Vehicle Reliability?*, Transportation Research Part A: Policy and Practice, Vol. 94, December 2016, pp. 182–193.

[36] Nees, M., "Safer Than the Average Human Driver (Who Is Less Safe Than Me)? Examining a Popular Safety Benchmark for Self-Driving Cars," *Journal of Safety Research*, Vol. 69, June, pp. 61–68, 2019.

[37] Schwall, M., et al., "Waymo Public Road Safety Performance Data," Waymo White Paper, 2021.

[38] Guerci, J. R., *Cognitive Radar: The Knowledge-Aided Fully Adaptive Approach*, Norwood, MA: Artech House, 2010.

[39] Mobileye, *True Redundancy—The Realistic Path to Deploying AVs at Scale*, 2020, https://www.mobileye.com/true-redundancy.

[40] Vargas, J., et. al., "An Overview of Autonomous Vehicles Sensors and Their Vulnerability to Weather Conditions," *Sensors*, Vol. 21, No. 16, pp. 5397-5419, 2021.

[41] Kuutti, S., et al., "A Survey of the State-of-the-Art Localization Techniques and Their Potentials for Autonomous Vehicle Application," *IEEE Internet of Things Journal*, Vol. 5, Issue 2, April, pp. 829–846, 2018.

[42] Fayyad, J., et al., "Deep Learning Sensor Fusion for Autonomous Vehicle Perception and Localization: A Review," *Sensors*, Vol. 20, No. 15, pp. 4220–4254, 2020.

[43] Monaco, C. D., and S. N. Brennan, "RADARODO: Ego-Motion Estimation from Doppler and Spatial Data in RADAR Images," *IEEE Transactions on Intelligent Vehicle*, Vol. 5, Issue 3, September 2020, pp. 475–484.

[44] Schlager, B., et al., "State-of-the-Art Sensor Models for Virtual Testing of Advanced Driver Assistance Systems/Autonomous Driving Function," *SAE International Journal of Connected and Automated Vehicles*, Vol. 3. Issue 3, pp. 23–26, 2020.

[45] Kaur, P., et al., "A Survey on Simulators for Testing Self-Driving Cars," *Fourth International Conference on Connected and Autonomous Driving (MetroCAD)*, Detroit, MI, April 28–29, 2021, pp. 62–70.

[46] Best, A., et al., "AutonoVi-Sim: Autonomous Vehicle Simulation Platform with Weather, Sensing, and Traffic Control," *Conference on Computer Vision and Pattern Recognition*, Salt Lake City, UT, June 18–22, 2018.

CHAPTER 11

Testing Automotive Radars

Leander Humbert

11.1 Introduction: Why Is Testing Necessary?

ADAS and AD functions in modern vehicles are becoming increasingly complex and integrated. In addition, vehicles utilizing those technologies are spreading into mass markets. Since they are safety critical, the implementation of these driving functions is especially dependent on technological progress in electronics and sensor technologies. Testing and evaluation of installed performance of novel electronic control and sensing systems poses a specific challenge in this context since development cycles are short and much of the functionality is software and signal processing driven.

The requirements for test instruments and systems vary depending on the tier of the automotive ecosystem where the testing is performed. There are several steps during the development process to cover the full value chain: classical use of instruments like vector network analyzers, oscilloscopes, and waver probes are used for the transition from simulation to prototyping of a chipset, validation of signal processing is accomplished by electromagnetic stimulation of the sensor, and finally the fusion of all subsystems is validated in the fully integrated vehicle. Traditionally at this point drive testing over the course of several millions of kilometers on real roads are performed. However, these test drives do not provide a reproducible or controlled environment, and the costs and time effort associated with them are tremendous.

The automotive market requires high performance and reliable radar sensors at an affordable price, forcing research, development, and production to be efficient in terms of cost. It follows that test and measurement of these radar sensors needs to be fast, reliable, cost-effective, and straightforward. Aside from the internal demand for testing of the respective manufacturers (e.g., chipset, radar module, full vehicle), there are external stakeholders that bring up additional requirements.

11.1.1 Verification and Validation of System Performance

When having a closer look at the V-model for an ADAS function (whether this model is actually used) the interdependencies among the supply chain are perceivable.

Specification begins for the original equipment manufacturers (OEMs) on an abstract functional level before it translates into increasingly technical specifications for lower tiers. Among OEMs, Tier 1 and Tier 2 suppliers it is necessary for common methods of validation and verification (Figure 11.1) to ensure quality and that design goals are met.

11.1.2 Conformance to Legal Regulations and Industrial Standards

Automotive radar sensors are used in complex and safety-critical applications such as ADAS and autonomous driving and have to operate in a dense spectral environment that is crowded by different services. Performance of the sensors is therefore a concern for both manufacturers and general public interest. Therefore, compliance with relevant regulations and standards, both under nominal and adverse conditions, must be verified. There are also liability considerations for OEMs and radar-module suppliers, requiring robust test and measurement procedures and trusted testing instruments to minimize risk and ensure an efficient approval and certification process. The structure of involved bodies usually is international, very complex, and interconnected, making harmonization of different standards and regulations mandatory.

The topics covered by standards and regulation range from specifying limits for mutual electromagnetic interference (differentiated among devices and services) to describing the exact test methodologies to be used, to advised best practices. Also, the degree of legal binding is complex, as sometimes standards might be incorporated into national law and therefore be part of a regulation or homologation (certificate that allows a product to enter a market).

11.1.3 Safety Performance Assessment

In contrast to the strict conformity required for standards, regulations, and homologation, in programs like the new car assessment program (NCAP) a baseline for a minimum compulsory performance of an electrical system or full vehicle is defined, with the intention that these voluntary assessment programs can be used as a benchmark for the safety of vehicles. Therefore, the performance requirements for those assessments are usually higher than for conformance of the same functions. Using a rating system, it is possible to compare vehicles more easily. This safety rating

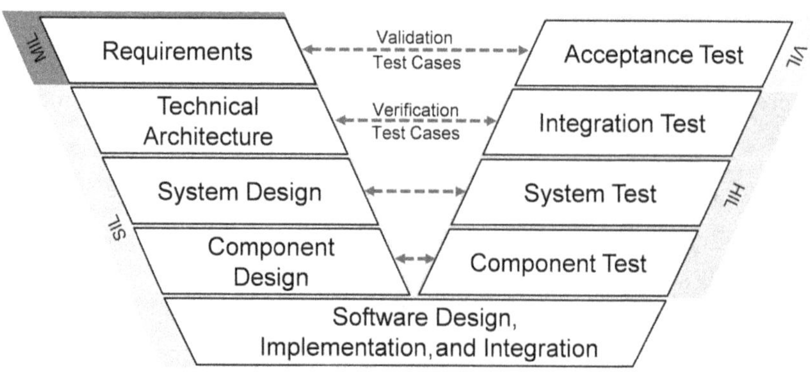

Figure 11.1 The complexity of modern vehicles is broken down into several components and/or subsystems developed by different suppliers. The classical V-model is an approach to organize the necessary verification and validation activities.

is determined from a series of vehicle tests, designed and carried out by consortia of public and private stakeholders. Test results are commonly presented by motor press, and in turn, greatly influence consumer demand for a vehicle [1].

This provides a continuing incentive for manufacturers to regularly enhance the assessment procedures and integrate new active and passive safety systems in the benchmarking process. For example, Euro NCAP recently introduced testing of scenarios for active braking or steering intervention as those functions reduce the collision speed and the severity of accidents. A shift in focus toward sensor (especially camera and radar) based ADAS functions has happened over the last few years.

Several comparable international NCAPs and organizations exist like Euro NCAP, US NCAP, or C-NCAP [2], so again there is not a single harmonized entity for definition of test cases.

11.2 Measurable Parameters: From Sensor Level to Vehicle Integration

The next section provides answers to the questions surrounding what parameters of an automotive radar can be measured, which of those make sense to be measured, and what should be measured but is not possible or only with an uneconomical high effort due to integration on chip.

Always keeping in mind *why* a measurement is necessary, the level of abstraction increases from testing of physical RF parameters for research and development (R&D) beginning on module level, up to full vehicle integration testing for homologation.

One final remark is that this excludes chip-level testing, as there is not much uniqueness of testing a radar chip compared to other integrated circuit (IC) or microwave monolithic integrated circuit (MMICs) wavers. The parameters that are unique to a radar chip are the same as tested on module level, only that measurements performed are conducted (using a waver probe) and not over-the-air.

11.2.1 Transmitter Tests

11.2.1.1 Operating Frequency Range

The general definition of the operating frequency range of transmitting equipment normally is a tool of regulatory bodies, like the International Telecommunication Union (ITU) or the Federal Communications Commission (FCC), to ensure interoperability of radio services. The operating frequency range of a radar under test (RUT) is derived from its occupied bandwidth, spanning from the lower f_L to the upper f_H frequency limits, where the mean powers emitted are each equal to a specified percentage of the total mean power of a given emission [3]. Unless otherwise specified, this percentage is 0.5%, so that the occupied bandwidth (OBW) contains 99% of the total integrated power of the transmitted spectrum centered on the center frequency f_c.

To ensure that the transmitted power that is not inside the operating frequency range also does not interfere with other RF services or devices, limits for unwanted emissions also apply. Normally there is differentiation between out-of-band (OOB) emissions on frequencies immediately outside the OBW that normally results from the modulation process and spurious emissions that include harmonic, parasitic emissions, intermodulation products, and frequency conversion products that normally have no informational content of the signal. The definition of these bands is up to the respective regulatory body, but can be translated to a spectrum mask, a definition of maximal effective isotropic radiated power (EIRP) values versus frequency centered around the OBW the RUT (see Figure 11.2).

11.2.1.2 Modulation-Related Measurements

Whatever radar waveform is used for an automotive sensor, whether FMCW, phase-coded FMCW (PC-FMCW), PMCW, or orthogonal frequency division multiplex (OFDM), there is some form of modulation involved. For a FMCW radar important chirp signal parameters include

- Chirp bandwidth;
- Chirp duration;
- Chirp rate or frequency slope;
- CPI or frame times.

Those parameters, as shown on Figure 11.3, have a direct impact on the sensor performance for range and Doppler accuracy and resolution and therefore validation of the defined specification is needed during the design phase, with these well-known relations for FMCW range and Doppler resolution:

$$\Delta R = \frac{c}{2 \cdot B} \tag{11.1}$$

$$\Delta v = \frac{c}{2 \cdot f_{Tx} \cdot T_{chirp}} \tag{11.2}$$

or in case of fast (multiple) chirps:

$$\Delta v = \frac{c}{2 \cdot N \cdot f_{Tx} \cdot T_{chirp}} \tag{11.3}$$

Additionally, when it comes to topics like chirp sequence waveforms, sequential operational modes (e.g., long- and short-range waveforms or calibration), or for some approaches in time domain for interference mitigation, a series of chirps is

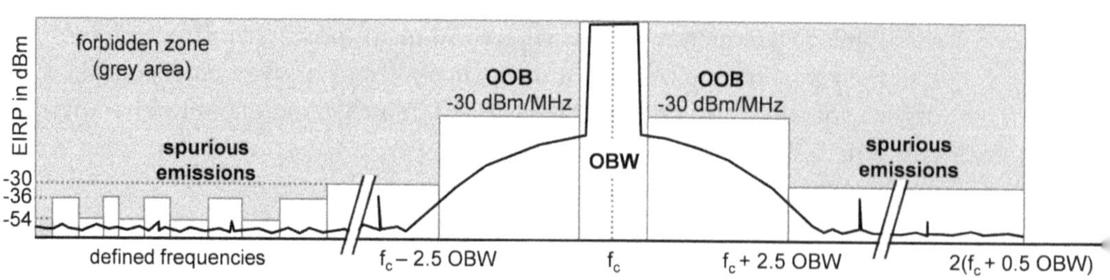

Figure 11.2 Spectrum mask based on values provided by ETSI EN 302 264, that defines OOB range as 2 · OBW and spurious emissions via the harmonic of the highest transmitted operation frequency.

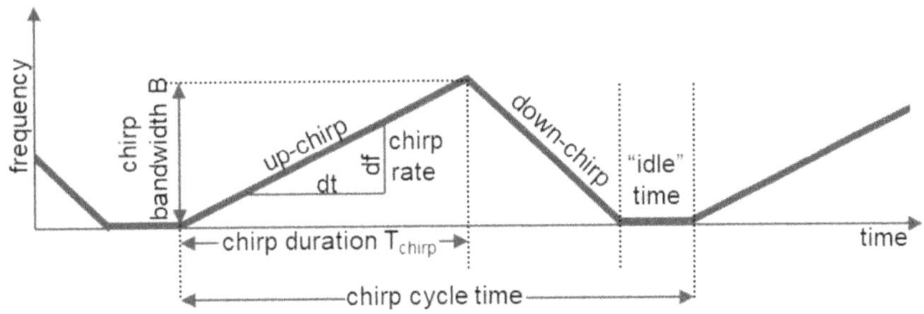

Figure 11.3 FMCW waveform parameters.

transmitted. Well-defined frames for coherent integration and transmission pauses for processing (and possible mitigation of collocated emitters) are defined as seen in Figure 11.4.

Some regulatory legislation relates to duty cycle measurements of a given transmitter. For example, the power access duty cycle is the percentage of active transmission over the total cycle time of an emitter and therefore is a metric (especially when communication channels could be influenced) for the occupancy time of that channel. Some examples of common methodologies and terms follow.

- *Power access duty cycle* is the ratio of the cumulated sum of transmission time per observation period. For a radar, the observation period normally is the frame interval time or pulse repetition interval.
- *Dwell time* describes the time interval for which a certain frequency range is used. Not to be confused with the time that the target of a scanning radar is inside the 3 dB beamwidth.
- *Spectrum access duty cycle* is defined for a particular spectral band and indicates the percentage of the observation time the emitter under test (EUT) has active transmissions in the observed band.

The measurement of those time and frequency-related parameters is normally quite straightforward and can be provided by different instrumentation (e.g., spectrum analyzer and zero span or oscilloscope including FFT analysis). The main differences are in the degree of automation or manual effort analyzing the corresponding values.

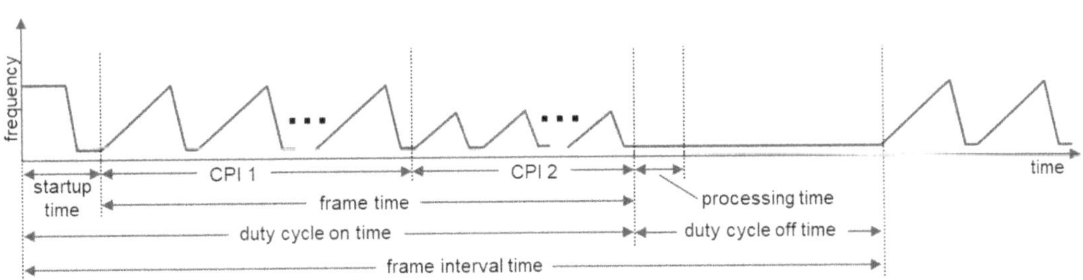

Figure 11.4 FMCW frame parameters for chirp sequences.

Dedicated interference mitigation techniques, the requirement for orthogonal waveforms for MIMO processing or information transmitted for cooperative radar are reasons for more complex digital modulation schemes like PM-FMCW, PMCW, or OFDM. In order to fully characterize digitally modulated signals, the signal must be demodulated. Key parameters here are the error vector magnitude (EVM) and the bit error rate (which is useful for determining that there is a problem, but not necessarily indicative of where the problem may exist), carrier frequency error, IQ-offset, IQ-imbalance, and quadrature error.

Many radar implementations today use a real mixer and real baseband and ADC chain with digitally obtained quadrature channels, as seen in Figure 11.5(a). This is partly motivated by the cost advantages obtained by not having to double some functional blocks in those architectures.

Alternatively, the many newer radar chips incorporate a quadrature mixer with complex-baseband processing (as seen in Figure 11.5(b)). Here, the received signal mixes with the LO and a 90° phase-shifted path in parallel. A duplicated IF filter chain and ADC for the in-phase (I) and quadrature (Q) channels provide the complex baseband signal [4].

This architecture brings several advantages, including an improvement in noise figure performance (up to 3 dB in theory) by eliminating the image band noise foldback to in-band. A possibly more important advantage is the implementation of advanced interference mitigation techniques. In a FMCW radar, the image band contains only noise and is free of any desired signal. Therefore, with a complex-baseband implementation, the image-band spectrum can be monitored to detect interference and/or estimate the thermal noise level accurately without clutter.

Both implementations come with an additional challenge in the form of IQ calibration. When utilizing a real valued baseband with digital transformation, this imbalance originates from effects like ADC clock modulation or inadequate isolation of the susceptible ADC connections (e.g., clock, power, ground) from other inadvertent signal emitters.

In a quadrature receiver, the real and imaginary part of the complex radar signal are sampled with two different physical channels. The nonidealities of the RF components, like low noise amplifier, phase shifter, mixer, and filter, as well as crosstalk between circuits, lead to an unwanted amplification and phase-shift in all

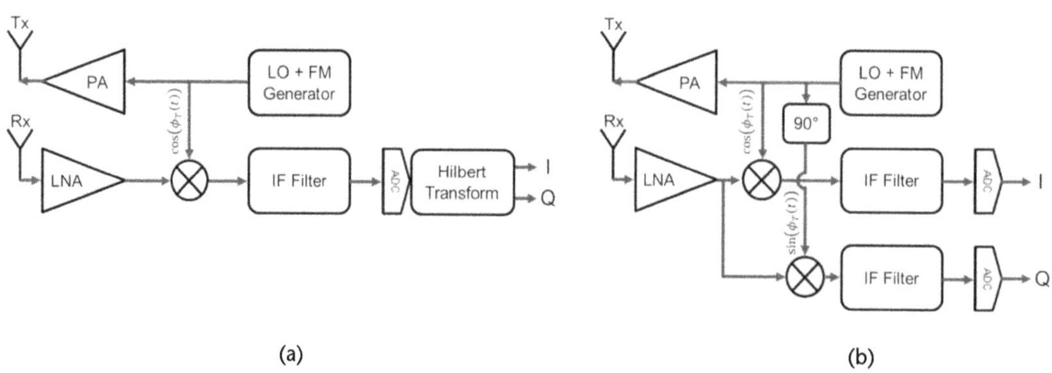

Figure 11.5 Digital (a) versus analog complex baseband (b) homodyne FMCW architecture.

I and Q channels. In a range-Doppler image, this will manifest as both an opposite range offset, and an opposite Doppler offset (i.e., appearing as a ghost image reflected through the zero-range and zero-Doppler point) [5].

In addition to the IQ imbalance of individual channels (single-channel IQ imbalance), an IQ mismatch between channels (array imbalance) is possible. Array imbalance can be caused by the same reasons as discussed, but also by antenna position errors, mutual coupling between different channels, and dielectric lens.

Another fundamentally different approach that requires an analog baseband processing is the class of PMCW radar as seen in Figure 11.6. In contrast to FMCW, a digital modulated and coded signal is transmitted that can be correlated via a matched filter in the receiver and provides direct time-of-flight information for range measurement without using a beat frequency. That poses additional requirements for the clock distribution and the full complex baseband has to be sampled compared the bandwidth of beat frequency sampling.

Regarding the modulations schemes in use for phase coding, the most relevant may be binary phase shift keying (BPSK) and Gaussian minimum shift keying (GMSK), both carrying only one bit per symbol and are robust to fluctuations of channel quality. Methods and metrology used in communication standards can be used to estimate the signal quality after modulation (see Figure 11.7 for an exemplary GMSK modulated signal).

The vector or polar diagram indicates the instantaneous value of the carrier signal at any point in time including the symbol-clock transition, as well as the transition between each decision point. The constellation diagram on the other hand focuses on the decision points defined by a modulation dependent pattern of phases and amplitudes in the complex plane. Therefore, both the ideal (reference) signal and the actual measured signal can be compared on the same plot. Typical effects that can be identified using a constellation diagram are:

- Gaussian noise causes the samples to land in a random ball about each constellation point;
- Noncoherent single frequency interference shows as samples making circles about each constellation point;
- Phase noise shows as constellation points spreading into arcs centered on the origin;

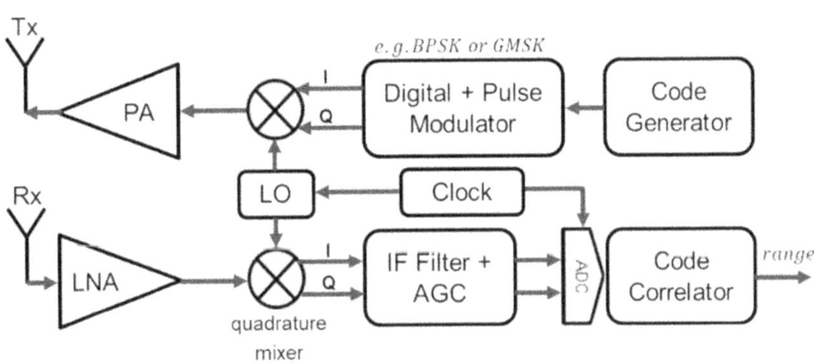

Figure 11.6 Simplified architecture of a PMCW architecture.

- Amplifier compression causes the corner points to move toward the center.

Closely related to the constellation diagram is the EVM, that is a single value indicator of the performance of a digital transmit or receive system. The error vector is the difference between the reference and measured data vectors, sometimes normalized to the peak signal level. From the captured set of error vectors, the RMS and peak EVM scalar values can be calculated either in dB or as a percentage:

$$EVM\ (\mathrm{dB}) = 10\log_{10}\left(\frac{P_{\mathrm{error}\ RMS}}{P_{\mathrm{reference}}}\right) \quad (11.4)$$

$$EVM\ (\%) = \sqrt{\frac{P_{\mathrm{error}\ RMS}}{P_{\mathrm{reference}}}} \cdot 100\% \quad (11.5)$$

Breaking EVM down into magnitude and phase components and comparing the dominant error provides a reference for error diagnosis. When average magnitude error is the dominant error mode, the undesired amplitude modulation could be a result of amplitude-to-amplitude (AM-AM) distortion, transistor biasing effects, thermal effects, and memory effects. Conversely, if the average phase error is the dominant error mode, the undesired phase modulation could be the result of amplitude-to-phase (AM-PM) distortion, noise, spurious and/or cross-coupling problems in frequency, or other frequency generating stages. The aforementioned IQ imbalance is indicated when the constellation is not "square," that is, when the Q-axis height does not equal the I-axis width. Quadrature error is seen in any "tilt" to the constellation.

Another tool to evaluate digital waveforms is the eye diagram as seen in Figure 11.8. The digital signal is repetitively sampled and applied to the vertical input, while the data rate is used to trigger the horizontal sweep. The triggering performed normally is a key to detect a problem. With the eye diagram it is possible to measure effects like noise or jitter, intersymbol interference (ISI), asymmetries in the waveforms, multipathing, impact of pulse shaping (e.g., root-raised-cosine or Gaussian), and more. It can be compared to spectral measurements masks, so violations of forbidden areas can be identified automatically.

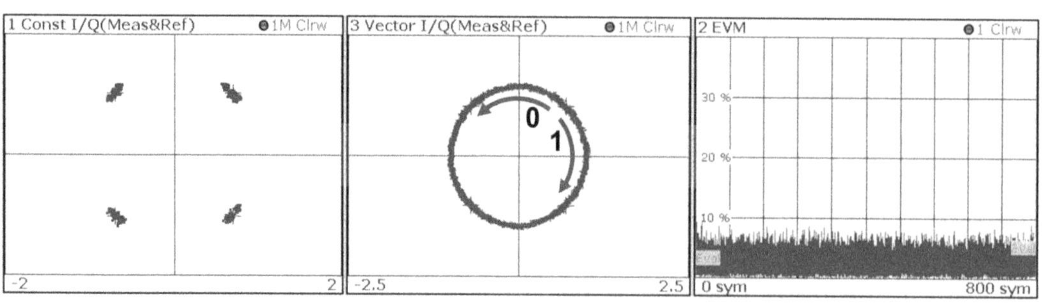

(a) (b) (c)

Figure 11.7 For setting up (a) the constellation diagram or (b) vector diagram of a GMSK modulated digital signal or (c) calculation of EVM several parameters of the modulation scheme, like symbol rate, transmit filter, and signal structure (burst, continuous, or pattern) have to be known.

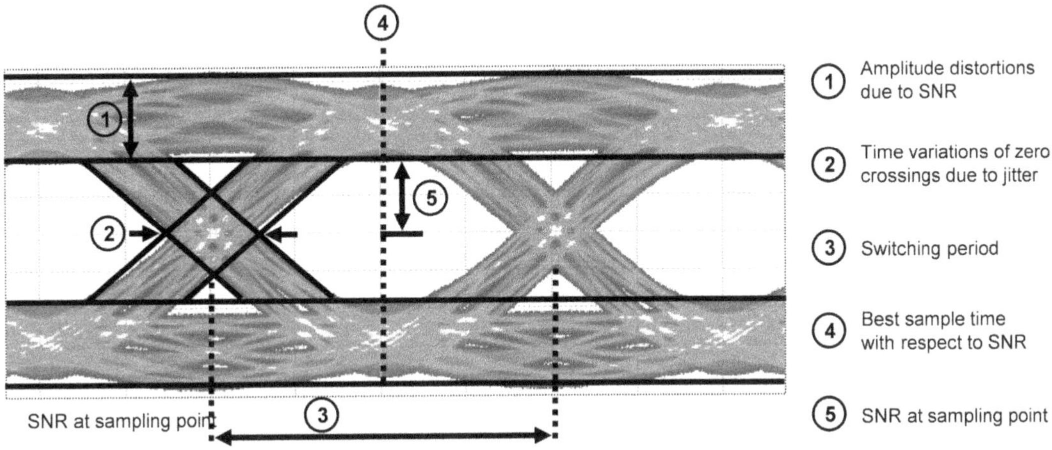

Figure 11.8 Eye diagram of a BPSK signal and some interpretation key points for measurement of binary-coded content.

11.2.1.3 Multiplexing-Related Measurements

Azimuth and elevation angles to a target are measured by using several transmitting and receiving antennas. To estimate the angle, a radar measures the phase difference of a received signal at multiple antennas, and by increasing the number of antenna elements, resolution and accuracy of the angular measurement improves. Today many automotive radars apply MIMO radar signal processing to improve angular resolution by using a virtual array to increase aperture size (see Chapter 4, Automotive MIMO Radar).

In order for the receivers to distinguish between the various transmitter signals or orthogonality in other words, several different approaches are used:

- TDM;
- FDM and DDM;
- CDM (both fast time and slow time);
- OFDM.

Each of these multiplexing techniques differ in their range-angle and range-Doppler coupling as well as their cross- and autocorrelation functions. It is necessary to test the transmit channels relative to each other in terms of time and frequency domain. For TDM and FDM the requirements are easy to understand, as orthogonality is achieved by simply shifting the signals of the different transmit channels in the respective domain. A test setup for TDM would require an oscilloscope, as only a single channel is transmitting at a time, and for FDM a spectrum analyzer can be used as each transmit channel uses an exclusive frequency band.

On the other hand, CDM and OFDM are orthogonal due to coded and correlated waveforms and separation of the transmit signals during measurement by frequency or timing is not possible. As there often is no connector on a radar module for a conducted measurement, it is either required to use RF probing or using a dedicated test mode of the MIMO radar, that only transmit sequentially through the channels.

Due to their highly flexible and statistical nature, OFDM signals cannot be measured like comparable analytical waveforms. But choosing the appropriate hardware design and OFDM parameters is crucial for several reasons:

- Deorthogonalized received signal due to Doppler shift by moving targets.
- Peak to average power ratio (PAPR) of the transmitted waveform imposes linearity demands on the signal chain.
- As a full-duplex system internal decoupling and crosstalk between antennas is an issue. An analog low pass filtering of the internal coupling, as used in FMCW radar, is not sufficient.
- Efficiency loss due to the guard interval, that determines the lower limit for the symbol repetition interval (SRI) and thus the unambiguously measurable velocity range.
- ISI from targets with a high RCS outside the distance range of interest.
- Time and frequency synchronization might impact receiver dynamic range and channel isolation [6].

Some transmit signal measurements require the instrumentation to demodulate the signal first. For this, a generic OFDM demodulator is required as there is no specific standard (like Wireless Local Area Network (WLAN) or Worldwide Interoperability for Microwave Access (WiMAX) in communication). Instead the OFDM demodulator has to be freely configurable (e.g., center frequency, level, guard interval, or FFT length have to be adjustable). Furthermore, in case of integrated data transmission, the structure of the preamble, the pilot, as well as position and modulation type of the data carriers must be defined to analyze parameters such as EVM versus carrier or constellation diagram.

11.2.1.4 Power Measurements

Power measurements are normally part of a regulatory framework that manages the use of the RF spectrum and ensures human safety. Therefore, often upper limits for peak or average power are defined.

All power measurements can be done with a signal and spectrum analyzer in spectrum mode and the time domain power measurement or a power meter. In the spectrum analyzer, the time domain power measurement is selected and configured similar to the previously discussed operating frequency range measurement. Some notes on how to set up different instruments for those measurements are given in Section 11.3.

It is critical to know the type of power measurement to be performed, in order to select the right equipment and respective settings. The following definition of terminology can help to make the selection process easier.

Instantaneous Power $P(t)$ is the signal power at any given instant of time. On a test instrument, it would look like a reproduction of the modulated waveform. Capturing instantaneous power requires instrumentation that samples a bandwidth limited signal fast enough to fulfill the Nyquist criterion. For an automotive radar with transmit frequencies in the microwave range, that approach normally is not

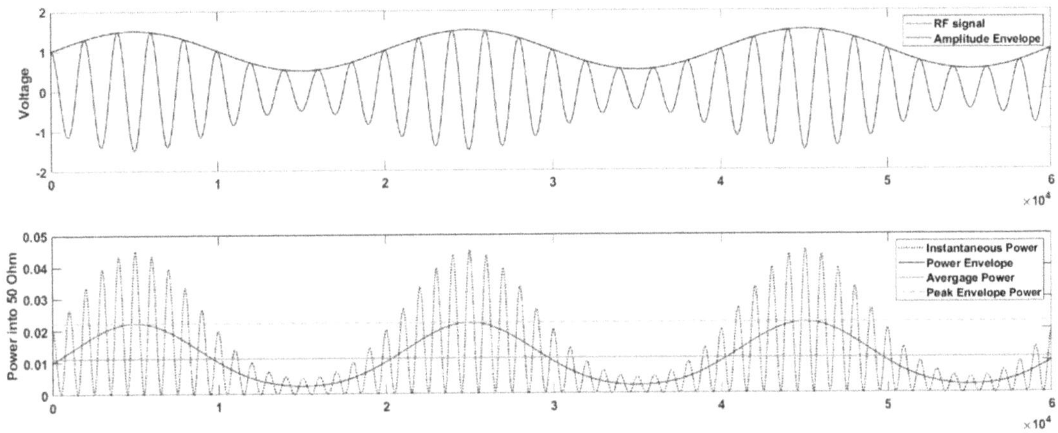

Figure 11.9 Shown are the (a) amplitude or voltage of an RF signal with (b) the correlated power parameters.

feasible, but more on the connection between dynamic range of an instrument and sample rate in Section 11.3.

The *average power* P_{Avg} of a modulated signal is determined by squaring the voltage trace $U_{RF}(t)$ (say, the according samples u_i) and then dividing it by the reference impedance. The value averaged over many periods of the modulated signal (measurement time) yields the average power of the signal. The example in Figure 11.9 illustrates an AM modulated RF signal with a modulation depth of $m = 0.5$ and the respective sinusoidal amplitude envelope. Additionally, the squared trace referenced to 50 Ω (dotted) and the average power (solid light) are shown.

$$U_{RMS} = \sqrt{\frac{1}{N} \cdot \sum_{i=1}^{N} u_i^2} \tag{11.6}$$

$$P_{Avg} = \frac{U_{RMS}^2}{Z_{REF}} \tag{11.7}$$

When continuously measuring the average power of a signal over a single RF period, the resulting trace is the *power envelope* $P_{env}(t)$ depicted with solid dark line. $Pe(t)$ describes the effect a modulation or transient condition have on a signal's amplitude, without examining details of the RF carrier waveform.

The *peak envelope power* P_{PEP} is the average power that the transmitter outputs during one period of the RF signal at the maximum value of the envelope. For a signal with a constant envelope (e.g., a pure FMCW signal), the peak envelope power is equal to the average power. In most cases a "peak" power measurement refers to peak envelope power and not the peak instantaneous power.

11.2.1.5 Effective Isotropic and Total Radiated Power

Effective isotropic radiated power (EIRP), also called the equivalent isotropic radiated power, is the measured radiated instantaneous power in a given direction (θ,φ) relative to the emitter under test (i.e., the transmit power supplied to the antenna multiplied by the antenna gain in a given direction relative to an isotropic antenna) (see Figure 11.10). When all angles of an antenna are measured, this provides the antenna radiation pattern as discussed under Section 11.2.3. For regulatory-related measurements normally a single value of EIRP is given: the maximum value of the EIRP(θ,φ) over all measured angles in the permitted range of operating frequencies. Depending on the instrument settings, measurements of peak or average power or the power spectral density can be related to the directional effect of the transmit antenna.

Total radiated power (TRP) describes the sum of all power radiated by an antenna connected to a transmitter. The main purpose of TRP measurement is to evaluate the performance of an antenna in a real environment and is closely related to the efficiency of the antenna, the only difference being that is not a function of the antenna only, but also of the transmitter and the connection in between. The TRP would be equal with the conducted power (antenna input power) if an ideal isotropic antenna was utilized, but when using a real antenna, the TRP is weaker than the conducted power and depending on the antenna efficiency. Therefore, it is also relevant for the PA specification to provide enough output power to overcome mismatch loss. Because TRP performance is impacted by the antenna and the PA, the PA must include compensation circuitry that counteracts the effects of load-impedance variation presented by the antenna [7].

The relevant power density entity for power radiating out of a surface is the component of the Poynting vector that is perpendicular to the surface. This entity is denoted power density:

$$TRP = \int_0^{2\pi}\int_0^{\pi} P_D(r,\theta,\phi) r^2 \sin(\theta)\, d\theta\, d\phi \tag{11.8}$$

Equivalent isotropic radiated power (EIRP) is defined at infinite distance and is related to the power density, given that r is beyond the far-field distance, with [8]:

$$TRP = \frac{1}{4\pi}\iint_0^{2\pi} EIRP(\theta,\varphi)\sin(\theta)\, d\theta\, d\varphi \tag{11.9}$$

With discrete measurement points distributed over an equidistant surface area that translates to:

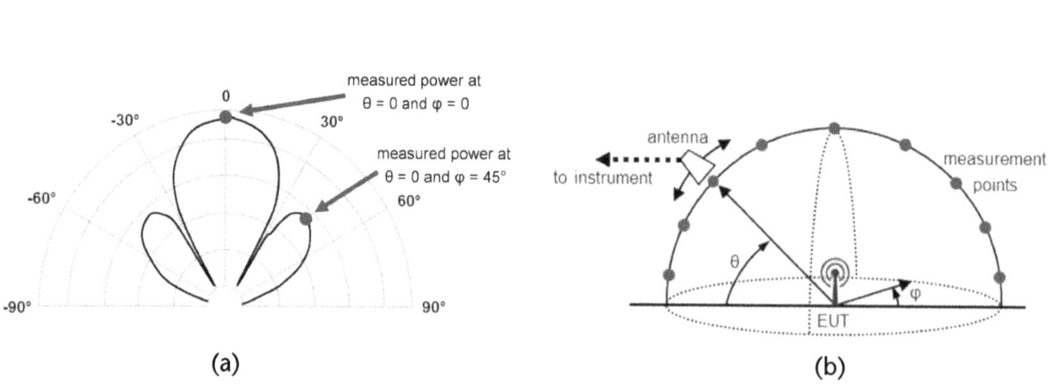

Figure 11.10 (a) EIRP as a function of angle dependence of the antenna diagram. (b) Definition of TRP with spherical measurement coordinates around EUT.

$$TRP = \frac{1}{N_{th}N_{ph}} \sum_{i=0}^{N_{th}-1} \sum_{j=0}^{N_{ph}-1} EIRP(i,j) \tag{11.10}$$

In practice the actual measurement is a combination of antenna radiation pattern (Section 11.2.3) and average EIRP measurements.

11.2.1.6 Crest Factor, Complementary Cumulative Distribution Function, Peak Envelope Power (PEP), and Peak Average Power Ratios

Power testing is important for regulatory and reasons, especially when designing complex radar waveforms. Where a pure FMCW modulated signal has a more or less a constant power envelope, novel modulation schemes (e.g., OFDM [9]), similar to communication signals, have increasingly higher peak average power ratios (PAPR). This ratio relates to the crest factor (CF) (the linear peak to RMS amplitude ratio, for example, using voltage x_{RMS}) of a signal and might pose some challenges in system design as the requirements in terms of dynamic range of ADC and linearity for amplifiers increases accordingly and might have an impact on overall system power efficiency (envelope tracking).

$$PAPR_{\mathrm{dB}} = CF_{\mathrm{dB}} = 20\log_{10}\left(\frac{|x_{\mathrm{peak}}|}{x_{RMS}}\right) = 10\log_{10}\left(\frac{P_{\mathrm{peak}}}{P_{avg}}\right) \tag{11.11}$$

Therefore, efforts in waveform design to reduce CF or PAPR (e.g., like clipping, special coding or selected mapping) have to be tested after simulation in terms of performance.

With periodic signals it is possible to determine the CF by measuring the peak and average or RMS values of the voltage or power using a state-of-the-art power meter or oscilloscope. If the signal is measured for the duration of one period, it is possible to completely measure the peak value as well as the RMS value, assuming the test equipment has sufficient speed for this task. This is not the case with signals with random modulation such as an OFDM signal. While the average value can be determined precisely in only a few seconds using a thermal power meter, the magnitude of the peak value is highly dependent on the measurement time. With high CFs, the signal peaks are rarer, so it is necessary to specify the CF that was determined along with the measurement interval or the number of samples. The CF determined is representative for the specified measurement interval only and does generally valid represent the entire turn-on interval.

A statistical way of quantifying those effects is the complementary cumulative distribution function (CCDF), which provides information on the probability of the signal being at or exceeding a given power level. This is expressed in either dB relative to the average power or as an absolute value in dBm (Figure 11.11(a)). The CCDF measurement is the integral of the distribution function versus the level (integration of the observed level to infinity) and therefore emphasizes the maximum, or peak power values. Therefore, in Figure 11.11(b), the more the CCDF curve moves to the right, the more frequently the signal takes on values that are significantly above the average value.

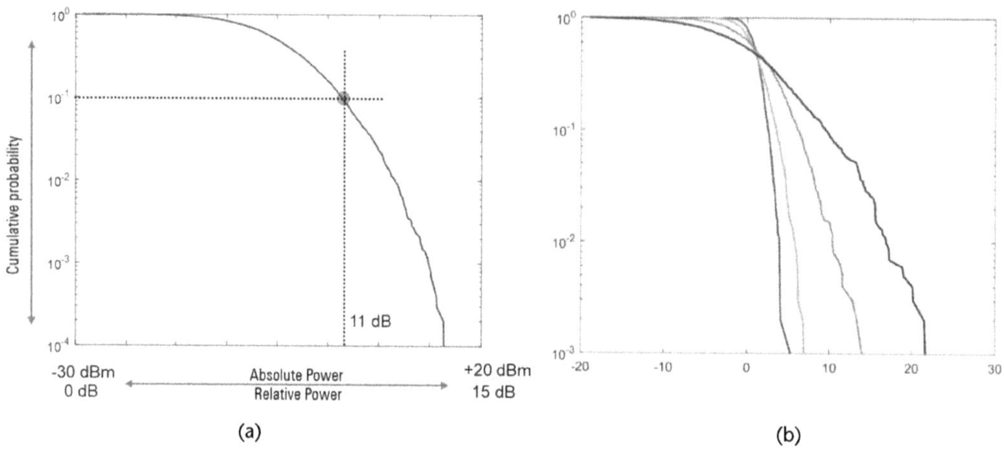

Figure 11.11 Shown are (a) definition of the CCDF and (b) examples of signals with different PAPRs.

In the example depicted in Figure 11.12, amplifier linearity is determined by applying a modulated signal and measuring the CFs of both the input and output signals. The crest factors or PAPR of the initial graph Figure 11.12(a) are very close, indicating linearity. However, the resultant graph Figure 11.12(b) of increasing the input power to the amplifier causes a discrepancy between the ratios, indicating nonlinearity. Accordingly, the same can be seen when comparing the CCDF of the input and output signals. If the amplifier output is distorted by compression, the CCDF curves will be increasingly different. This same methodology can be used with many other active and passive components.

11.2.1.7 FMCW Signal Quality

Radar performance relies on accurate and reliable target detection. A FMCW radar uses a linearly swept frequency sources to generate the transmit signal. One design choice for the FMCW is the transmit signal source and the kind of signal generation.

Typical sweeper implementations include:

- Open loop VCO;
- VCO and frequency discriminator.

VCO implementations typically use some form of signal corrections, such as a look up table (LUT) and with precalibrated and stored control for the data DAC.

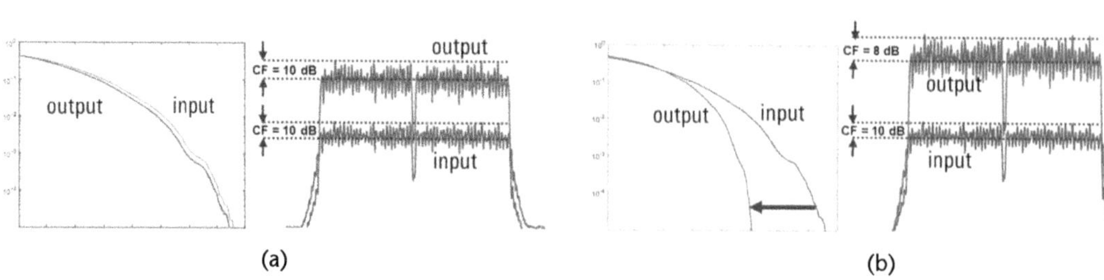

Figure 11.12 CCDF and CF of a power amplifier's input and output signal. In (a) the PA operates in linear region, and in (b) input power drives PA into saturation and nonlinearity.

However, the calibration is often not static over time and temperature changes limit the achievable linearity. VCOs with a frequency discriminator generate a voltage output signal proportional to frequency which is fed back as a closed loop correction. Unfortunately, these analogue frequency discriminators have limited performance, which is why the required signal performance is often not achieved in wide-band radar systems.

Synthesizer subsystems include:

- Phase locked loop (PLL) synthesizer;
- Direct digital synthesizer (DDS).

PLLs generate an output signal with phase related to an input signal. This output signal is fed back over a frequency divider and then compared to a reference signal using a phase detector (or a time-to-digital converters if using a purely digital approach) to drive the VCO. One drawback of PLLs is the increase of phase noise, which should be kept as low as possible in radar systems.

By contrast, a DDS creates the output signal by reconstruction based on a look-up table (which contains amplitude values as a function of the phase) and a digital to analog converter. A typical disadvantage of the DDS approach is the increased the level of spurious emissions due to frequency multiplication. To limit these emissions, the multiplication factor should be kept as low as possible [10].

For slow to moderate chirp rates, the resulting baseband signal falls into the close-in phase noise region of the PLL (Figure 11.13). To reliably detect a target, a certain signal-to-noise ratio must be given at the mixer output, that is, the signal power of the beat frequency associated with the target must rise above the phase noise (the "skirt") level by a certain amount. Low close-in phase noise is therefore imperative to ensure high detection sensitivity of the radar. Furthermore, detection reliability can be improved by generating fast chirp rates because this results in higher beat frequencies for the same radar scenario and potentially pushes them out of the 1/*f* noise and up toward lower phase noise, increasing the baseband signal-to-noise ratio as well. Faster chirps, however, require fast PLL settling and thus again a wide loop bandwidth [11]. Unfortunately, achieving low close-in phase noise with wide PLL bandwidths presents a fundamental design trade-off with conventional

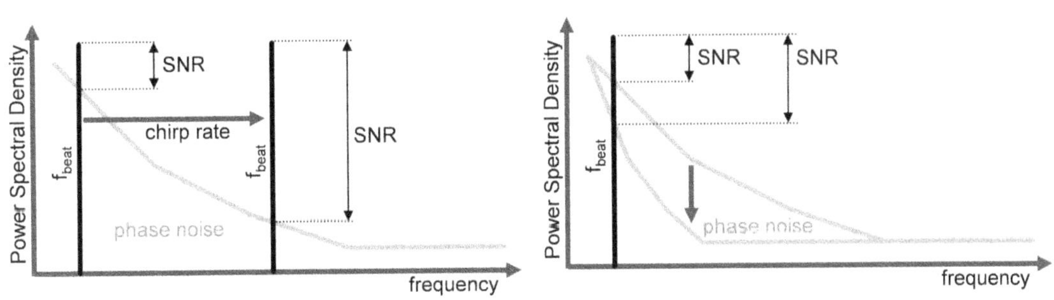

Figure 11.13 To improve SNR and target detection reliability (a) a faster chirp rate or (b) an RF design with lower phase noise can be used.

PLLs—a wide loop bandwidth allows more noise from the phase detector to contribute to the output phase noise.

Phase noise describes the effects of the frequency synthesizer at a constant frequency, but when it comes to the accuracy and resolution of the radar sensor, linearity and transient observation of the signal are important as well. Any unintentional nonlinearity in the receive signal that is mixed with the nonlinear transmit signal may cause the frequency bin to become broader than expected (compare Figure 11.14). For example, when saturating at a higher frequency, long range targets are more affected than short range targets because the radar mixes more linear parts with nonlinear parts of the signals. This is not the case for close range targets [10].

Slow frequency deviation from a perfect linear signal slope over a certain bandwidth may occur as depicted in Figure 11.14(a). Due to down-conversion of the receive signal with the instantaneous transmit frequency, the beat frequency will exhibit a trend. Hence, the Fourier transformed signal will result in a broader frequency peak. This decreases range and radial velocity parameter estimation accuracy and resolution, as the beat frequency measurement is less accurate.

The linearity of the transmit signal becomes even more important for targets that are located at longer ranges, as the receive signal may be down-converted with another frequency than expected. This could even result in false range estimation, as the wrong beat frequency is measured.

Another effect on transmit signals are ripples, as illustrated in Figure 11.14(b). This frequency deviation affects the accuracy of the beat frequency measurement and causes unwanted side-lobes to appear in the IF signal spectrum. The beat frequency measured by down-conversion and Fourier transformation will result in

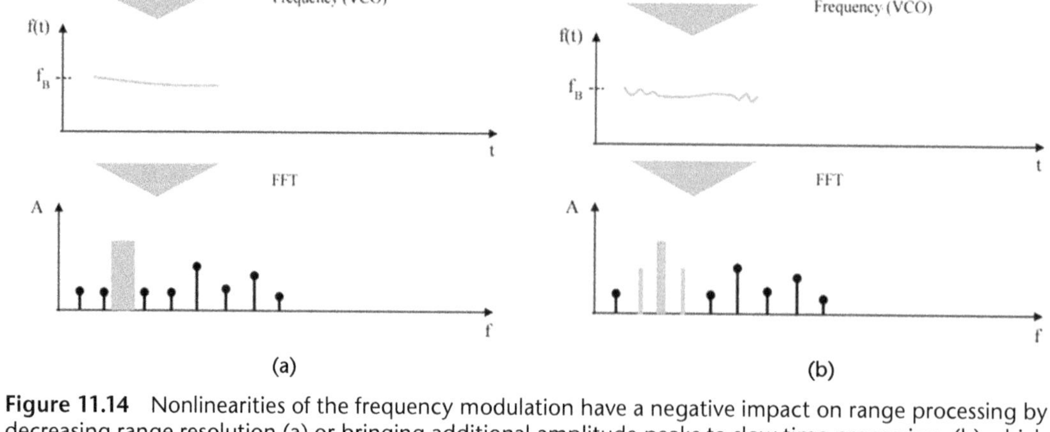

Figure 11.14 Nonlinearities of the frequency modulation have a negative impact on range processing by decreasing range resolution (a) or bringing additional amplitude peaks to slow time processing, (b) which may induce errors in processing.

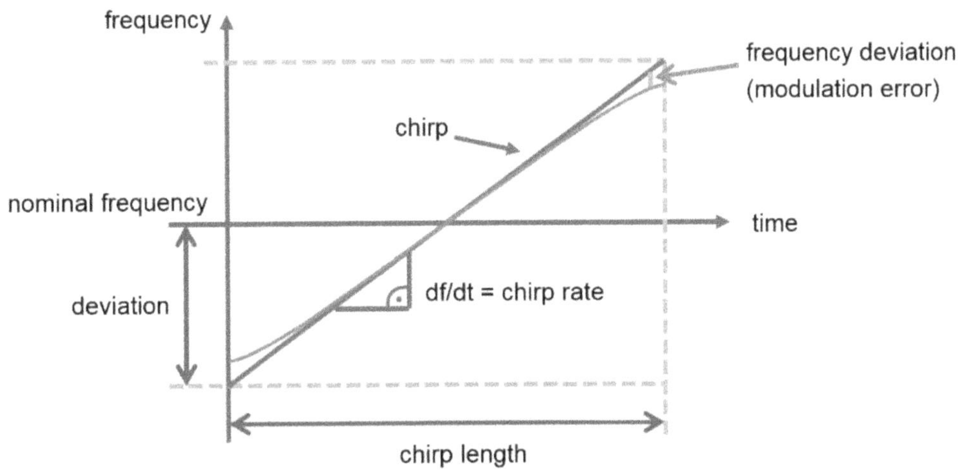

Figure 11.15 Linearity of an FMCW chirp describes the deviation of the real signal from the nominal instantaneous frequency.

a wider frequency peak in the Fourier spectrum compared to the transmission of perfectly linear ramps. Hence resolution in both domains (range resolution, radial velocity resolution) and accuracy are degraded during the FMCW signal processing.

The most intuitive way to measure a performance indicator for FM linearity is to reference the signal to a nominal frequency and determine the deviation as seen in Figure 11.15.

Note that the term *deviation* is unfortunately used with two different meanings: Deviation from nominal frequency due to modulation and deviation of the actual modulation frequency from the intended ideal modulation frequency (i.e., modulation error). Also, deviation is sometimes defined from minimum chirp frequency to nominal frequency as shown here and sometimes from minimum chirp frequency to maximum chirp frequency (factor of two).

Another parameter of interest is defined by FM/PM settling time or how long the FM/PM signal takes to remain within a specified tolerance around the nominal frequency (Figure 11.16) after a transition. The limitation here is that this only represents the effect of one frequency change at a given measurement time and therefore is most suitable for hopping signals.

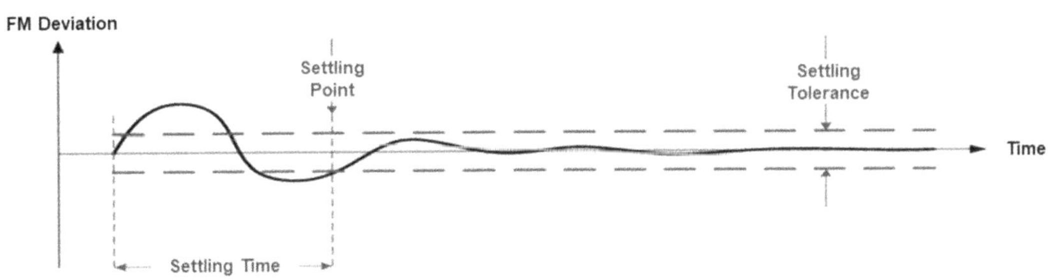

Figure 11.16 Definition of settling time, for example, after a change of an operating frequency.

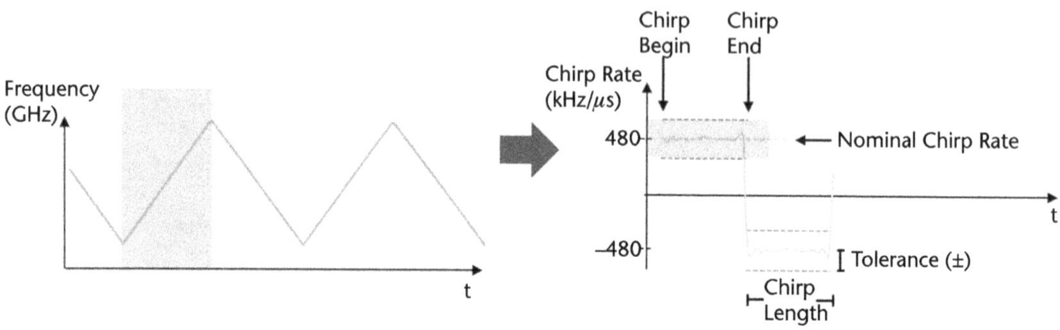

Figure 11.17 For FMCW normally transient quality is of importance. In this example, the chirp rate is compared to a reference value of the duration of a given chirp. This requires that the chirp length and the nominal chirp rates are known.

With a FMCW signal it is more interesting to observe the linear behavior of the signals over the full CPI of a chirp or frame. Here it is beneficial to use the deviation of the chirp rate over time as an indirect indication of the signals linearity (Figure 11.17). Any deviation beyond the tolerance is directly attributable to a certain time/frequency of the FM ramp.

11.2.2 Receiver Test

Testing integrated radar chip receivers involves two separate dimensions, as seen in Figure 11.18. There are analog effects and specifications of the RF frontend, e.g. noise figure or phase noise, intermodulation or mutual coupling of the RF channels. Unfortunately, testing of those parameters are normally only possible at chip level requiring dedicated probing on waver level or relying on the self-monitoring capabilities of the chip directly.

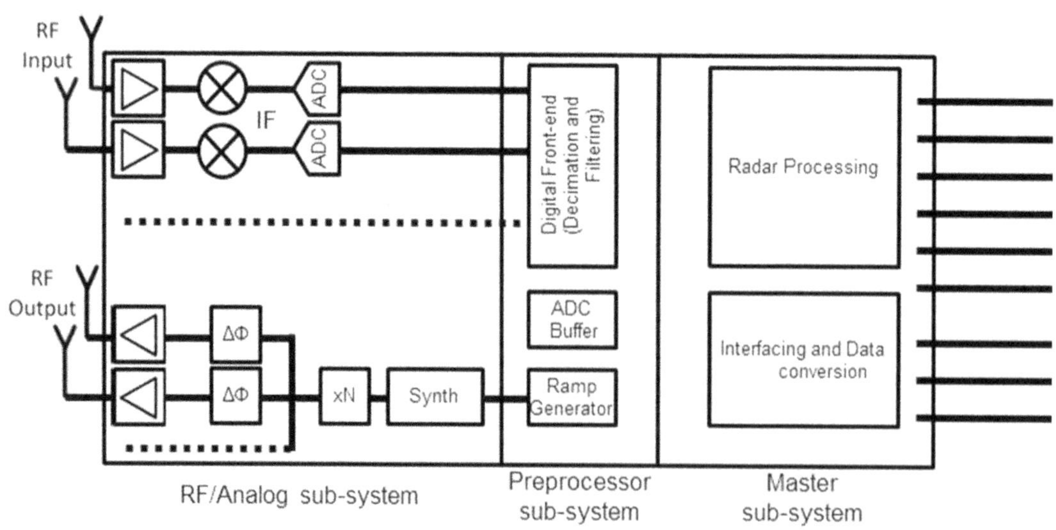

Figure 11.18 Simplified block diagram of an automotive radar sensor with integrated DSP, MCU, and radar accelerator [12].

On the other hand, the receiver can be tested on a quantified functional level by simulating a defined input signal under different conditions and monitoring of the processed (or if possible raw) radar data.

11.2.2.1 Receiver Spurious Emissions

A radar receiver might act as an emitter for spurious emissions; this can be the result from the interaction of the receiver with the coupling transmitter's signal. As part of a regulatory framework receiver spurious emissions normally should be tested in a mode other than the transmit mode. This is important because certain receiver stages, such as low noise amplifiers, can become unstable and tend to oscillate, causing radiated emissions outside the operational frequency of the radar. A receiver spurious emissions power measures the power emissions generated or amplified in a receiver that is transmitted via the RX antenna. The detection of spurious components normally involves measurements at different frequencies, angles, and polarizations. Some spurious emissions tests require a quasi-peak detector of the measurement receiver or signal analyzer.

11.2.2.2 Receiver Sensitivity, Saturation, Dynamic Range

To quantify the analog performance of the receiving path of an automotive radar, sensitivity is a parameter that to a certain degree can be evaluated independently from signal processing. By using an approach to simulate an artificial radar echo (see Section 11.3.2) with a well-defined RCS and range, the minimum signal level detectable by the radar sensor can be evaluated indirectly. The performance could therefore be expressed as a minimum detectable RCS at a given distance as often specified in relevant test scenarios (see Section 11.4.4).

At the other extreme the analog frontend has to handle high signal levels, such as a high RCS target in close vicinity to the sensor. At certain levels harmonics or intermodulation products of multitone targets (targets with different Doppler shifts) due to the radar's receiver mixing stage, may provide enough SNR at the ADC to be detected as ghost targets. This can also be evaluated using a radar target generator (RTG) and increasing RCS of the simulated target to the point where ghost targets appear in the object list after processing.

The MDS and the maximum level before spurious signals appear define the DR of the receiver, which is of utmost importance as it defines the capability of the receiver to handle strong signals (like a truck close to the sensor) in parallel with signals with low SNR (e.g., a bicycle 200m away). Aside from inducing spurious signals, a strong reflector (or a strong sum signal at the radars ADC) might also drive the receiver into saturation and therefore reduce the sensitivity directly.

11.2.2.3 Receiver Interfering Signal Handling

With respect to electromagnetic susceptibility or EMS testing (see Section 11.2.5) two fundamentally different effects have to be evaluated. One effect is how the receiver of the automotive radar handles interfering signals inside its operational frequencies. Using a strong CW signal to evaluate a saturated receiver has a limited

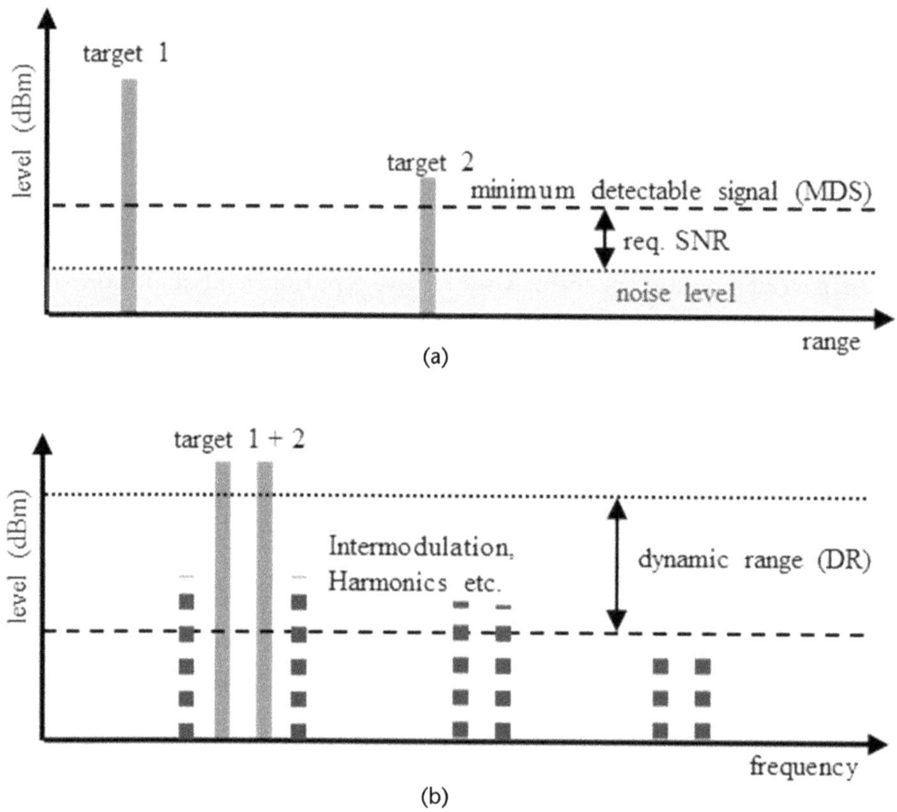

Figure 11.19 (a) For detecting a target the received power needs to be above the noise level by a defined SNR. The sensitivity is a mix of hardware (noise level) and processing (e.g., CFAR) parameters. (b) Definition of DR as the span between MDS and target input highest level before spurious signals are detectable.

relevance in real driving scenarios (nevertheless this is often part of a regulatory framework). Currently, in-band interference comes mostly from other automotive radars, and testing those is described in Section 11.2.4 in more detail.

On the other side high power out-of-band and remote-band emitters, like digital video broadcasting-terrestrial (DVB-T) broadcast, long-term evolution (LTE) base stations, or air traffic control (ATC) radar might impact the radar performance as well via radiative coupling. The conductive lines of the RUT act as antennas for frequencies of those strong emitters and therefore receive interfering signals. A validation of the correct operation of the sensor under these conditions therefore has to be validated.

For testing this type of interference handling, a signal with a defined field strength is transmitted at the RUT in parallel to a reference object (e.g., an RTG). Dependent on the underlying standard, the frequencies that have to be tested range from a few MHz up to several GHz. The combination of different aspect angles and antenna polarizations make that kind of measurement very time consuming, as for all these iterations, the correct function of the RUT has to be tested.

Possible effects or sensor behavior that can be observed when under harsh EMI conditions are shown on Figure 11.20.

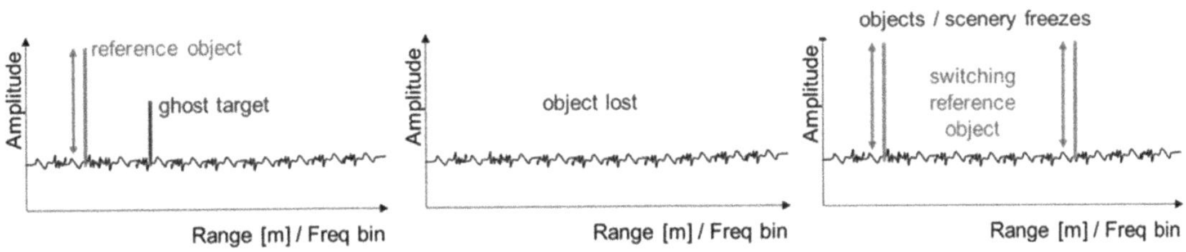

Figure 11.20 Possible errors due to external signals interfering with the radar receiver, like detection of ghost targets, drop of targets, and freeze of the complete radar module (this most likely happens with high power interferers).

Due to saturation of the radar receiver and a decreased SNR or other effects in signal processing, targets might be dropped (false negatives) or additional ghost targets might appear. Both effects can be evaluated by comparing the reference object(s) with the radar's object list. More severe effects impact the overall integrity of the radar module, potentially requiring a hard reset of the sensor. Scene freezing—where a physical change in the target does not update in signal processing—is one manifestation. This can be tested by switching between reference objects that are separated in range and comparing with the object list also.

11.2.3 Antenna and Radome Test

11.2.3.1 Antenna Pattern

The antenna radiation pattern describes the distribution of radiated energy into space under far-field condition as a function of direction. Key characteristics include half power beam width (HPBW), peak-to-sidelobe ratio (PSLR), antenna gain (TX), and sensitivity (RX), as can be seen in Figure 11.21. These parameters influence the FOV, unwanted multipath effects, and the required transmit power for realizing a maximum detection range. Regarding power, S_{11}-Paramenters or VSWR versus frequency are of interest too, as they provide information of the power added efficiency (PAE) of the radar when operating a given frequency.

The measuring setup described in Section 11.4.1 shows the basic principle of having a reference object with a determined RCS, a known and compensated air

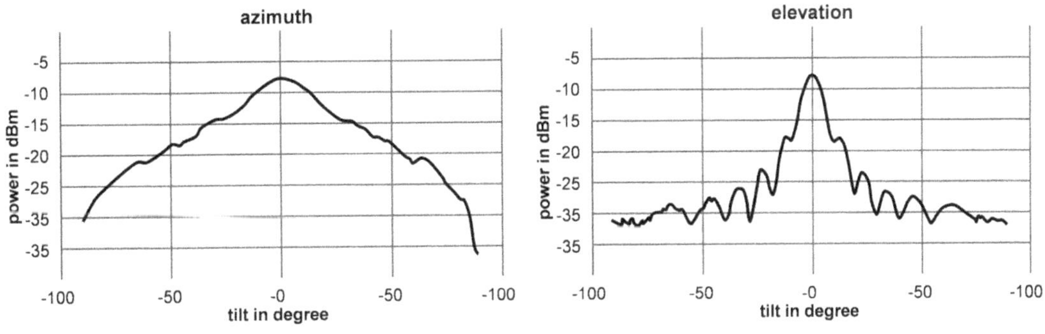

Figure 11.21 Exemplary antenna diagram of a radar transmitter (Note that EIRP and not gain on ordinate).

gap and a positioner that turns the RUT in azimuth and elevation. Some peculiarities exist in antenna measurements:

- One aspect is that the radar normally doesn't transmit a constant CW signal; the measured signal at the input of the power meter is time variant. This must be configured in the detector and poses additional challenges, when using complex waveforms like OFDM.
- The measured pattern actually is not the antenna diagram, but rather a "module EIRP." Additionally, with MIMO architectures, a geometric sum of all transmit antennas is measured.
- The measurement of the voltage standing wave ratio (VSWR) normally is not possible on module level. The antenna integration (the extreme would be antenna-on-package) doesn't allow a network analyzer to be connected without unrealistic effort. In that regard, the antenna is unity with the module.
- For the last two reasons, it is not possible to measure the receive antenna diagram. Neither directly nor using reciprocity of TX antenna (which would work only if identical anyway).

The module EIRP, that is the additive antenna diagram of the transmit group antenna times the transmit, is the often the only available figure of merit.

11.2.3.2 Radome Tests for Integration of Radar Sensors

One nontechnical aspect that makes radar sensors attractive for automotive applications is that they can be hidden behind nontransparent surfaces, like emblems or bumpers, maintaining a clean design of the vehicle. The drawback is depicted in Figure 11.22 as the materials used in manufacturing a radome are forming layers of different permittivity. This causes disturbances to the planar wave front that induce phase errors in between the elements of a receiver antenna array. Additionally,

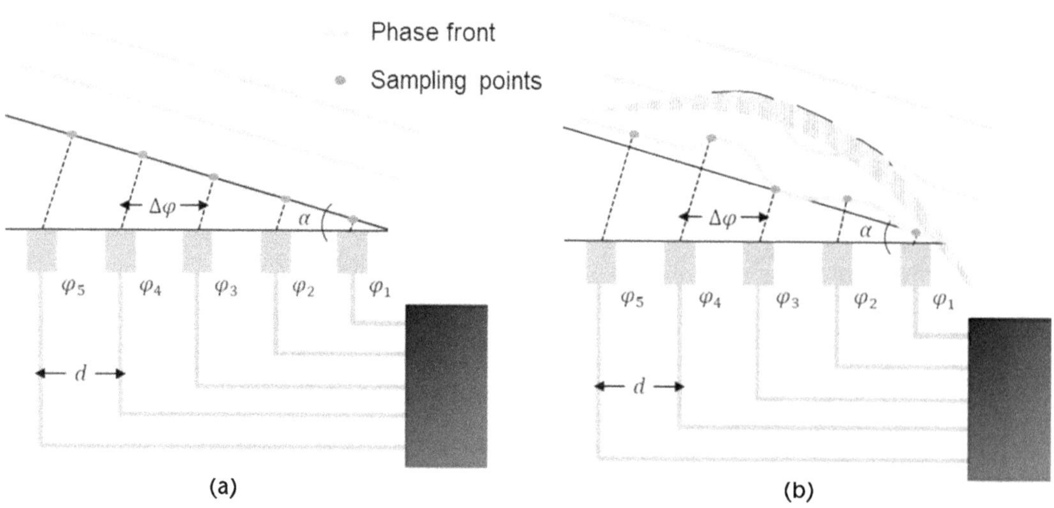

Figure 11.22 The incoming undisturbed wave front (a) is influenced in phase by bumper or facia material and (b) inducing errors in AOA processing.

unwanted microstructures in the radome material due to manufacturing, like air gaps or glue, can influence the radar. If the structures are systematic, sensor integration can be optimized by determining suitable positions.

No mathematical correlation is known to the author about how the radome's homogeneity precisely influences angular errors in the radar. This applies for both reflection and phase measurements. As every radar processes data differently, thresholds on angular information only apply to specific radar/radome combinations and all measured values only apply to the specific combination.

Key parameters for specifying a radome are the measurement of the materials transmission loss and the reflectivity. The most basic kind of radome is a sheet of a dielectric material with thickness d and permittivity ε_r. More complex radomes consist of multiple layers with different physical dimensions and material parameters, as seen in Figure 11.23.

The influence that the radomes transmission has on phase estimation leads to an additional error in the magnitude of up to 1° in AOA estimation. At long target distances this error can be enough for a lane mismatch and misinterpretation of the environment. The direct impact on the AOA processing cannot be determined by measuring the radome only—it depends on both the radar and radome. The reason for this is basically the different spatial antenna configurations of modules being influenced by different geometrical portions of the radome. Just because a radome is appropriate for a certain radar does not necessarily mean that it will also work well with another.

There are different approaches to characterize a radome's attributes:

- The most basic method is the use of several reference corner reflectors with a defined RCS spatially distributed in the FOV of the sensor. A first measurement using the radar as instrumentation without a radome acts as a reference. The second measurement is taken with the radome mounted. The results from these two measurements are subtracted from each other to determine the influence of the radome on radar performance. This intuitive method provides basic

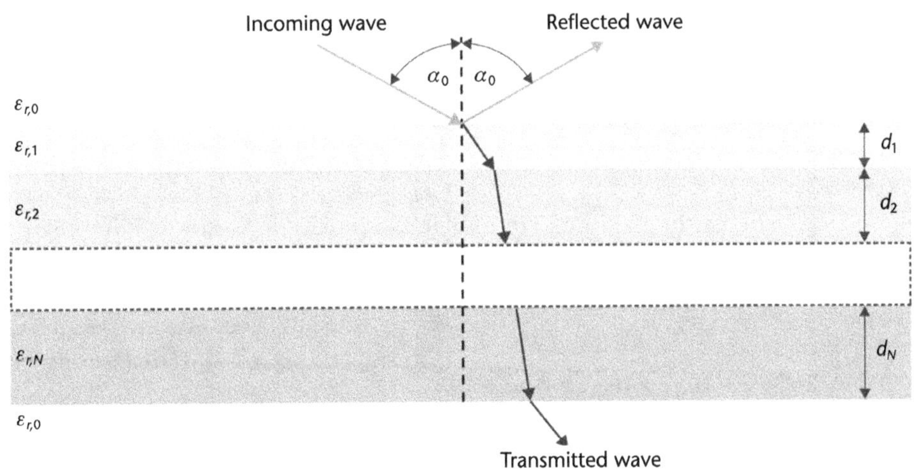

Figure 11.23 Effects on a transmitted electromagnetic wave due to a radome consisting of a multiple layer compound material.

information about radome homogeneity as transmission loss is evaluated at two to four random points (depending on the number of reflectors), whereas reflection is not measurable at all.

- Network analyzers produce a full set of S-parameters (reflection and transmission loss measured from both sides of the DUT). Network analyzers are the best instruments for obtaining the electrical parameters for a raw material. The analyzer operates over a wide frequency range and offers precise measurement results. Generating a spatially resolved picture of the radome usually takes a huge amount of time, with the measurement of one radome taking up to four to six hours depending on the required resolution.
- Imaging methods combine the advantages of measuring with a network analyzer and corner reflectors. With a large panel that consists of a massive number of transmit and receive antennas (and processing similar to a MIMO radar) a high-resolution image of the radomes properties is available. An external transmit antenna is required for transmission loss measurements. The intuitive results provide a quick and easy interpretation of the measurement of the radome under test.

11.2.4 Performance and Functional Tests

After bringing together all necessary components and processing into a radar module, two central questions arise. The first question is if the module's performance is meeting its specification, in terms of key performance indicators (KPIs) like dynamic, accuracy, and resolution in all measurable degrees of freedom.

From a design point of view, this would be described first, as it basically defines the module, but from a testing perspective it is normally from the simple to the complex, following the development process. Second question is whether the related vehicle function in its entirety is working as intended. Testing functions can be realized at different levels (e.g., interference mitigation can be tested at the chip, module, or architectural level), so sometimes it is fluid whether functional testing should be performed on a module level or as part of integration. The focus for this section is

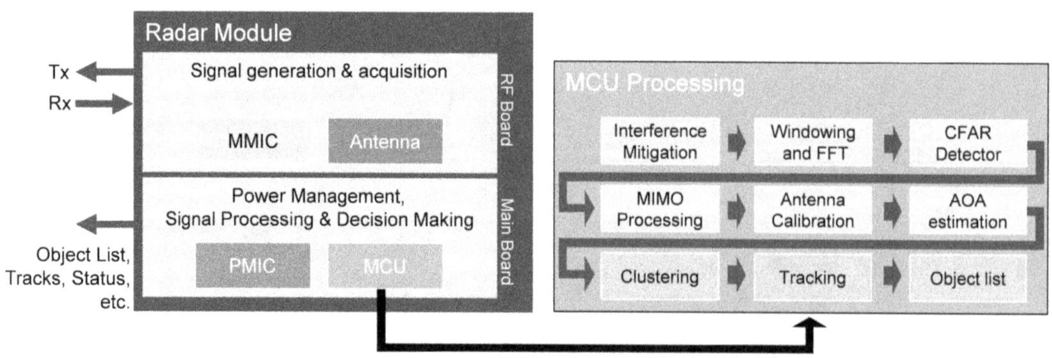

Figure 11.24 Generic architecture of a complete radar module. Module-based HiL testing utilizes instrumentation on the RF side and digital processed references (e.g., in the form of the object list on the other).

on module-based functions and module hardware-in-the-loop testing (HIL); higher level ADAS and vehicle integration testing are described later in this chapter.

11.2.4.1 Performance Testing of Radar Modules

Specifications provide the key performance indicators of a radar module and how to measure them. In contrast to functional testing, performance testing normally is not referenced to an object list (the result of several processing steps), but on the raw data of the sensor only after detection and Doppler and AOA processing. Before measuring the performance of a radar module, it has to be calibrated to compensate systematic imbalances and offsets (see Section 11.2.5).

The test setup for receiver DR applies, as described in Section 11.2.2, to determining the operational minimum and maximum detection range. With an RTG and targets at defined distances and a defined RCS, the limits of sensitivity, saturation, and dynamic range can be evaluated.

For evaluation of the range accuracy, the reference signal (e.g., RTG or a corner reflector) with a known distance is compared with the measured distance of the radar module. This should be repeated over the complete instrumented range, as often there is a relationship between accuracy of measurement and the distance of the target (e.g., by lower SNR). Range resolution is measured by using two simulated targets in the same AoA separated by a given distance and evaluating the object list.

For Doppler-related measurements (accuracy and resolution), the test setup is similar but brings up some additional requirements. Obviously, a static corner reflector is not able to induce a defined Doppler shift. Waveforms used in FMCW radars often have a strong range-Doppler coupling and therefore integrate a compensation of the induced range error. This makes it necessary to evaluate both parameters in parallel especially as targets that offer a Doppler shift without an associated range rate (or vice versa) might be dropped in processing for being implausible.

The angular accuracy of a sensor module can be evaluated with the help of a high precision positioner or turntable that changes the bearing of the module or the relative angle a reference object. Combined with an RTG this allows simulated

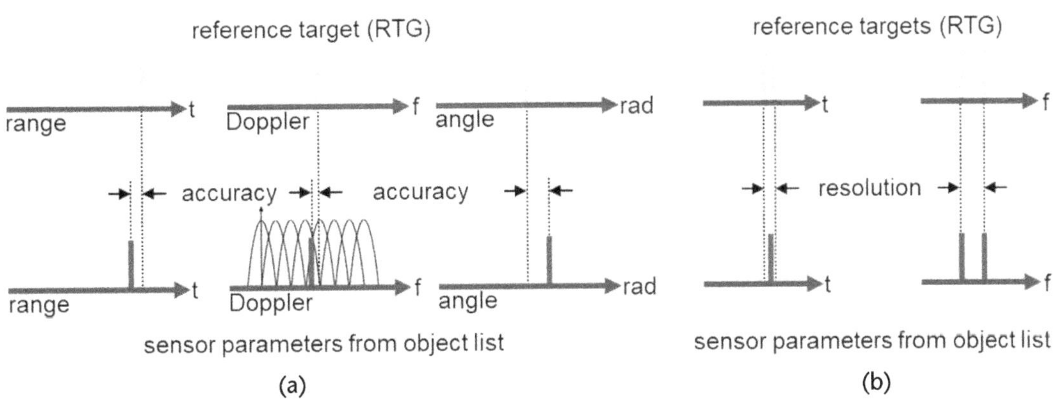

Figure 11.25 (a) By comparing the reference target in range, Doppler, and angle with the processed data from the radar module, the accuracy can be measured. For measuring the resolution, (i.e., capability to separate close targets) (b), it is necessary to provide two targets in the respective dimension.

objects at different distances, with different RCS at different AOA. Therefore, not only angular resolution of the sensor over the full FOV can be measured, but also if there are any blind spots. Measurements in spatial resolution require at least two RTG and separated frontends for the artificial targets. This can be achieved by the use of separate radial positioners that move physically or an electrically switched antenna array. Using positioners might lead to additional issues with vibration during the testing and both approaches are hard to integrate into a small-scale shielded environment.

11.2.4.2 Testing of Radar Interference Mitigation Techniques

There are two possible classes of in-band interference that influence automotive radar operation. The first relates to classical EMI topics (i.e., electrical devices that unintentionally operate or produce spurious signals in the same frequency range as the victim radar). For a 24 GHz radar that certainly is an issue, especially since rollout of 5G. The second—transmission from other vehicle radars—is much more likely in the 76–81 GHz band due to increasing use of ADAS in small-size, middle-class, and upper-class cars, as well as the increasing number of radar-based systems per car that are both wideband and limited to a common band in spectrum.

Various problems occur when interfering signals are received by the radar sensor together with the real echo of the object. The interferers often raise the noise floor, leading to a significantly degraded detection probability of objects that produce echoes with small amplitude (e.g., from pedestrians). Depending on the road scenario (Figure 11.26) and the actual combination of the radar systems and modes that are present in this scenario, the victim radar can experience a decrease in sensitivity, corrupted measurements, and ghost targets. This clearly shows the need to the performance of the automotive radar sensor's interferer mitigation techniques (see Chapter 8, Radar Interference).

Testing the robustness of automotive radar sensors against interferers is important due to its safety-relevant nature. To avoid an overwhelming amount of real-world testing, different scenarios can be evaluated by simulating interfering radars. Though different approaches are possible here, from using actual radar modules or recorded live signals, a signal generator offers the most flexible solution (see Section 11.4.4).

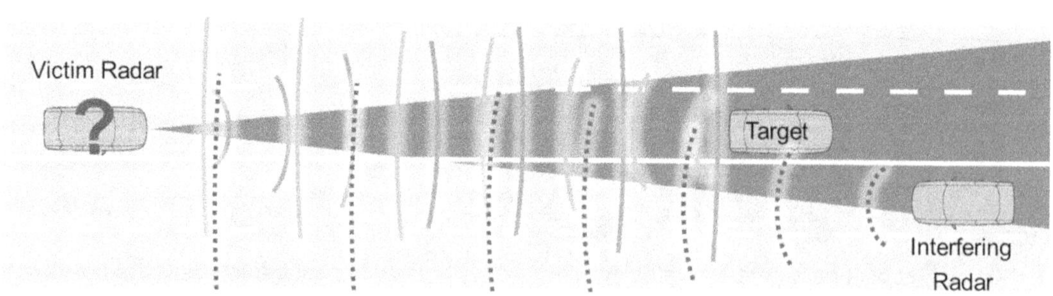

Figure 11.26 The victim radar (left side) should detect the target vehicle on the same lane while being under the influence of the interfering radar (right side). The question is, do the mitigation techniques work as intended for all possible interferers?

11.2.4.3 Tracking/Software/Classification Tests

As radar usage transitions from simple ADAS to full autonomy there is often a parallel desire to move from just identification of the presence of an object to its classification (e.g., can the radar distinguish between a standing bicyclist and a guard rail). In addition to attributes like RCS and speed of the center of reflection, many radars utilize more sophisticated information like Micro-Doppler and scattering of the different structural components of an object that are mostly related to the sensor's angular resolution. Software simulation and direct injection are typically the first approaches used testing this machine learning-aided processing step, but later in the development chain and during sensor integration, testing with a simulated target becomes more attractive, leading to additional requirements to radar target generators (see Section 11.3.2). As the simulated target must be modeled via several spatially separated artificial echoes [13] an increased number of microwave frontends are required each with an individual delay and Doppler shift. Such an approach would be limited by numbers and the geometry of scatter centers that can be simulated, as this translates to the density of the individual radiating frontends of an RTG.

Another requirement for testing software processing arises due to the implemented tracking algorithms. After detection of a target from the received baseband signal, the relevant data is mapped into a digital format for the tracking system. The data structure is usually in the form of an array of detections for a given time and together with additional information such as RCS or target classification sorts the detection into existing tracks, new detections, and clutter. The tracking filter (most prominent are Kalman and alpha-beta filter) establishes a position-velocity (acceleration) kinematic model that describes the motion of the tracked radar target. This model is the key enabler for functions like evasive steering. Testing tracking on an integration level requires the RTG setup to simulate the complex spatial movement of a real target in more than the range domain (simulated angular movement), as well as providing a granularity of the steps in simulated range and angle that match the tracking filter.

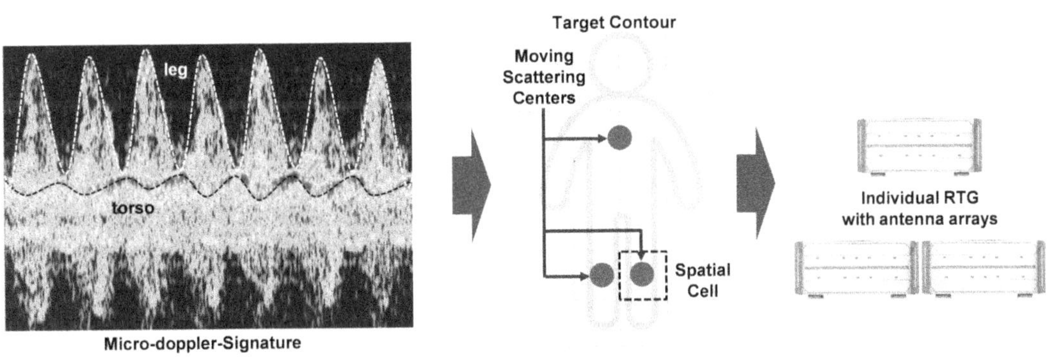

Figure 11.27 The Micro-Doppler profile of a pedestrian originates on different spectrally overlapping scatter centers of the human body. In this simple example only torso and legs are taken into account, which can be mapped to RTGs for testing purposes.

11.2.5 Integration Testing

The processing capability, chip integration, and interfaces to sensors have increased drastically in the last years, while the manufacturing costs have decreased, leading to new possibilities of driver assistance and driving autonomy. In addition to progress in data processing and sensor technology, the complexity of such systems is increasing due to a wider range of function and deeper integration into the overall vehicle system. Additionally, the increasing ratio of software testing to hardware testing has led to longer test times to ensure safe and reliable behavior of the function can increase exponentially.

To streamline the development and validation of new driver assistance systems, test environments can be realized on different levels. The classical approach is drive testing on real road or proving ground, which is relatively cheap (in terms of investment, not the time involved) and offers realistic vehicle behavior with the whole vehicle but is less reproducible and limited in flexibility. Furthermore, a roadworthy prototype is required, limiting the testing capacity to the end of the development cycle.

An attractive supplement to real road testing is hardware-in-the-loop (HiL) test beds, which offer developers an easily accessible and reproducible but abstract and model driven environment. Here all ADAS relevant subsystems are physically present as hardware and integrated into the overarching software, but with an artificial environmental stimulus.

Vehicle-in-the-loop (ViL) testing is in between the HiL and the proving ground (see Figure 11.28). It combines the ease, flexibility, and reproducibility of HiL testing with the benefits of the complete vehicle testing at the proving ground. This involves test driving of a real vehicle in a virtual open space, while test scenarios are simulated by embedding real components in this virtual environment.

Other relevant integration aspects include calibration of the module, including all effects that originate from sensor mounting and obscuring (e.g., by bumpers). Finally, integration testing should evaluate how the integrated sensor functions in a congested spectral environment with high power out-of-band signals being present.

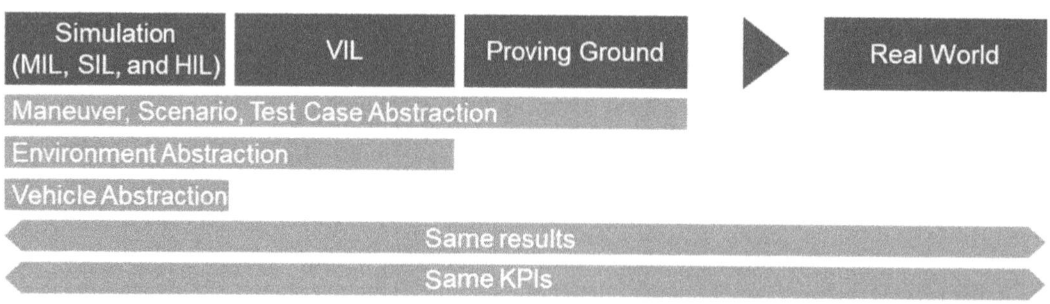

Figure 11.28 With decreasing levels of abstraction during testing (i.e., de-embedding of the vehicle under test [VUT]), the evaluation results and technical KPIs must be maintained to have a valid transition to the "real world."

11.2.5.1 Sensor Mounting and Calibration

Digital beamforming and the angle of arrival estimation require compensation of the differences of systematic errors in between the different transmit and receive channels of an antenna array. Main sources for time variant errors in an active group antenna are the frontends of the transmit and receive chains and the respective A/D and digital-to-analog (D/A) conversions, but due to mutual coupling, the antenna array itself usually is the dominant contributor for the static errors. The sum of all these errors decorrelate the channels or elements in respect to their amplitude and phase response [14].

Calibration is therefore necessary to ensure a defined manifold of the antenna array, but the methods for error correction must be differentiated. Calibration to compensate *time variant* errors must be performed regularly, often on startup of the radar module using internal calibration ports that bypass the actual antenna group. This is a module or chip internal process to compensate errors (e.g., of temperature drift).

Calibration of the *static errors* in the antenna array normally only happens once, based on an external reference target, and should be performed under anechoic conditions to avoid corruption from other objects in the scene. The reference target is measured at different positions to estimate an accurate correction or calibration matrix C for that given radar module.

$$C = (V \cdot K)^{-1} \tag{11.12}$$

This form of calibration combines all coupling effects within the matrix K and the phase and amplitude errors of the transmit and receive chains in the diagonal matrix V. In practice, the number of positions of the reference object is limited for the sake of economical test duration.

With this matrix, the range bias and RX channel gain offset will compensate for the inherent offset due to signal chain processing delays and board delays, while the phase measurement offset compensates for the phase delay between receivers to give a linear cross-channel phase response.

Although calibration of a radar module is typically performed in the production factory, it is necessary to have additional calibration after the sensor is integrated in a vehicle, since the as-installed performance can be affected by mounting, reflections, and transmittivity of radomes. Depending on the type of mounting of the radar sensor, manufacturing errors can happen when installing the module into the car or bumper, requiring end-of-line tests to make sure that the radar sensor is mounted correctly and the radar sensor only looks through the RF optimized parts of the radome.

An incorrectly mounted sensor may be the source of several errors during operation that cannot be corrected via calibration. Figure 11.29 indicates possible effects that may occur.

There are different approaches to ensure correct mounting of the sensors that rely on computer aided design (CAD) data of the radar sensor and the bumper to enable detection algorithms to align with surface features.

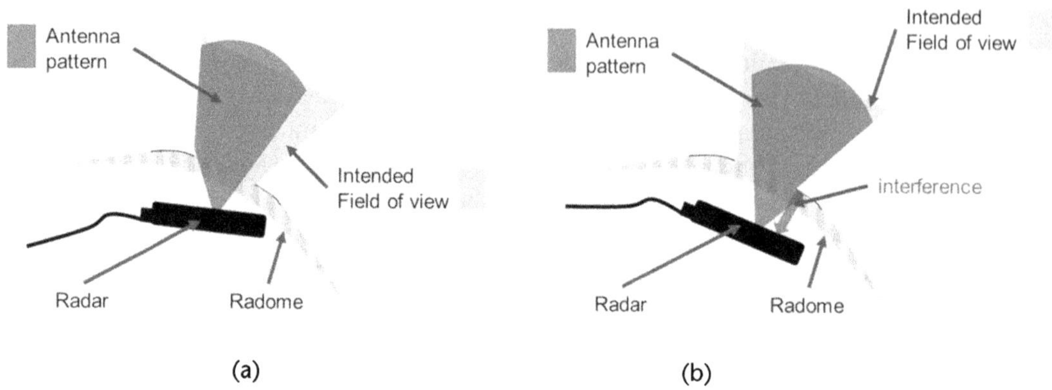

Figure 11.29 Possible effects of a misaligned sensor module: (a) intended FOV is not illuminated due to a tilt of the transmit antenna, and (b) reduced sensitivity of the sensor by having a strong interfering reflection directly at the receiver.

- Videometric alignment checks geometry with a stereo camera setup, but only works without radome obscuring visible light.
- Imaging methods, as described in Section 11.2.3, provide a high-resolution image of the mounting including radome properties as well.

11.2.5.2 HiL Testing

ECU functions can be tested in a HiL setup by simulating a small-scale reference scenario and feedback loop of the sensors object list. However, several sensors typically contribute to the situational awareness needed for ADAS functions. The fusion of the individual sensor data in the respective ECU has to be tested in conjunction as well. In order to test sensor fusion systems in HiL environments, the scenario models for the vehicle environment need to precisely convey the vehicle's surfaces and inertia, road situations as well as traffic environments and the behavior by others in traffic.

It is possible to simulate the entire sensor system used for the vehicles ADAS functions on the electronic and physical interfaces of HiL systems. Either the sensors are electronically simulated in real-time systems on the interface level via digital injection or the actual sensors are physically stimulated by target simulations. The latter requires a test bed that can simulate a driving scenario in multiple domains. For example, over the course of a given scenario simultaneous emission of radar signals, simulated targets for lidar sensors, and projecting images for cameras must be provided synchronously. This is required as the sensor fusion unit needs to perceive the same scenario in real-time from all sensors.

All this information is collected in real-time from the simulated car surroundings and the reactions by the sensor fusion control device are synchronously returned to adapt the simulation accordingly and to achieve closed loop scenarios. Test systems for ECU testing require scalable input/output (I/O) and bus/network interfaces, restbus simulation on the behavior and protocol level and manipulation and test options for buses and networks. Fault insertion on a physical, protocol, and model level allows testing of complex errors and the according systemic reaction.

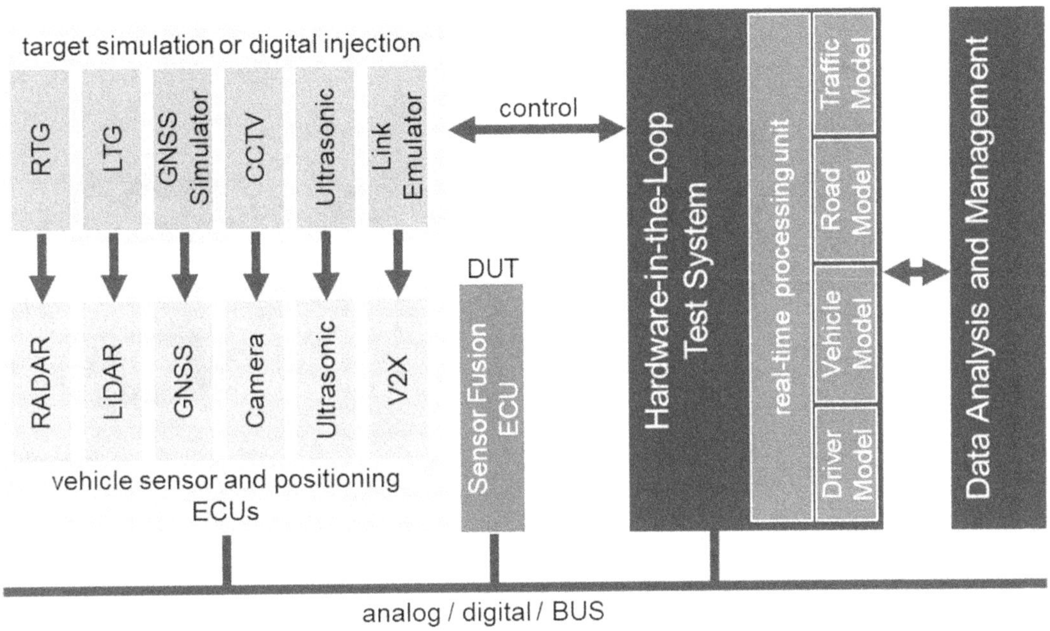

Figure 11.30 In ADAS HiL testing the DUT is normally an ECU combining data from different sensor systems. Embedded into simulated vehicle and environment models and connected via physical interfaces and gateways.

Examples are fail-safe and fail-operational tests as well as vehicle network testing, such as sleep or wake-up tests.

11.2.5.3 ViL Testing

A step further in the development chain is homologation: the process of validation and certification of vehicle features. With Level 4 of autonomy the responsibility for the driving task is transferred from the driver to the system. A major challenge is to ensure and prove the autonomous driving functionality and safety, increasing the scope of testing required. Road tests are time consuming, expensive, and not always sufficient to proof ADAS functionality, increasing the need for testing in a controlled environment so that challenging, complex, and risky driving scenarios can be generated and tested reproducibly. Test setups that simulate the complete driving behavior of a ready-to-drive car are the purpose of ViL testing.

ViL test setups should focus on simulating a complete vehicle level and avoid interference with the vehicle's internal systems to keep the interface development effort at a minimum. The stimulation of the vehicle's internal sensors is done by using feasible actuators, stimulators, and injectors. While this is comparable to HiL testing, it involves having a (virtual) test driver in a real vehicle maneuver in an open simulated space and therefore requires additional integration effort as seen in Figure 11.31, because it is not a single ECU that is tested but an overarching function.

In this example the software simulation of the environment, the vehicle dynamics, and a (virtual) driver must be translated into actual synchronized physical stimuli. The test bed provides instrumentation of the relevant vehicle in terms of

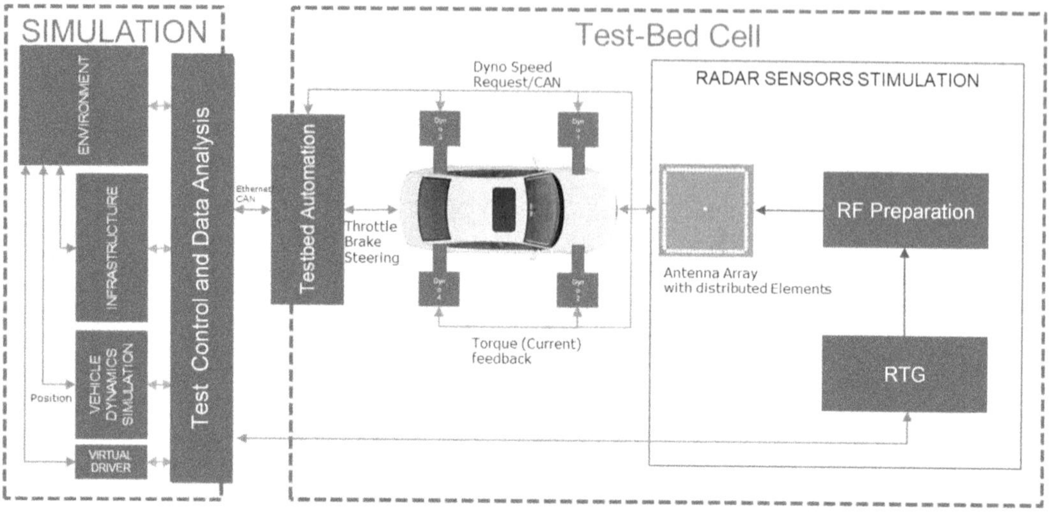

Figure 11.31 Example for an ViL test bed. This could be used for ADAS functions that only rely on a radar sensor, like adaptive cruise control. When functions rely on sensor fusion, stimulus, and integration for other sensors, like camera and/or lidar, are required.

Ethernet and CAN-Buses and actuator subsystems like a dynamometer for information about torque and speed, as well as control of brake, throttle, and steering of the embedded vehicle.

The fundamental performance indicator that can be measured with such a ViL test bed is whether the actuators show the anticipated reaction to a determined stimulus of the sensors. An example could be the question if autonomous emergency braking brings a vehicle to halt in a given time, when sensors detect a crossing pedestrian. The ADAS function to be tested determines the scenarios, which in turn dictates the details of the setup of the test bed, as shown in Section 11.4.7. When comparing rear collision warning and lane keeping assist functions for example, it becomes obvious that the involved actuators and their behavior, and the sensors and their respective FOVs are different in most aspects.

11.2.5.4 Electromagnetic Compliance Testing

Electromagnetic compliance (EMC) aims to ensure the correct operation of different equipment in a common electromagnetic environment, by either avoiding unintended radiation (electromagnetic interference or EMI) of electrical equipment or by reducing the vulnerability of electrical equipment (electromagnetic susceptibility or EMS) against interfering spectral emissions. EMI is categorized into the two areas of conducted and radiated emissions with their specific compliance standards and limits. Conductive, inductive, and capacitive coupling are primarily a board-level and mixed signal challenge that requires proper stackup design, layout, routing, and grounding techniques. The introduction of new RF technologies such as automotive radar systems and 5G connectivity require verification for compatibility with each other and existing car systems.

As an example, the International Special Committee on Radio Interference standard CISPR 25 details limits and procedures for the measurement of on-board radio disturbances in the range of 150 kHz to 6 GHz to protect inboard services, like GNSS, entertainment, and connectivity. Here, testing on component and full vehicle level apply and not only over the air measurements are performed, but also conducted emissions under open lab conditions.

Similarly, CISPR 12 deals with the protection of off-board receivers and is described for emitted frequencies between 30 MHz and 1 GHz, while CISPR 36 adds further requirements for electrical vehicles (due to DC to AC power inverters or battery management and charging) for frequencies below 30 MHz. As the measurements only involve the radiative coupling with external devices, an open area test site (OATS) or anechoic chamber is required to eliminate spurious signals. Normally this testing is performed at the full vehicle level under all relevant operating conditions, not at the component level. EMI measurements and conformity therefore is all about standards, but those normally apply not only to a specific vehicle component, like the radar module, but rather the full integration.

EMC compliance includes testing susceptibility to outside interference from electromagnetic fields as well; it is the degree to which electronic equipment malfunctions or breaks down when subjected to varying levels of EMI. It is different to a purely OOB receiver interfering testing as described in Section 11.2.2 in terms of the coupling components of the RUT and the fields strengths used for testing.

Depending on the used standards and the frequency, the maximum field strength the radar is exposed to while operating correctly can be 200 V/m. The most important standards for that are ISO 11451, which describes vehicle test methods for electrical disturbances from narrowband radiated electromagnetic energy and the corresponding electrical and electronic components located within the vehicle as described in ISO 11452. The main differences here are in the interfering frequencies to be covered (both involve test up to 6 GHz, but start from 200 MHz for components like a radar module and from 100 kHz for the full vehicle tests), the respective field strengths, and the test methodologies. Often manufacturers derive their own house standard from these international standards as well.

The component-level tests are performed on a grounded EMC test bench and involve a fully functional test of the radar on object list level. The tests on vehicle level occur after all components have already been EMC tested and therefore assume functional safety. They are performed in a shielded EMC measurement chamber and in contrast to EMI normally the vehicle is exposed from the front (possibly the sides) only and only the status and error memory of the components are monitored. There are trends indicating a change from today's pure EMS testing, using CW and AM signals, to coexistence testing using real signals including complex modulation and signal payload.

11.3 Test Equipment

In the following section, the instruments for measuring radar-related parameters will be described at a high level. As most measurements can be done via using several different instruments, one focus will be on how to select the optimal equipment for a given setup. As an example, most power-related measurements can be done using a spectrum analyzer, a power meter, or an oscilloscope. This will not consider perspectives like differences in cost or versatility of an instrument, but only relate to a specific task.

The other focus will be on the relevant instrumentation parameters that influence a given measurement. The intention is not about providing exact values for instrumentation settings, but instead to develop an understanding of what a measurement parameter does, serving as a starting point for setup and troubleshooting.

Often a priori knowledge of signal or a structured order of measurements is required to do a meaningful configuration of the instrument. An example would be the measurement of the average power of a transmitted radar waveform. Normally waveforms are cyclic and have inherent pauses, therefore before measuring average power the duty cycle of signal is required. For a signal analyzer, sweep time must be chosen in a way that a full cycle is captured and averaged, for a power meter this information is used directly.

If antenna connectors are accessible, a vector network analyzer can be used for characterizing antenna arrays, the most common application would be S-parameter measurements (both transmission and reflection coefficients). Unfortunately, an antenna array in mm-wave frequency range will not provide antenna connectors due to added complexity and cost, physical size limitations, and resulting insertion loss. Consequently, over-the-air (OTA) tests are required for most setups. This raises some additional requirements to support a controlled environment (in terms of electromagnetic shielding, temperature, and vibration) and to calibrate a reference plane (impact of the air gap on frequency response, group delay).

The emphasis in the next sections is on dedicated test equipment for radar measurements. More general purpose test equipment will not be explained in detail but related to automotive radar only.

11.3.1 General Test Equipment

11.3.1.1 Spectrum and Signal Analyzers

Modern spectrum analyzers are essentially an FFT analyzer utilizing sampled time-domain signals and a traditional spectrum analyzer, working based on the superheterodyne principle. The advantages include (1) the input frequency range is independent of the A/D converter rate and a preselection to suppress harmonics and other undesired spectral components, and (2) high measurement speed at low resolution bandwidths and preservation of the signals phase information [15].

This is the defining difference to a signal analyzer (SA): by maintaining IQ information, demodulation of digitally modulated signal is possible. Both terms (signal and spectrum analyzer) will be used interchangeably hereafter. Signal sampling without sweeping the center frequency sweep (so called *zero span*) offers a way the SA can display a time domain signal. The main difference from an oscilloscope is that the SA data has been down-converted, so the displayed data is relative to the center frequency of the measurement.

A SA used for OTA measurements captures the radar signal on RF level and does not benefit from the radar's processing gain. The analysis bandwidth of the instrument must be at least 4 GHz when measuring modern radar systems in the 79 GHz band.

A SA can be used for all frequency domain-related measurements, but as those normally display magnitude or power in one dimension, some considerations have to be taken. Correct setup of resolution bandwidth (RBW) and video bandwidth (VBW) are crucial to avoid or compensate "pulse" or rather cycle time desensitizing, especially as chirp durations are getting shorter.

General parameters in time domain like chirp duration and idle times can be measured via zero span and provide the basis for setting up power measurements of a radar signals. With known time structure, it is possible to set the SA's sweep times and detectors correctly and accurately to measure parameters such as OBW and EIRP [16].

The capability of a SA to demodulate a complex signal is the enabler to analyze novel waveforms like OFDM or PC-FMCW with measurement methods like PAPR, EVM, and so on. To reduce testing effort and required test time, most of the described steps and functions can be implemented as dedicated functions of a SA.

11.3.1.2 Oscilloscopes

Modern digital oscilloscopes support a multitude of specific measurements and applications for troubleshooting circuits or checking the quality of acquired signals. Examples are serial bus trigger and decode for debug and compliance tests, jitter measurements, frequency response analysis using Bode plot functions, power electronics measurements, digital signal analysis for mixed signal designs, EMI analysis, and automotive radar debugging.

In a digital oscilloscope, the analog signal at the input (e.g., via a probe), after attenuation or amplification and preliminary filtering, is directly converted to digital domain with a given sampling rate. Each channel has a separate ADC, each with an external clock. Therefore, in contrast to a SA, an oscilloscope is not a tuned receiver. But with a SA introducing dedicated time domain measurements, the same holds true for oscilloscopes with FFT processing (providing a frequency domain measurements) and ADCs that offer sampling rates in the magnitude of several hundred GS/s. The differences in those two instruments seem to be decreasing.

Assume a typical radar signal in the 79 GHz band with a chirp bandwidth of 4 GHz. There are two ways an oscilloscope is able to capture this signal: either it has a bandwidth that covers all of the signal directly (DC to 81 GHz) or an external downconverter is used to bring the signal into the oscilloscope's bandwidth. The first approach collects 81 GHz BW of noise, as it has a wide-open frontend and normally instruments beyond 20–30 GHz of BW are usually more expensive than a comparable SA. The second brings in the disadvantage of introducing images of the down conversion processes. And both lack due to another general difference between SA and oscilloscopes that is the data resolution. Typical spectrum analyzers sample data at 14 or 16 bits, whereas typical oscilloscopes are usually 8-bit devices and especially for OTA testing this translates to a reduction of the dynamic range.

11.3.1.3 Power Meter

Power meters can be more accurate, easier to use, and less expensive when compared to a SA and oscilloscopes for pure power measurements. There are different sensor technologies that can be used for this task. The most accurate are thermal power sensors (thermistor or thermocouple) that absorb RF power, turn it to heat, which in turn translates into a proportional resistance or voltage change. However, even if they offer high accuracy for an average power measurement independent of the signal's modulation and can integrate frequencies well beyond 100 GHz, they are normally not suitable for automotive radar measurements. With a low dynamic range and a sensitivity of less than –30 dBm, the transmit power of a radar module often is too small, additionally the necessary measurement time is quite long.

Another approach are diode-based power sensors that convert an AC signal to a DC signal via rectification. In the square law region, diode output voltage is linearly proportional to the square of the input voltage. When the input voltage crosses a certain threshold, the diode response moves into the linear detection region, where output voltage is proportional to input voltage. At that point, several measures are possible to avoid errors. Correction factors compensate for the deviation from square law region and can be used to extend the dynamic range of a power sensor. These correction factors are typically stored within the sensor itself. Alternatively, the sensor uses multiple overlapping paths (one diode per path) with different input attenuation and routes the signal to a path such that the diode is always in the square law region. Overall diode-based power sensors offer the sensitivity, DR, and fast measurement time that is required for radar signal.

For example, diode-based power sensors can be used to measure the antenna diagram of a radar module as shown in Section 11.2.3, but especially when measuring in signals in the microwave range, test system properties such as the lower test limit and measurement accuracy are affected by mismatches, attenuation losses, reflections, and frequency dependent antenna gain. To reduce and compensate those errors, the length of RF cables connecting antenna and detector diode should be minimized and the RCS of the power meters antenna should be as small as possible. With such a setup, it is possible to measure relative power with an uncertainty of < 0.1 dB and absolute power values down to –76 dBm.

Diode-based power sensors are also often used in EMC applications as probes for calibration and reference (see Section 11.4.4). Normally multipath average power sensors are used covering the relevant frequency range from 8 kHz to 18 GHz.

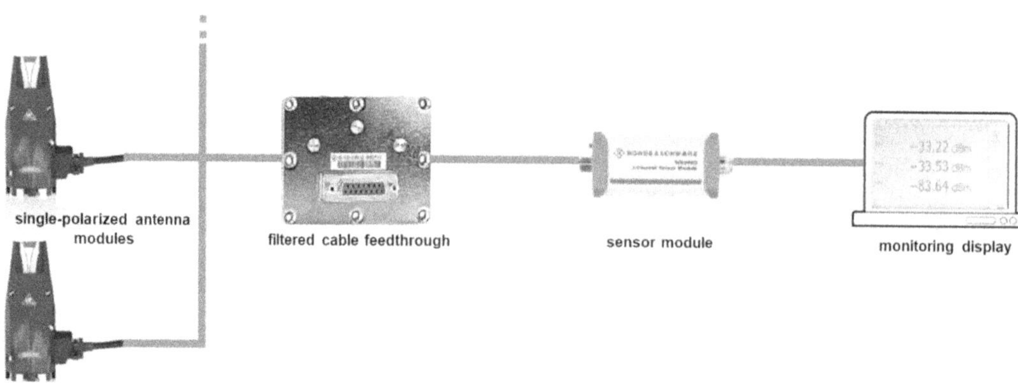

Figure 11.32 Setup with diode-based powermeters. To reduce cable losses of mm-wave signals the detector is integrated directly in the antenna module. With multiple parallel modules, it is possible to speed up measurements (e.g., by combining horizontal and vertical polarized antennnas).

11.3.1.4 Vector Network Analyzer and Imaging

A vector network analyzer (VNA) includes both a calibrated source signal and a coherent receiver. The receivers detect changes to the output signal from a device (or network) compared to the source signals input to the device. By comparing the ways in which current and voltage are affected by the device, a VNA can measure both the amplitude and phase responses caused by the device. The results are transmission measurements (transmission coefficient, insertion loss, gain), reflection measurements (reflection coefficient, VSWR, return loss), impedance, and the S-parameters (scattering parameters) S11, S12, S21, and S22.

Aside from the more classical applications in conjunction with (waver) probes for measuring printed circuited board (PCB) or chips, there are some other areas where VNAs can be used. One area is the characterization of radar radomes (see Section 11.2.5). Another area is generation of high-resolution RCS models of different traffic participants (e.g., car models, bicycles, pedestrians) for target classification [17]. Measurement techniques for large targets generally require measuring the RCS in azimuth by rotating the target on a turntable and can be used to detect specific scatter centers of the object.

11.3.1.5 EMI Test Receiver

EMI measurements require a different approach than other types of general RF tests. In EMI, engineers never quite know what signals may be present. Two instruments are used for EMI testing: spectrum analyzers and test receivers. In summary, spectrum analyzers can be used for EMI, but various work around methods must be used in order for the results to be accurate and are prone to error in consequence.

Test receivers are designed for compliance testing, which translates to:

- Higher number of measurement points (e.g., tens of thousands) to improve accuracy.
- Automatic control of attenuation and preselection filtering, as well as setting of preamplification and RBW and step size.
- EMI type RBW filters are narrower, rolling off at 6 dB per decade rather than 3 dB. Common CISPR RBW filters are 200 Hz, 9 KHz, 120 KHz, and 1 MHz.
- Display of field strength (in contrast to power) as most EMI standards and limit lines are specified in field strength at some distance from the DUT.
- Some of the OOB signals encountered in EMI testing are high amplitude and could cause an overload of the instrument. Test receiver use preselection filters to attenuate any signals not being displayed on the screen. Not to be confused with preselection in SA that have the purpose of image rejection.

11.3.1.6 Phase Noise Analyzer

While a SA could be used for measuring the phase noise of PA or VCO, radar applications require a very low phase noise, because it has a direct influence on many performance parameters. For signals that have very low level of phase noise, the phase noise of the oscillator within the analyzer will add to that of the signal under test and will distort the result, therefore the phase noise performance of an instrument should be at least 10 dB better than that of the DUT.

That is the main benefit a dedicated phase noise analyzer offers. It features very high sensitivity due to extremely low-noise internal sources and cross-correlation of multiple internal LOs to suppress their phase noise compared to the DUTs. Often it is possible to measure residual phase noise (i.e., additive phase noise of a DUT only, without contributions from the signal source) and not only the absolute value.

11.3.2 Radar Echo Generators

A radar echo generator is needed as a reference target for performance testing and validation of a radar sensor. It provides the sensor with a defined target, simulating all relevant attributes that a real target on the road would also possess, like RCS, distance, speed or angular position. There are different approaches to realizing a radar target simulator/radar echo cenerator (RTS/REG) solutions that range from simple reflectors to digital test equipment for simulation of complex scenarios. This section describes several types of equipment and some of their respective advantages and disadvantages.

11.3.2.1 Analog

Analog target simulators apply physical delay lines to delay the incoming electromagnetic wave from the radar and simulates an object in a certain range R_{sim} from the radar. The simulated range is the sum of the speed of the electromagnetic wave in air/vacuum (speed of light c_0) divided by 2 and in a certain medium (with speed v_1) multiplied by the time of flight t_{tof} and time within the medium t_{proc} respectively. This includes the time for passing physical distance to the RTS/REG forth and back as well as the internal signal processing time.

$$R_{sim} = \frac{c_0 t_{tof}}{2} + v_1 t_{proc} \tag{11.13}$$

- R_{sim} simulated distance of a target (in m);
- c_0 speed of light (in m/s);
- t_{tof} time of flight (in s);
- v_1 speed of electromagnetic wave in medium 1 (in m/s);
- t_{proc} RTS/REG processing time (in s).

To delay the signal fiber optical delay lines (FODLs) are often used as they are relatively flexible, phase coherent, and can create small systems that convert the RF signal of the radar to optical and delay of a fixed duration. The number of delay lines determines the steps that the translated range can be switched and architectures with electrical switching of the signal or parallel optical switches are possible. The number of discrete steps is limited though, as every switch induces additional unwanted attenuation and reflections in the RTS. Often an IF is used because low frequency signal handling is easier and creates less loss. After delay, the signal is then reconverted to RF and retransmitted to the radar. Some systems are also able to introduce Doppler frequency shift by an additional IF mixer after signal delay.

FODLs offer constant delay versus frequency, are immune to vibration, are largely resistant to electromagnetic interference, and fiber delays do not radiate energy. Repeatability of simulation, low system cost, and time-savings are key advantages. However, FODLs cannot generate time-variant range-Doppler targets, nor do they offer continuous range settings or arbitrary signal attenuation and gain.

RCS can be simulated by attenuating the radar signal. After considering free space loss and antenna gain (DUT and RTS/REG) the attenuated signal represents a target with a certain RCS.

11.3.2.2 Digital

Digital target simulators rely on digital signal processing. The target generator receives an electromagnetic wave from the radar and modifies it digitally to picture the desired scenario. Adding delay on the signal simulates a distance to an object, frequency shift simulates a Doppler shift indicating the velocity of an object, and signal attenuation indicates a certain RCS. In addition, it is possible to add multiple delays, frequency shifts, and attenuation simulating multiple targets.

Often digital radio frequency memory (DRFM) is used to manipulate the radar signal digitally—down-converting, filtering, and digitizing the received RF signal before storing and modifying it. Signals are then reconverted to analog and mixed to RF frequency using the same LO used for down-conversion. This is important to reduce phase noise. The minimum delay introduced by a DRFM is mainly limited by its ADC and DAC. In addition, signal processing adds several processing cycles to the radar echo signal. Typical minimum range delays range from below 100 ns to below 1 µs. A further consideration is how the analog RF signal is represented in the digital domain (amplitude, phase, I/Q) and the number of bits, because this is what mainly determines the DRFM's signal fidelity.

11.3.2.3 Differences and Hybrid Architectures

The analog target simulator offers the possibility to simulate short distances while a longer signal processing delay in the digital units results in a greater minimum range to an object. On the other hand, the digital target simulator often offers

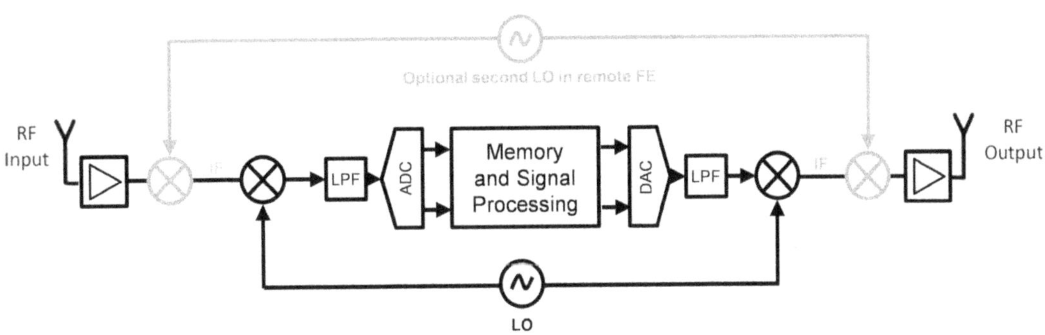

Figure 11.33 General architecture of a digital radio frequency memory. A second stage of the superheterodyne architecture in a remote frontend can be used to reduce cable losses by converting to a lower intermediate frequency.

more flexibility to simulate complex scenarios with multiple targets. Analog target simulators exist with one fixed distance, multiple fixed distances, or variable distance using a switch matrix to toggle between different fiber-optic lengths. Faster switching of the distances can simulate a dynamic scenario of a target but phase jumps and attenuation challenge the fidelity of this. Depending on the update rate of the radar, the required phase coherence and RCS switching between different lengths back and forth could also simulate multiple targets.

Hybrid target generators combine analog and digital target simulator architectures and reduce the disadvantages of each individual approach. Usually, the analog technique is used to simulate short distances which is more complex when using a digital target simulator since the required signal processing time limits the distance minimum that can be simulated. On the other hand, the benefit of the high flexibility of digital target simulators can be used to simulate more complex schemes with multiple targets. Generating a scenario using analogue and digital simulation techniques requires precise synchronization and calibration between both modes.

11.3.3 Measurement Antennas

Depending on the measurement task different antennas or antenna configuration are required for a proper setup. RTS/REG may support bistatic, monostatic, or both antenna configurations, normally consisting of waveguide antennas due to their low-loss transmission line (Figure 11.34).

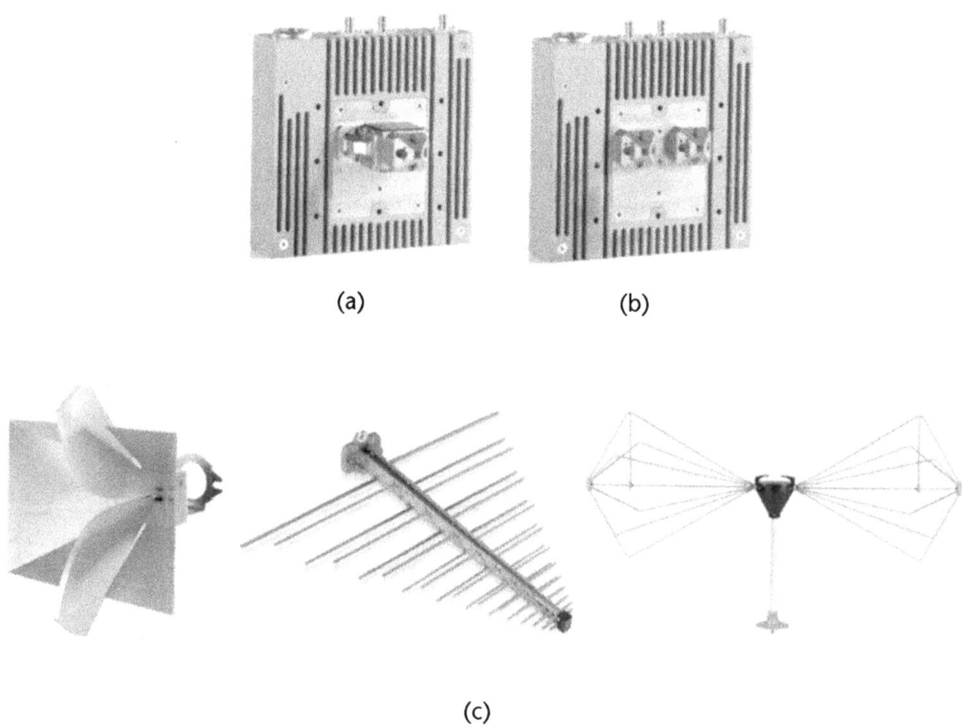

Figure 11.34 Examples for antennas for radar target simulation: (a) monostatic, (b) bistatic, and (c) various antennas for EMI/EMS testing.

A bistatic configuration uses independent antennas for receive and transmit paths providing the higher isolation needed to handle very high receive power levels. A monostatic configuration uses a single antenna with an integrated circulator providing superior performance to test MIMO radars, whose performance suffers when RTS/ EG transmit and receive antennas aren't collocated, but their TX-RX isolation is limited by circulator performance.

It is often useful for the RTS/REG to use a remote frontend module detached of the simulator base unit, so that the position of the antenna in the setup is more flexible and unwanted delays and path loss are minimized.

A full set of antennas is necessary for EMC testing, because a very wide frequency range has to be covered. Typically, antennas that are used are biconical and logarithmic periodic measurement antennas covering the frequencies from several MHz up to the lower GHz range. For frequencies beyond 1 GHz normally

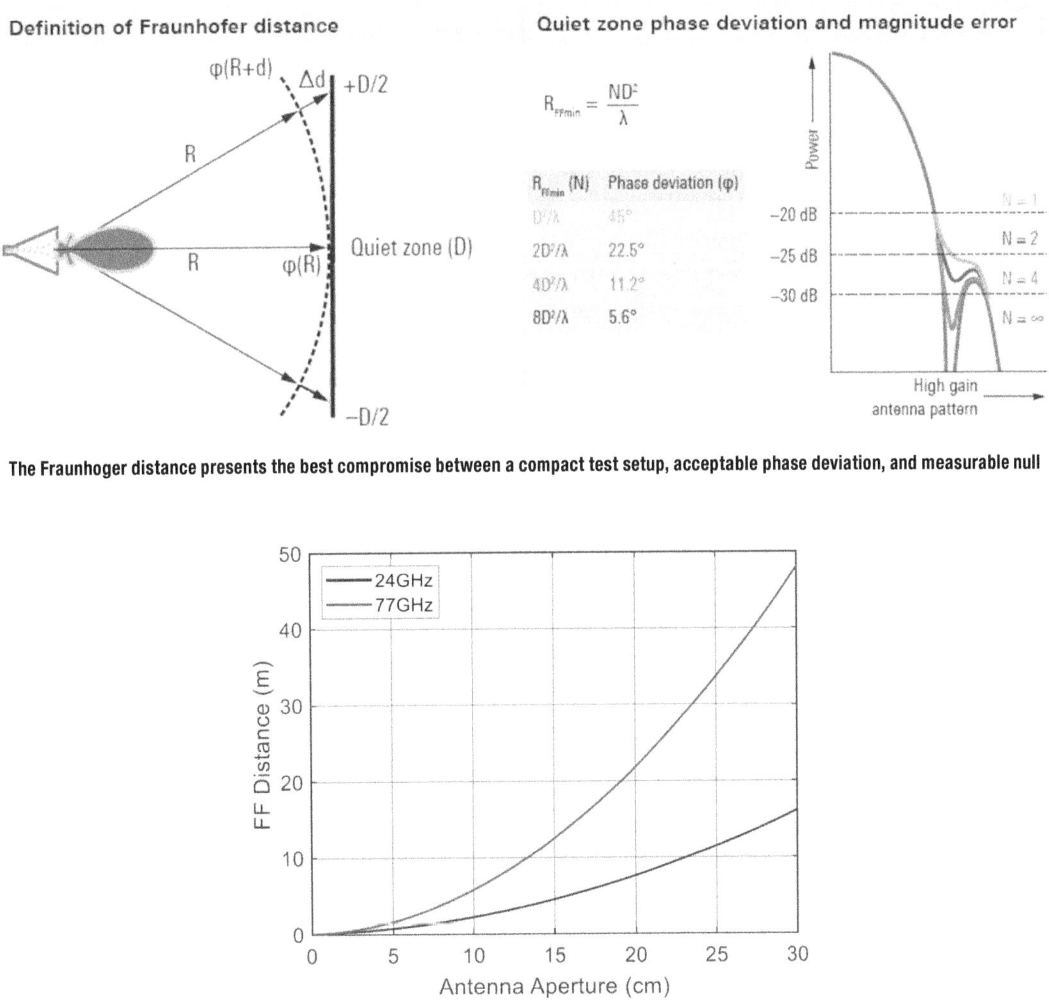

Figure 11.35 Definition of Fraunhofer distance and quiet zone in terms of wavelength and antenna aperture.

horn antennas are common, as well as stripline antennas (not to be confused with microstrip lines) for frequencies down to kHz.

11.3.4 Anechoic Chambers

An anechoic chamber for provides a quiet RF environment whereas the DUT only detects the objects being simulated and repetitive test results are possible. Accurate and repeatable RF measurements require the so-called *Fraunhofer distance* (see Figure 11.35), the far-field distance where plane wave propagation can be achieved. Typical amplitude tapers of less than 1.5 dB and phase taper of 22.5° are achieved within the quiet zone.

The required higher angular resolution for radars is typically achieved through increased number of TX and RX antenna arrays, which increases both the far-field distance and the required quiet zone for the measurements. For example, a far-field distance of more than 20m is required for measurements of an EUT of 20 cm antenna aperture size at 77 GHz. Higher measurement distances lead to higher paths losses, decreased dynamic range for measurements, and higher minimum distance of objects that can be simulated.

Some test methods, like ETSI EN 303 396 which utilizes traditional test equipment (power meter, spectrum analyzer, signal generator), allow the use of reduced measurement distances. Depending on the radar itself (e.g., applying MIMO) the recommended far-field distance can become impractical even at mm-waves for larger antenna apertures.

For functional test of a radar using RTS/REG, depending on the type of signals and modulation used, the correct EUT measurement may not be possible if they are done in the near-field range and they directly depend on the phase or amplitude measurement results. For example, range and radial velocity measurements are possible for linear FMCW radars in the near-field range of the RTS/REG because they depend on the beat frequency between the transmit and receive signal. On the other hand, RCS measurements in the near-field is not possible because RCS depends on the amplitude of the signals and the errors in the near-field are too high.

The use of high-performance shielding effectiveness (> 60 dB typical) and high-quality absorbers that minimize unwanted reflections (ghost targets) from chamber multipaths or installed equipment enables reliable test results. If unwanted reflections exist, they may be avoided by time gating, or simply tolerated if their level is at least 20 dB below the level of the intended object being simulated throughout the FOV.

11.3.4.1 Direct Far-Field Setup

Direct far-field setups typically require very large measurement distances and have smaller quiet zones than compact antenna test range setups. Nevertheless, whenever other approaches are not technically possible, then either direct far-field testing with a very large chamber or relaxed compliance limits with higher measurement uncertainties for practical measurement distances in near-field ranges will need to be employed. This is the case for the measurements of unwanted emissions in the OOB and spurious domains, which are difficult to be covered by other setups.

According to the far-field calculation formula the minimum distance R_{min} is:

$$R_{min} = \frac{ND^2}{\lambda} \tag{11.14}$$

Where R_{min} is the minimum distance in meter and D is the antenna aperture in meter; typically $N = 2$ is used.

11.3.4.2 Near-Field/Far-Field Transforming Setup

There are different approaches to perform a near-field/far-field transformation of incident waves. Some require sequential measurement of phase and magnitude of near-field (E-field) surfaces and use software-based Fourier transformers to generate a synthetic far-field signal. This approach has a strong time dependency and therefore is normally not suitable for anything but CW-fed antennas.

Hardware-based approaches like (Figure 11.36) are better suited for a dynamic signaling system, like radar. But Fresnel lenses are very hard to manufacture and often too narrow band, whereas plane wave converters, with their requirement for high numbers of phase shifter and necessary calibration effort are not economical and lack the dynamic range for radar applications in the microwave range.

Compact antenna test ranges (CATR chambers) on the other hand, provide real-time near-field/far-field transformation on hardware without the need to post-processing (postprocessing is typically impractical for functional testing) and are comparatively easy to manufacture. The parabolic geometry of a mirror/reflector transforms near-field spherical waves into far-field plane waves, without the free

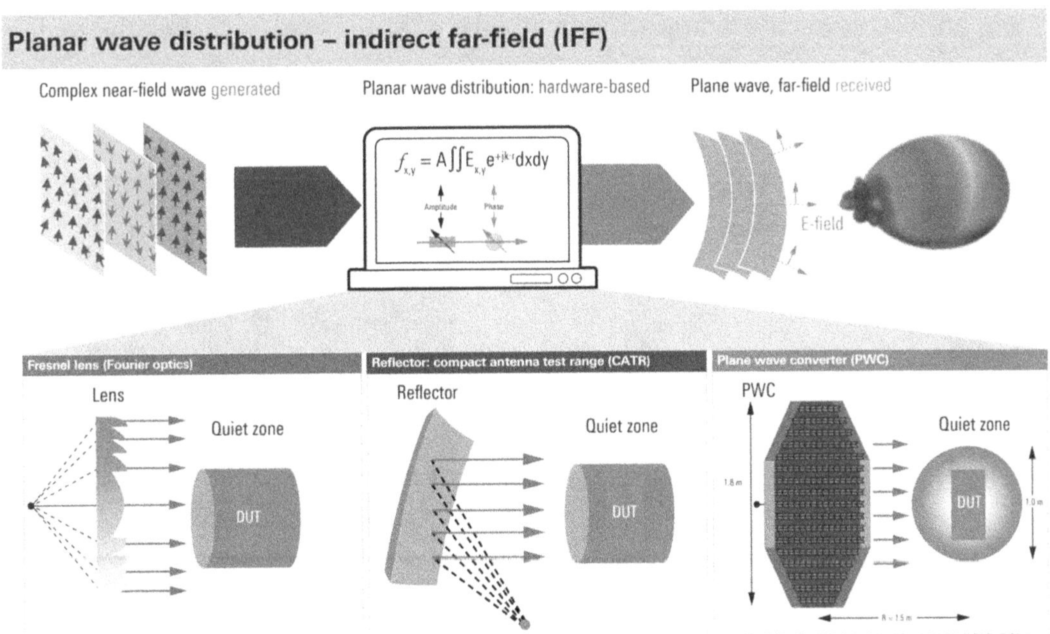

Figure 11.36 Hardware-based near-field to far-field transformation.

space requirements of a direct far-field approach. Smaller setups for bidirectional transmit and receive measurements translate directly to shorter cables lengths and smaller total path loss that is especially important for mm-wave test setups. The feed antenna(s) from the RTS/RTG are positioned at the reflector focal point, the setup therefore is reciprocal in terms of wave geometry. The resulting quiet zone of a CATR setup is cylindrical and there is close to zero path loss between the reflector and the EUT making the total free space path loss (FSPL) for the measurements lower and making the measurement accuracy of the system independent from the accuracy of DUT positioning system in one of the axes (whether vertical or horizontal depends on the CATR setup). The size of the quiet zone is typically 25% to 70% of the reflector size, basically dependent on the quality of the reflector edge treatment and the feeding arrangement.

In summary, a complete functional test of a radar requires measurements to be done at a far-field distance. If this distance is too high for a certain frequency and antenna aperture, then either methods to achieve near-field to far-field transformation in real-time, like a CATR setup, shall be applied, or some of the functional tests will not be possible.

11.3.4.3 Impact of MIMO Architectures

Depending on the physical distance of a target to the radar antenna, the propagation of the electromagnetic waves is under far-field or near-field condition (see Section 11.3.4). Following the MIMO configuration of the radars group antenna, the processing has to compensate for spherical wave fronts (close targets in near-field) or not (long range targets in far-field) to estimate AOA correctly. Normally, a radar makes a range estimation before AOA processing to choose the correct compensation.

This is the reason why an inherent and systematic error may arise when using an RTG to simulate targets. Normally the RTG frontend, including the feed, is

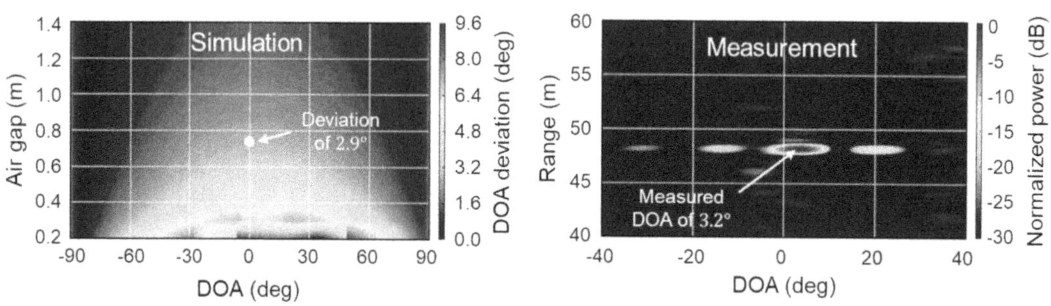

Scenario parameters:	
Air gap:	75 cm
Direction of arrival:	0°
Simulated range:	47.2 m

Measurement results (near-field deviation):			
	True location	Simulated	Measured
Range:	47.95 m	47.95 m	48.2 m
DOA:	0°	2.9°	3.2°

Figure 11.37 Effect of a simulated target in far-field range with the actual antenna being in near-field. The target, that should be on boresight inherits an error of about 3° in angle estimation.

quite close to the sensor. As the physical target echo source in this case is always in the near-field, but can simulate a virtual target in the far-field by delay of the signal, the models for compensation of the spherical waveform in radar processing is wrong and induces a phase and therefore angle error [18].

For a measurement in a CATR chamber this is compensated (but in contrast simulation of a close target might induce an error), but for functional testing without a near-to-far-field transformation, this has to be kept in mind. A similar issue emerges, when bistatic TX and RX antennas of the RTG's frontend (see Figure 11.34) are physically separated. The deviation of the physical AOA of the TX antenna and the directions of the artificial objects increase linearly with the separation of the TX and RX antennas in the angular domain. Those deviations cannot be compensated, as the separation of the TX and RX antennas introduce additional phase shifts to the received signal of the radar.

11.3.4.4 Environmental Conditions

Sometimes OTA measurements must be taken in extreme conditions, such as very high and low temperatures. However, it is not feasible to change the temperature in the whole OTA chamber or room; the wear on absorber materials, fixtures, and the positioner system goes up unnecessarily, increasing maintenance and risking possible failure of the test range. In addition, it takes much more time and energy than necessary to apply temperature changes to the large volume contained in chambers.

Instead, RF "invisible" material is used to contain the region of controlled temperature in a small space around the RUT. This keeps all other equipment in the chamber untouched by the extreme conditions and increases the efficiency for temperature changes. The challenge lies in accurate temperature and humidity control as well as using the right materials to contain the controlled region while not disturbing the traveling waves in a way that degrades the measurement data.

Figure 11.38 Shielded antenna test chamber with DUT under a "temperature bubble."

11.3.5 Positioners

Many measurements involve rotating the DUT in several degrees of freedom. Most automotive radars are singular polarized and the positioners must keep the polarization alignment while having enough accuracy for the azimuth and elevation angular measurements. Accuracy of angular measurements requires that positioner movements keep the RUT always accurately and repeatedly positioned at the rotation center. Today's imaging radars with MIMO aim for very high angular resolution. Positioners should have a higher angular resolution than the radar under test with an active feedback of the position. Typical high-performance positioners have an accuracy of 0.04°.

If the test range features positioning devices, the phase center of the DUT must be placed in the rotation center of the positioning setup. This way, the distance between the DUT and probe antenna stays constant, independent of the measurement angle. Therefore, path loss calibration only has to be conducted for a single direction, as all directions exhibit the same path loss and phase shift. With increasing frequency, the correct positioning in the phase center becomes more important as the wavelengths get shorter. The mechanical setup will add measurement uncertainty depending on the accuracy of positioning and the used frequency range [19].

The mechanical alignment of the DUT and positioner setup is especially important when measuring phase information such as a near-field to far-field transformation. At short wavelengths, a few millimeters of mechanical misalignment can have a significant impact on the phase deviation in the measurement.

The alignment of both rotation axes to meet in the center of rotation must be verified during site acceptance, often using laser guidance or automated positioning. When the phase center and rotation center are aligned, the accuracy along the elevation axis of conical-cut positioners can be confirmed electrically by performing a flip-test, as seen in Figure 11.39. During this test, multiple elevation cuts are measured over the elevation in the positive and negative direction for pairs of azimuth orientations, offset by 180°. For each pair, the amplitude measured at positive elevation with the first azimuth orientation is identical to the amplitude measured at negative elevation of the second orientation, and vice versa. For example, a measurement at

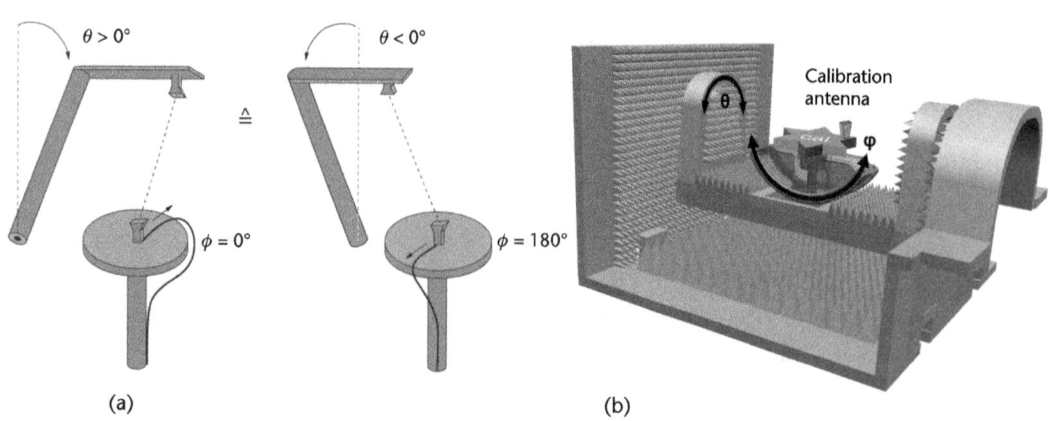

Figure 11.39 Calibration of (a) a distributed axis system (conical cut method) and (b) a 3-D tilt-tilt positioner.

0° azimuth, 10° elevation is identical to 180° azimuth, −10° elevation. The phase will experience a 180° shift. If the azimuth and elevation rotation center are not aligned perfectly, differences will occur in the measurement. Alternative setups using a smaller 3-D tilt-tilt positioner require a calibration disc mounted with an antenna at different positions to ensure correct alignment of the setup.

11.4 Example Test Setups

The previous sections described the reasons why testing is necessary, what the relevant parameters are, and the methods for testing them, as well as a description of the required test equipment. We now bring those together in some example test setups.

Antenna design relies heavily on software tools for simulating the complex field distribution (based, for example, on finite element modeling). For automotive radar modules where an array of microwave patch antennas is used, it is necessary to validate the simulated models against measurements of the finished antenna. The radiation pattern of the transmit antenna is of particular interest as it fundamentally defines the FOV of the sensor. In terms of internal EMC, the side- and backlobes of the beam also can be of interest, especially for highly integrated antenna in package.

For the reasons discussed in Section 11.2.3, it is not possible to directly measure the antenna diagram for every transmitting element, but rather a geometric sum of all transmitters EIRP. The parameters for the power sensor, as shown in Figure 11.40, are dependent on RUT signaling. Aperture time and duty cycle need to be assigned with respect to the transmitted waveform. For complex waveforms, like OFDM, simple peak, and average power measurements have to be replaced with statistical metrologies like CCFD.

The proposed setup includes an RTG connected to a frontend for providing a reference target so that the RUT is continuously transmitting. The receive antenna

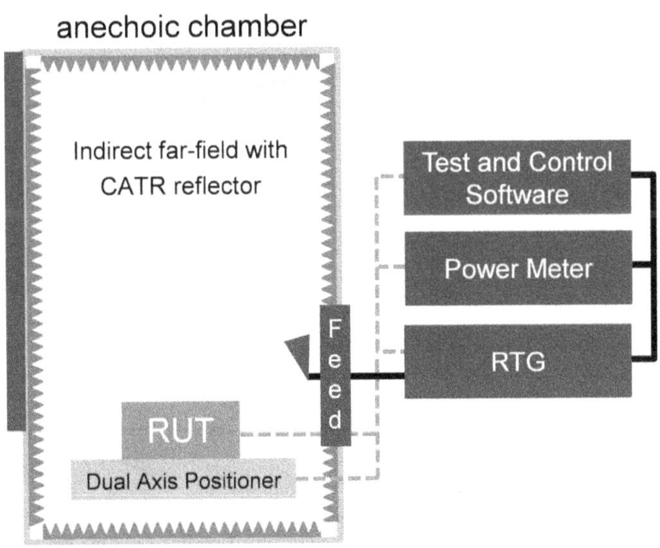

Figure 11.40 Setup for measuring antenna characteristic of a radar module, like EIRP.

of the frontend is also connected to the power sensor and is therefore recommended to use a bistatic configuration to avoid coupling influence on the measurement.

Over the course of the measurement, the RUT is shifted in azimuth and elevation by the positioner, with the angle step size determining the granularity of the resulting antenna diagram. The measurement is performed for horizontal and vertical polarizations by turning the RUT axis by 90°. Granularity, maximal angular deviation (only FOV, sidelobes, 360° × 180°) and aperture, and averaging time of the power meter determine test time for a single RUT.

11.4.1 Transmitter Test Setup

Transmitter testing involves in-band transmissions to validate operational parameters, and unintentional OOB emissions that may concern EMC or other regulation. Figure 11.41 shows a typical setup for measuring radar transmitter parameters that uses an RTG to provide a steady and predictable transmission to the radar. The main difference from the setup described in Section 11.4 is that a spectrum or signal analyzer—not power sensors—are used for instrumentation. The reason for this is the necessary measurements in frequency domain, as described in Sections 11.2.1.1 and 11.2.1.3. For valid measurements, correct configuration of the analyzers' sweep and averaging times, as well as RBW and VBW are crucial. Those settings are dependent on the actual waveform and require knowledge of the transmitted frame signal. For details on setting up spectrum analysis for automotive radars, refer to [15], [16], and [20].

Note that by using the according detectors (RMS or peak) of the analyzer, this test setup can be used for power and antenna measurements as well. When operating

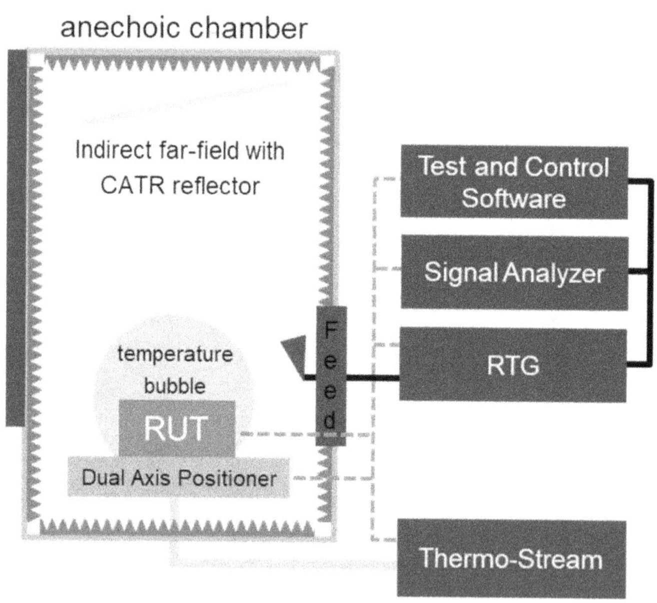

Figure 11.41 The integrated RTG in this setup ensures a defined mode of operation of the radar module, so that the measurement of the actual transmitter parameters using the signal analyzer is valid.

the spectrum analyzer in zero span mode, time domain related parameters like chirp duration, cycle and idle times or even the frame structure are possible. This versatility makes a signal or spectrum analyzer a good choice, but with the required analysis bandwidth (e.g., more than 4 GHz) other more limited instruments might be more economical or offer faster throughput for end-of-line testing.

11.4.2 Setup for Sensor Calibration and Performance Tests

Figure 11.42 depicts two test setups that can be used for calibration of the radar module and validation of the core performance parameters in range, Doppler, and spatial domain. Here no receiving instrument (e.g., signal or spectrum analyzer) is necessary, as the measured data are captured by the radar module itself. The setup only has to simulate artificial radar targets with high accuracy in all possible degrees of freedom (RCS, AOA, range, and speed). This can be realized by using an RTG and high accuracy dual axis positioner.

The first step before testing the specification of a RUT is calibration of the module to eliminate systematic errors of the antenna network and normalize the respective manifold. For calibration the RUT is positioned at boresight in respect to the simulated reference target and module estimates the range bias and the phase offset for a linear cross-channel response, then the RUT is swept through different angles using the positioner to repeat this step for several AOA. The angular granularity of this process is a compromise between error reduction and required test time.

After calibration the actual radar performance measurement as described in Section 11.2.4 can be performed. A single RTG that provides several artificial targets of defined range and speed for accuracy from a single AOA is enough to cover most of the test cases. The limitation here is that for validation of angular resolution more than one RTG is necessary with respective feeds and antennas. The easiest setup would be mounting the feeds with a gap determined by the angular resolution of the sensor. This is valid for that simulated position only, as the capability to resolve

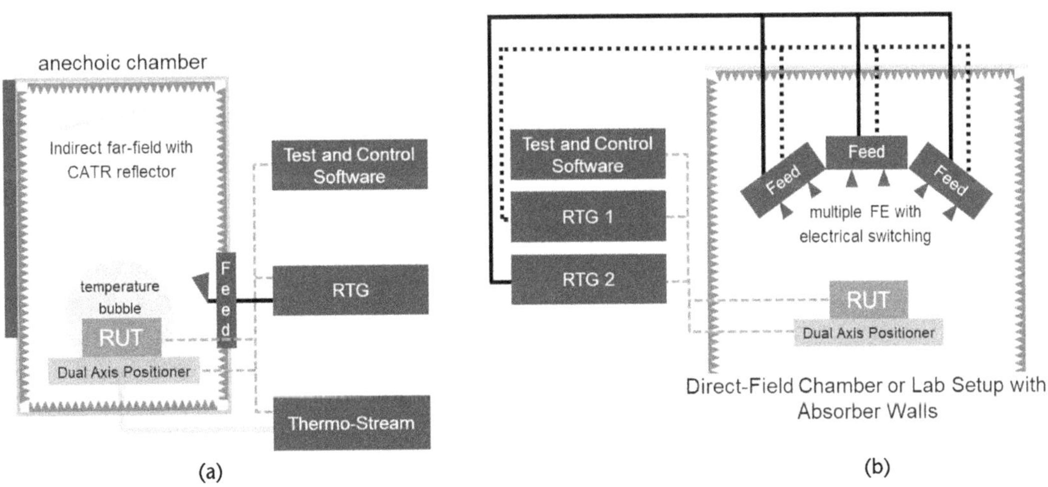

Figure 11.42 Proposed setup (a) for calibration and range and Doppler performance of a radar module and (b) for angular resolution and accuracy (azimuth only).

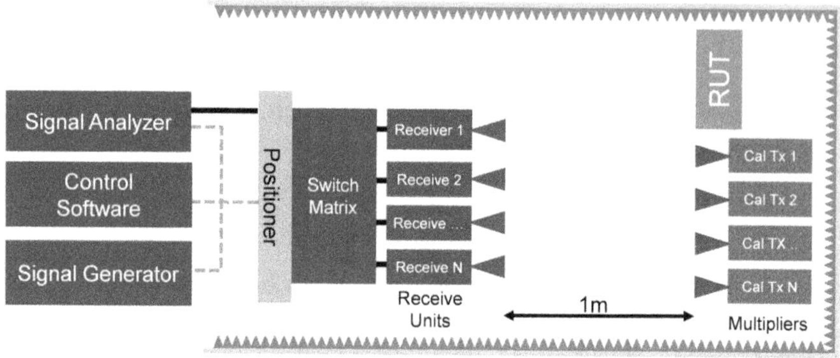

Figure 11.43 Direct-field chamber or lab setup with absorber walls for performing calibrated EMI (pre-) compliance measurements.

targets in angle is dependent on the targets position. This might include some blind spots as well and therefore a fast and accurate way of changing the position of the feeds in the sensors FOV is necessary.

The proposed setup avoids rails with feeds moving physically to avoid effects due to vibration and electronically switches between a high number of transmitting elements to simulate different AOA. By using two RTGs this provides two targets that can set independently to a generic spatial position. This is an enabler to more than pure validation of a specification but allows for a more detailed analysis in terms of sensor capabilities.

The drawback of this multifeed approach is the size requirement for spatial distribution of the feeds to cover the full FOV of the RUT. Due to that, an anechoic chamber normally isn't an option, but the measurement has to be performed with a direct far-field setup in a measurement chamber with comparatively limited shielding.

11.4.3 Setups for EMC and OOB Testing

Figure 11.43 shows a test setup than can be used for two kinds of tests: EMI testing (see Section 11.2.5) and for measuring OOB emissions (see Section 11.2.1).

The core of this setup includes several receiver chains consisting of a measurement antenna suitable for that frequency band, a receive unit performing signal amplification, and frequency conversion to an IF. Following the conversion is a spectrum analyzer (for precompliance) or EMI receiver (full compliance) with high sensitivity and dynamic range.

The aim of the measurement is to obtain information about the emitted field strength over a wide frequency range, but two considerations are required for a correct result. First, the physical output of an antenna is a voltage, thus a correction curve, the so-called *antenna correction*, must be multiplied, as field strength (V/m) = measured amplitude (V) × antenna correction factor ($1/m$). Second, in addition to the antenna correction, the attenuation of signals due to cabling must be considered. In general, the higher the frequency of the signal, the greater the attenuation that occurs, which is the main reason to bring the receive units as close as possible to the antenna. The cable attenuation is a unitless quantity that must also be considered in the result of the measuring receiver. In an automated test environment, the correction of the antenna and the cable attenuation is part of calibration. It is often possible to determine the cable attenuation with the devices and enter it directly into the setup. It should be noted that two factors are usually specified for the antenna correction, depending on the orientation of the antenna. A distinction is made between the horizontal and vertical polarization of the antenna.

To make sure that the test setup is operational and accurate, an external signal generator connected to an antenna at the test position can be used for calibration. To keep cable attenuation of the signal (transforming into a reference field strength) low, conversion from IF happens at the antenna via frequency multiplier directly.

So far this setup would be viable for any DUT, but when it comes to radar, several other requirements have to be considered. To ensure operation throughout the power range, the RUT should transmit with maximum specified EIRP as this has an impact of parasitic radiation of the module. However, this might have an effect on some measurements, as the EMI receivers are optimized for sensitivity and active radar signal 1 meter away might drive them into saturation. It could therefore be necessary to add filtering into the EMI receiver chain, to attenuate the actual radar signal.

Figure 11.44 depicts a possible setup required for EMS testing on vehicle level. EMI is simulated by using a signal generator, broadband amplifiers, and transmission using a measurement antenna for that given frequency band. The signal that is transmitted normally is a either a pure CW carrier or inherits an additional amplitude modulation with a 1 kHz sinusoid. Although not an official standard, some use wideband interfering transmissions like OFDM as well.

For calibration of the field strength at the measurement position, this system uses field probes at that position, without a vehicle under test (VUT) in place, that are connected to a measurement receiver. After calibration, those probes can be used for verification as well.

When an RTG is used for providing reference targets, it must be ensured that the correct operation of the RTG is not corrupted by the high field-strengths present

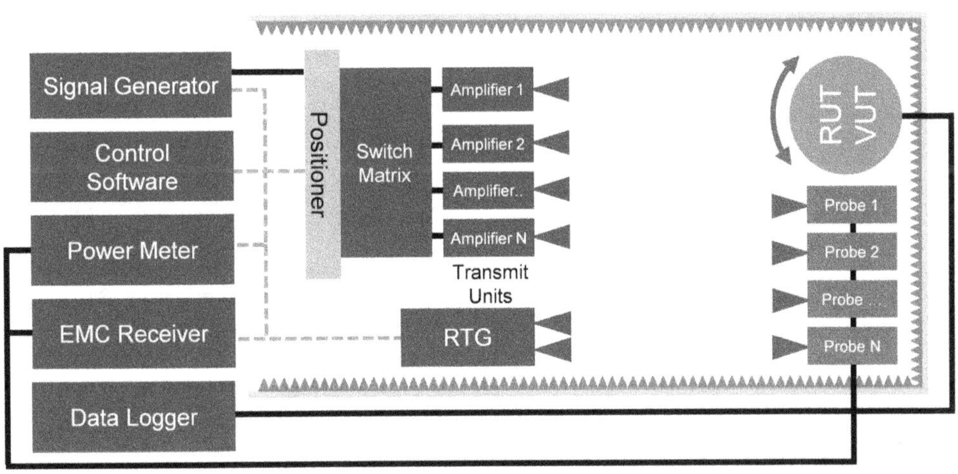

Figure 11.44 The signal generator provides an amplified test signal, while the RTG acts as the reference for proper function of the integrated radar module under interfering conditions.

in the vicinity of the EMC test antenna (e.g., by geometrical separation or additional filtering).

11.4.4 Simulating Interference from Other Automotive Radar Transmitters

Testing the effectiveness of RFI mitigation techniques may use an RTG for simulating a reference target (Figure 11.45), and a vector signal generator to represent the interfering signal, which is asynchronous with the victim radar. With an according control software for scenario generation, this interfering emitter can also simulate dynamic behavior, like nonlateral movement (range only) or changes antenna pattern due to steering.

The interference in this example is generated as an IF signal and uses the RTG for upconversion to RF. In this setup the interferer always occupies the same angular cell as the reference target but maintains a size of the instrumentation that fits an anechoic chamber and minimizes environmental influence on the setup. Testing of interference mitigation measurements and algorithms requires a lot of different interference signals for pairing with reference target echoes, therefore a general purpose signal generator is advised.

11.4.5 Exemplary Test Scenario

Testing an integrated radar module requires reference scenarios that cover most relevant corner cases for the function under test. Those functional requirements are only indirectly connected to the performance specification of the individual involved sensors, but stem from scenarios originating from regulation or standardization

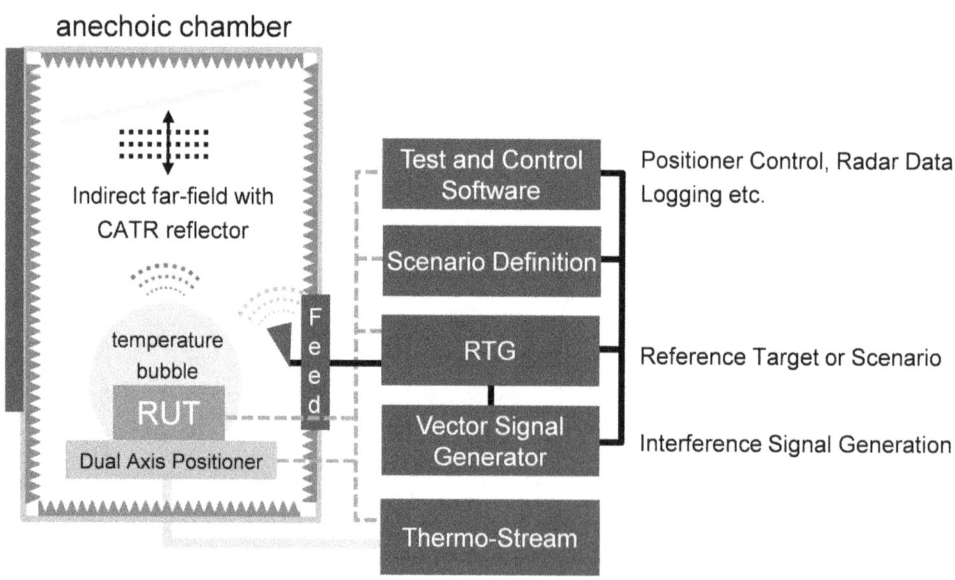

Figure 11.45 The RTG and the vector signal generator are controlled according to a predefined scenario to provide a reference target and the interference signal in parallel.

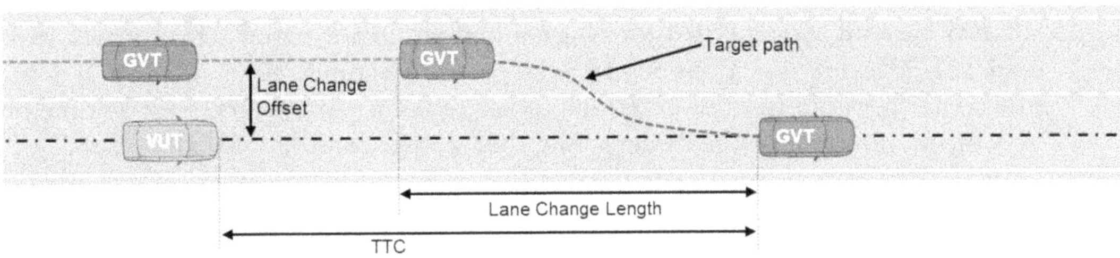

Figure 11.46 Cut-in reference scenario with VUT driving at 70 km/h and the cutting in GVT driving with 120 km/h. The target path during lane change is determined by lateral acceleration, change length, and radius of turning segments.

(house or industry). As an example of how to translate a generic scenario into a modified test setup comes from the Euro NCAP test and assessment protocol for highway assist systems [21].

Here we examine the performance of the ACC function. This includes the reaction to other vehicles during operation but excludes other functions like AEB. For highway assist systems, only car-to-car performance is assessed. Our example scenario (Figure 11.46) is a cut-in where a global vehicle target (GVT) on the adjacent lane will perform a full lane change (3.5m lateral displacement) into the lane of the VUT.

By determining the behavior of the GVT, the reaction due to the ACC function of the VUT can be evaluated. The criteria by which such an assessment is made are of less of concern here, as the key question is what instrumentation is required to simulate the scenario. For this, it is necessary to have a closer look at the different phases of the scenario from sensor perspective, as in Figure 11.47.

Even for this simple scenario, it is obvious that at least three to four different sensors (front and side radars and cameras) of the VUT are involved over the course of events, each requiring an adequate and synchronized stimulus. For the radar modules this could be separate RTGs, with defined handover between phase 3 and 4. Additionally, the targets for the side radar in phase 1 and 2 appear at very close distances, requiring an additional low latency target generator like an FODL in the RTG.

To fully realize this scenario via an integrated test bed at least three RTGs are necessary. An DRFM based RTG for stimulating the front radar providing full flexibility in range and Doppler simulation and for the side-looking radar a

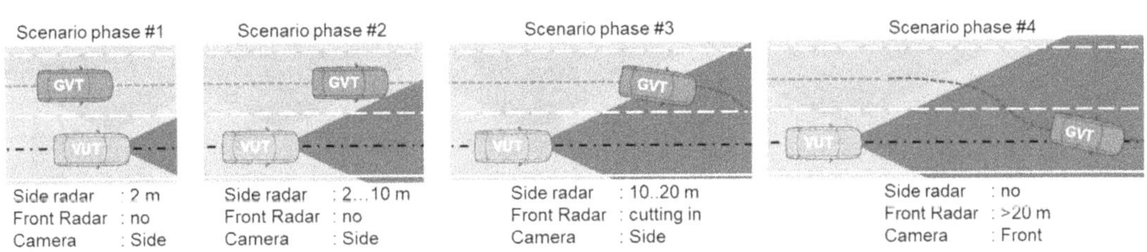

Figure 11.47 The reference scenario from Figure 11.46 separated into different phases for mapping the hardware requirements of a test bed to the involved sensors and actuators of the function under test.

DRFM-based RTG is needed for phase 3 and an OFDL-based RTG for phases 1 and 2. The next two sections describe the complete setup to simulate this. Of note, the scenario would need to determine other relevant parameters, such as time of day or sun position for the cameras and the weather condition for all sensors.

11.4.6 ADAS Integration Test Bed

ADAS and AD functions normally are implemented and distributed across multiple ECUs, requiring testing of the overarching software functionality. The communication infrastructure between the individual ECUs must be managed by the test system, including simulation of the vehicles remaining bus. Sensor (and actuator) data in the HiL-setup can be simulated using digital data injection or direct stimulus of the hardware. All those different data streams of the test bed need to be synchronized precisely.

Figure 11.48 shows a HiL setup that could be used to test the proposed cut-in scenario and evaluation of ACC, implemented on the ADAS domain controller. In this example, actuators are only software simulated and not shown to focus on sensor integration but could be integrated as hardware if required. Communication with relevant sensors and actuators require dedicated physical interfaces. Depending on the sensors used, up to three different data streams are generally available from a single sensor. This could be raw data (e.g., radar reflections and video images, object lists of recognized detections) or internal sensor data (e.g., status messages).

Embedded in this setup are three sensors contributing to the ACC function, namely a side-looking SRR (RUT 1), a forward-looking LRR (RUT 2) and a camera-based sensor. To provide a stimulus to the sensors that compares to the driving scenario two radar target generators are necessary. The frontends must be able to simulate the movement of the GVT, so either electronic switching is required or mechanical movement of RTG's feed.

The virtual ego driver drives along the flat road without making any steering maneuvers during simulation with an initial speed of 70 km/h, the accelerator pedal is depressed by 50%. At the start of the test drive, another GVT is approaching from 10m behind on a side lane with a lateral displacement of 3.5m and an initial speed

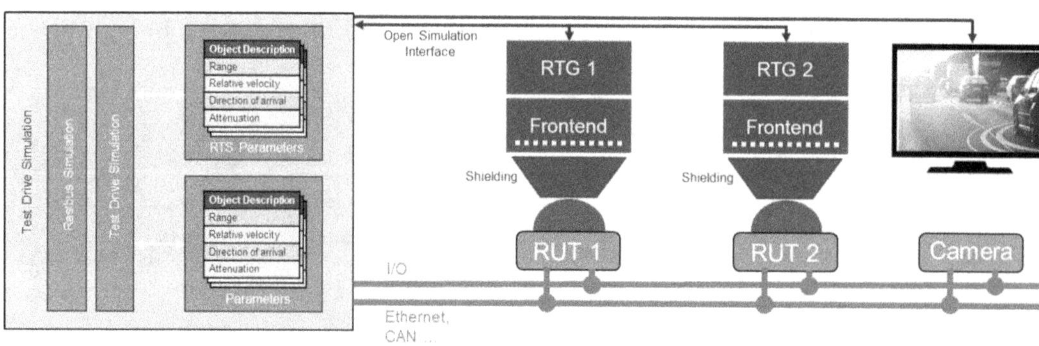

Figure 11.48 Depending on the phase of the test scenario, the simulated GVT must be injected to different sensors in parallel with synchronized model behavior.

of 120 km/h. As soon as the GVT enters the FOV of the ego vehicle's sensors phase 1 applies, ergo RTG 1 (using FODL) and the CCTV must synchronously provide the bypassing target with correct distance, direction, and speed. In phase 2 the GCT is still outside the FOV of RUT 2, but distance has increased to the point where an analog delay is no longer necessary and RTG 2 can switch to DRFM mode. In phase 3 the GVT is in the overlapping FOV of both radar sensors, therefore both RTG have to simulate the same target. This is the point where the cut-in maneuver of the GVT can be anticipated by the ADAS ECU and the ACC function should start decreasing set throttle or handover to driver. In phase 4 GVT finished cut-in and therefore is visible on RUT 2 and the front view of the camera. Ego vehicle should now maintain a distance and speed that ACC function defines.

11.4.7 ViL Test

The functional principle of ViL is similar to the HiL methods, because the same sensor information is used. But in comparison, ViL uses a physical vehicle instead of a virtual prototype, eliminating the need for a simulation of the remaining bus. The simulated environment is injected to the vehicle's sensor and the ADAS function in turn sends its own information to the physical vehicle. The loop is closed by sending the position and status (speed, torque, CAN) of the physical vehicle to the simulation environment.

Continuing with the same scenario but with a ViL test bed (Figure 11.49), we see that from a pure stimulus of sensors perspective it is very comparable to Figure 11.48, including the described phases over the runtime.

The main difference is that the exact position of the real vehicle has to be determined via instrumentation (not via modeling) and the resulting position, movement, and location have to be transferred to the simulation. This is the enabler for including real vehicle dynamic influences and reduce the need for real test driving. For the exemplary NCAP scenario, the function under test should ideally behave like described in HiL testing with the ego vehicle maintaining a distance and speed that ACC function defines.

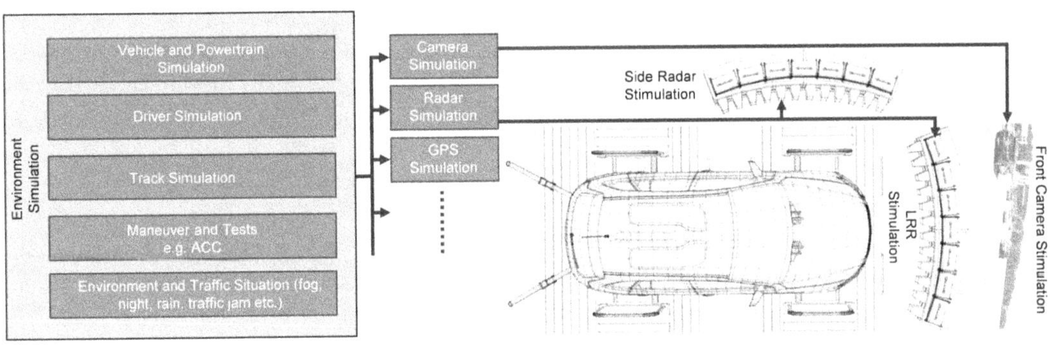

Figure 11.49 Proposed setup for a ViL testbed that can be used to test the ACC scenario.

References

[1] Euro NCAP, "Euro NCAP 20th anniversary," https://www.euroncap.com/en/about-euro-ncap/timeline/euro-ncap-20th-anniversary/. [Access at 07/21/2021].

[2] Wikipedia, "New Car Assessment Program," Available online: https://en.wikipedia.org/wiki/New_Car_Assessment_Program. [Access at 07/15/2021].

[3] International Telecommunications Union, "Recommendation ITU-R SM.443-4: Bandwidth Measurement at Monitoring Stations," Geneva, 2018.

[4] Ramasubramanian, K., "Using a Complex-Baseband Architecture in FMCW Radar System," [White Paper], Dallas, TX: Texas Instruments Incorporated, 2017.

[5] Doerry, A. W., "Balancing I/Q Data in Radar Range-Doppler Images," *SPIE Defense and Security Symposium*, Vol. 9461, 2015.

[6] Sturm, C., Sit, Y. L., Braun, M., and T. Zwick, "Spectrally Interleaved Multi-carrier Signals for Radar Network Applications and Multi-input Multi-output Radar," *IET Radar, Sonar and Navigation*, 2011.

[7] Hanaf, B., *Design of Silicon Power Amplifiers and Arrays for Millimeter Wave*, San Diego, CA: University of California, PhD Thesis, 2014.

[8] J. Fridén, J., Razavi, A., and E. Pucci, "Over-the-Air Testing for Total Radiated Power," *IEEE Conference on Antenna Measurements and Applications (CAMA)*, Västerås, Sweden, 2018.

[9] Lim, D.-W., Heo, S.-J., and J.-S. No, "An Overview of Peak-to-Average Power Ratio Reduction Schemes for OFDM Signals," *Journal of Communications and Networks*, Vol. 11, 20 June, No. 3, 09, pp. 229–239.

[10] Heuel, S., "Automated Measurements of 77 GHz FMCW Radar Signals," Rohde & Schwarz GmbH & Co. KG, Munich, 2014.

[11] Weyer, D. J., *Design of Digital FMCW Chirp Synthesizer PLLs Using Continuous-Time Delta-Sigma Time-to-Digital Converters*, MI: University of Michigan, 2018.

[12] Texas Instruments Incorporated, "AWR1843 Single-Chip 77- to 79-GHz FMCW Radar Sensor Datasheet (Rev. B)," https://www.ti.com/product/AWR1843. [Access at 08/02/2021].

[13] Khomchuk, P., "Pedestrian Motion Direction Estimation Using Simulated Automotive MIMO Radar," *IEEE Transactions on Aerospace and Electronic Systems*, Vol. 52, No. 3, June 2016, pp. 1132–1145.

[14] Freese, J., "Untersuchung von Gruppenantennen für den Einsatz in aktiven Antennensystemen," Darmstadt: Darmstädter Dissertation, 2007.

[15] Rauscher, C., *Fundamentals of Spectrum Analysis*, Munich: Rohde & Schwarz, 2016.

[16] Sanders, K.-U., "Peak and Mean Power Measurements on Wideband FMCW Radar Signals," Munich, Germany: Rohde & Schwarz, 2019.

[17] Kamel, E. B., Peden, A., and P. Pajusco, "RCS Modeling and Measurements for Automotive Radar Applications in the W Band," *IEEE*, Paris, France, 2017.

[18] Späth, S., "Investigation of the Simulation Capabilities of the Angular Directions of Objects by Radar Test Systems," Munich, Germany: Technical University of Munich, 2018.

[19] Nähring, A., "Demystifying Over-the-Air (OTA) Testing," Munich, Germany: Rohde & Schwarz, 2019.

[20] Ramian, F., "Speeding Up Spectrum Analyzer Measurements," Rohde & Schwarz, Munich, Germany, 2014.

[21] Euro NCAP, "Automated Driving Highway Assist Systems," Euro NCAP, Leuven, 2020.

[22] Will, C., Vaishnav, P., Chakraborty, A., and A. Santra, "Human Target Detection, Tracking, and Classification Using 24-GHz FMCW Radar," *IEEE Sensors Journal*, Vol. 19, No. 17, September 1, 2019.

[23] Murray, S., "The Perils of Spectrum Analyzer Power Averaging," Keithley Instruments, Inc, 2006.

[24] International Telecommunication Union, "Recommendation ITU-R SM.1754-0: Measurement Techniques of Ultra-Wideband Transmissions," Geneva, 2006.

[25] Dudek, M., Nasr, I., Kissinger, D., and R. Weigel, "The Impact of Phase Noise Parameters on Target Signal Detection in FMCW-Radar System Simulations for Automotive Applications," *Proceedings of 2011 IEEE CIE International Conference on Radar*, Chengdu, 2011.

[26] Pfeffer, C., Feger, R., Jahn, M., and A. Stelzer, "A 77-GHz Software Defined OFDM Radar," *15th International Radar Symposium (IRS)*, 2014.

List of Acronyms

ACC	Adaptive cruise control
ADAS	Advanced driver assistance systems
ADC	Analog to digital converter
ADS	Automated driving system
AEB	Automatic emergency braking
AoU	Assumption of use
APAS	Almost perfect autocorrelation sequence
APQP	Advanced product quality planning
ASPICE	Automotive software process improvement and capability determination
ASIL	Automotive safety integrity level
B	Modulation bandwidth
BPF	Band pass filter
BSD	Blind spot detection
BW	Bandwidth or beamwidth
c_0	Velocity of light
CDM	Code division multiplex
CDMA	Code division multiple accessing
CMMI	Capability maturity model integration
CMOS	Complementary metal-oxide-semiconductor
CPI	Coherent processing interval
CW	Continuous waveform
DAC	Digital-to-analog converter
DBF	Digital beamforming
DCM	Digital code modulation
DDM	Doppler division multiplex
DDT	Dynamic driving task
DIS	Draft international standard
DOA	Dircction of arrival
DOD	Direction of departure
DRAM	Dynamic random-access memory

ESPRIT	Estimation signal parameters via rational invariance techniques
f_C	Carrier frequency
FCM	Fast chirp modulation
f_d	Doppler frequency shift
FDM	Frequency division multiplex
FFT	Fast fourier transform
FMCW	Frequency modulated continuous wave
FMEA	Failure mode and effects analysis
FMEDA	Failure mode and effects diagnostics analysis
FOV	Field of view
FTA	Fault tree analysis
GPIO	General purpose input/output
IAA	Iterative adaptive approach
IATF	International Automotive Task Force
IEC	International Electrotechnical Commission
IEEE	Institute of Electrical and Electronics Engineers
IF	Intermediate frequency
INCOSE	International Council on Systems Engineering
IQ	In-phase- and quadrature principle
ISO	International Standards Organization
JTAG	Joint test action group (interface)
L	Number of objects in a radar scene
L_C	Spreading code length
LCA	Lane change assist
LNA	Low noise amplifier
LO	Local oscillator
LPF	Low pass filter
LRR	Long range radar
M	Number of accumulations
MIMO	Multiple input/multiple output
mmW	Millimeter-wave
MUSIC	Multiple signal classification
MRR	Medium range radar
N_R	Number of receiver
N_T	Number of transmitter
ODD	Operational design domain
OMP	Orthogonal matching pursuit
P	Ping

P	Power of a signal
PA	Power amplifier
PLL	Phase locked loop
PMCW	Phase modulated continuous wave
PRF	Pulse repetition frequency
PRI	Pulse repetition interval
PW	Pulse width
QM	Quality managed
RCS	Radar cross-section
RDM	Range-doppler map
RF	Radio frequency
RFI	Radio frequency interference
RoC	Radar on chip
R_r	Range resolution
R_u	Maximum unambiguous range
RX	Receiver, receive channel
SAE	Society of Automotive Engineers
SAR	Synthetic aperture radar
SEooC	Safety element out of context
SiGe	Silicon-germanium technology
SLA	Sparse linear array
SoC	System on chip
SPICE	Software process improvement and capability determination
SRAM	Static random-access memory
SRR	Short range radar
SS	Svensk Standard (Swedish Standards Authority)
T_C	Chip time
TDM	Time division multiplex
TX	Transmitter, transmit channel
ULA	Uniform linear array
VDA	Verband der Automobil Industrie (German Association of the Automotive Industry)
VGA	Variable gain amplifier
VRX	Virtual receive channel
Δv	Radial velocity
Δf	Beat frequency
v_u	Unambiguous velocity

About the Editor

Matt Markel started his radar career on his first day of work after college graduation (in 1990—before the internet) with a borrowed copy of Stimson's *Introduction to Airborne Radar* to prepare him for work on the Advanced Medium Range Air to Air Missile (AMRAAM) program. Today he is an engineering leader with more than 30 years of experience in the development of advanced technologies, specifically in the fields of radar and electronic warfare for both defense and commercial applications.

In the commercial autonomy space, Dr. Markel led the radar team at Waymo, formerly the Google Self-Driving Car Project, with his radar department designing and delivering multiple generations of indigenously designed tracking and imaging radars to Waymo's self-driving fleet. He developed the radar research and development framework that has provided numerous cutting-edge technologies and capabilities to Waymo's radar line. Currently he leads the radar division at Ghost Locomotion, an autonomous driving company using breakthroughs in collision avoidance technology to develop safe, attention-free self-driving for mass market consumer cars.

In the defense space, Dr. Markel was a Principal Engineering Fellow for Raytheon, and served as their Technical Director for Advanced Electronic Warfare (EW). He was directly involved in the transition of key radar and EW capabilities to the U.S. military and has been a Principal Investigator for multiple radar and EW research programs that advanced the state of capability, effectiveness, autonomy, and cognition in U.S. systems.

Dr. Markel received a bachelor's from the University of Missouri–Rolla (now Missouri Science and Technology) and a master's and PhD from the University of Florida, all in electrical engineering. He is the inventor on multiple patents from his time at Raytheon and Waymo.

About the Contributors

Murtaza Ali is VP of Systems Engineering at Uhnder, Inc., where he oversees development of radar systems. He has more than 25 years of experience in systems development. His current research focus is on radar signaling techniques, MIMO methods for imaging radars, and interference mitigation/avoidance. Murtaza holds a PhD degree in electrical engineering from University of Minnesota. He has 51 U.S. patents and has published over 40 papers in refereed and invited forums.

Tim Campbell joined the Google self-driving car team (now Waymo) in 2013 as a radar system designer. While there, he was inventor on over 45 patents worldwide. Prior to that Tim spent 10 years doing airborne radar system design and development in the aerospace industry.

Gary Clayton has 34 years of experience in the aerospace industry in radar system design and development for airborne and spaceborne radar, and as a consultant for surface-based radar development. Gary's leadership in the aerospace industry included developing industry radar technology roadmaps and as an industry consultant for technology investment for multiple customers. Gary joined Waymo in April 2020 as part of the radar leadership team and is the head of radar development.

Ryan Fan is a Radar Engineer at Ghost Locomotion. Previously, he was an engineer at Waymo, Raytheon, Electromagnetic Systems, and Boeing where he worked on a wide variety of radars. He received a master's degree in electrical engineering from UCLA in 2011 and a bachelor's degree in electrical engineering from University of Texas, Arlington in 2007. His areas of expertise include the design of radar systems and modes, electronic protection, signal processing, and radar systems analysis.

Michael Fina received a PhD in mechanical engineering from the University of California at Berkeley for research in Organic Light-Emitting Diodes in 2012. Following his graduate work, Michael worked at Raytheon for five years in Advanced Concepts and Technologies developing new hardware and algorithms. For the past three years, Michael has worked on developing automotive radar field tests at Waymo including tests aimed at studying and modeling radar in weather phenomenology.

Griffin Foster is a researcher at Zendar working on radar systems and algorithms. Previously he was a radio astronomer working on digital instrumentation, synthetic aperture imaging with interferometric arrays, SETI, and searching for rare transients. He was the project scientist for Breakthrough Listen on the MeerKAT telescope in South Africa. He has built instrumentation for telescopes in the United States, United Kingdom, Italy, Ireland, and South Africa. His DPhil work at the University of Oxford focused on low-frequency instrumentation for wide-field imaging.

Leander Humbert is a technology manager for radar-based measurement applications at Rohde & Schwarz. Leander earned his Diplom-Ingenieur degree focusing in Communications from the Helmut-Schmidt-University in Hamburg. He has 17 years of professional experience in designing radar and electronic warfare systems from different perspectives of the ecosystem. He worked in the German Air Force as EW Officer and Systems Engineer and later as a consultant for scientific and operational studies regarding sensors for air and missile defense systems.

Mike Keaveney received a bachelor's degree in electrical and electronics engineering from University College Dublin in 1986 and a masters in engineering from the University of Limerick in 1993. He has spent most of his career at Analog Devices in Limerick, Ireland, working on ICs for communications systems and automotive radar and became an ADI Fellow in 2012. Mike has served on the Technical Program Committees of the ISSCC and ESSCIRC international IEEE conferences and on the Technical Review Board of the Microelectronics Research Centre Ireland (MCCI). He has numerous peer refereed journal and conference papers and has been granted over a dozen U.S. patents. Mike joined Robert Bosch Ireland Ltd. in 2021 where he is Radar Architect and Limerick Design Center Manager.

Yaohui Liu is an application engineer at Uhnder and focuses on working with customers to bring digital radar solution to autonomous vehicles, Previously, he worked at Veoneer, NXP, Freescale semiconductor, and Nokia where he worked on RF, Microprocessors/SoCs, and signal processing algorithms targeting wireless networking and advanced driver assistance systems. Yaohui received his master's degrees in information science from Beijing Jiaotong University and licentiate degree in electrical engineering from Helsinki University of Technology.

Xiaoli Lu earned a PhD degree from the Department of Electrical Engineering, University of Victoria, Canada. She joined Nvidia in 2021 as principal engineer to lead the systems efforts for L3 automotive product. Prior to Nvidia, Xiaoli worked for Waymo and Cruise, two key players in the L4 AV industry, on sensing and perception from requirements, design, testing to functional and SOTIF safety. Xiaoli also worked for Apple on their special project after she served 10 years in Raytheon on both air traffic control radar and high frequency surface wave radar.

James McGinley is a principal engineer with Analog Devices. During the past 25 years, he has held roles in all areas of the product lifecycle, spanning systems, software, and hardware, across several industries. For over a decade he's specialized in the safe deployment of ADAS and AD technologies. James serves as head of delegation for Ireland for international standardization activities concerning automotive hardware and software safety. He holds a bachelor's degree in physics and master's degree in software.

Vinayak Nagpal co-founded Zendar to develop advanced imaging radars for autonomous systems. Zendar has demonstrated software defined imaging radar that matches lidar in resolution without increasing system cost. Zendar is backed by Tier-1 Silicon Valley venture capital firms. Previously, Vinayak developed instrumentation for radio astronomy imaging that enabled projects like the first submillimeter images of the black hole at the center of the Milky Way. Vinayak got his PhD in electric engineering from University of California (UC) Berkeley in 2012. He has held research positions at the Harvard Smithsonian Center for Astrophysics and Nokia Bell Labs.

Antonio Puglielli is Director of Systems Engineering at Zendar, a Berkeley-based startup developing high-resolution radar for autonomous systems at scale. At Zendar he leads hardware and algorithm development. Antonio received his PhD in electrical engineering at the University of California Berkeley, where he worked on chip, system, and signal processing development for mm-wave communication links.

Darsh Ranjan received his bachelor of arts degree in mathematics from Princeton University in 2005 and attended the graduate program in applied mathematics at UC Berkeley. Darsh has developed and implemented algorithms for processing automotive sensor data since 2010 and currently develops radar signal processing software at Zendar.

Ralf Reuter received a doctorate degree in electrical engineering in 1998 from the University Duisburg, Germany, and a diploma degree (master's) from the University Duisburg in high-frequency microwave and electrical engineering in 1992. In 1997, he joined the Fraunhofer Institute for applied semiconductor physics in Freiburg, Germany, and was leading the department for high-frequency devices, circuits, and technologies. In 2000 he joined Motorola and with Freescale in 2015, Ralf became a Fellow for Radar Systems and Applications. With NXP in 2016, he was the Fellow and Senior Director for Analog Radar frontends. Since July 2018, Ralf is working at Uhnder as Corporate Fellow and Senior Director of Customer and Applications Engineering. Ralf has authored or coauthored more than 110 technical papers in refereed journals and conferences and owns more than 30 patents in the field of mm-w circuits, system and sensor design.

Arunesh Roy is Uhnder's senior director for advanced applications and perception, leading the feature set definition and performance specifications for automotive radar signal processing algorithms. He and his team develop new radar system concepts and processing algorithms including advanced detection and estimation techniques, multitarget tracking, and multisensor fusion. He has developed automated methods for extrinsic calibration of sensors with dissimilar resolutions as well as sensor fusion and simultaneous localization and mapping (SLAM) algorithms for multisensor data. Prior to joining Uhnder, Arunesh held senior engineering positions with NXP and Oculii, where he focused on radar algorithm and signal processing architectures, approaches, and developmental techniques. Arunesh earned his master's and doctorate degrees in electrical engineering from Wright State University in Ohio. He holds several patents for technology related to radar architectures, signal processing and sensor fusion for cameras, and radar for multitarget tracking.

Donnie Smith is a radar engineer at Waymo and has been working on radar-based solutions for autonomous vehicles, on the ground and in the air, since 2013. Donnie is an inventor on over 15 patents worldwide. Previously, he worked at the Georgia Institute of Technology Research Institute, where he developed novel tracking solutions and simulation-based metrics for tracking and track fusion algorithms. Donnie holds master's and bachelor's degrees in electrical engineering from the Georgia Institute of Technology.

Mark Steigemann joined Philips Semiconductors System Laboratories in 1998. He has been working with NXP Semiconductors since 2006, where he is presently a Fellow and Chief Architect for ADAS products, which includes mm-wave transceiver, radar MCU, and one chips. He developed and industrialized the first RFCMOS-based software defined broadcast radio and steered the development of the first single-chip 76–81 GHz automotive radar one chip in 28nm CMOS technology. He holds a degree in electrical engineering from the University of Hamburg.

Shunqiao Sun is an assistant professor at The University of Alabama. He received his PhD degree in Electrical and Computer Engineering from Rutgers University in January 2016. Previously, he was with Aptiv, Technical Center Malibu, California, and led angle finding research and development efforts for short-range-radar with booking of 130-million units. He received the 2016 IEEE Aerospace and Electronic Systems Society Robert T. Hill Best Dissertation Award. He coauthored a paper that received the Best Student Paper Award at the 2020 IEEE Sensor Array and Multichannel Signal Processing Workshop.

Cicero S. Vaucher joined Philips Research Laboratories Eindhoven in 1990. He has been with NXP Semiconductors since 2006, where he is presently a Fellow and an Automotive Radar Product Architect. He developed and industrialized the first silicon-based receiver IC for 12-GHz satellite outdoor units and steered the development of the first single-chip 76–81 GHz automotive radar transceiver in 40 nm CMOS technology. He holds a doctorate degree in electrical engineering from the University of Twente and is presently a part-time professor of mm-wave radar frontends at TU Delft.

Jimmy Wang co-founded Zendar in May 2017 to develop advanced imaging radars for autonomous systems. Zendar has demonstrated software defined imaging radar that matches Lidar in resolution without increasing system cost. Prior to this, Jimmy was a Lead Research Engineer at HERE Technology (previously Earthmine, acquired by HERE Nokia in 2011), where he led the software and algorithm development for processing large scale 3D street level imageries. Jimmy obtained his PhD in Vision Science from UC Berkeley in 2010, focusing on computer vision and unsupervised machine learning.

Index

Artech House Radar Library

Joseph R. Guerci, Series Editor

Adaptive Antennas and Phased Arrays for Radar and Communications, Alan J. Fenn

Adaptive Signal Processing for Radar, Ramon Nitzberg

Advanced Techniques for Digital Receivers, Phillip E. Pace

Advances in Direction-of-Arrival Estimation, Sathish Chandran, editor

Airborne Early Warning System Concepts, Maurice W. Long

Airborne Pulsed Doppler Radar, Second Edition, Guy V. Morris and Linda Harkness, editors

Aspects of Modern Radar, Eli Brookner

Aspects of Radar Signal Processing, Frank F. Kretschmer, Bernard Lewis and Wesley W. Shelton

Basic Radar Analysis, Mervin C. Budge, Jr. and Shawn R. German

Basic Radar Tracking, Mervin C. Budge, Jr. and Shawn R. German

Bayesian Multiple Target Tracking, Second Edition, Lawrence D. Stone, Roy L. Streit, Thomas L. Corwin, and Kristine L Bell

Beyond the Kalman Filter: Particle Filters for Tracking Applications, Branko Ristic, Sanjeev Arulampalam, and Neil Gordon

Cognitive Radar: The Knowledge-Aided Fully Adaptive Approach, Second Edition, Joseph R. Guerci

Coherent Radar Performance Estimation, James A. Scheer

Computer Simulation of Aerial Target Radar Scattering, Recognition, Detection, and Tracking, Yakov D. Shirman, editor

Control Engineering in Development Projects, Olis Rubin

Countermeasures for Aerial Drones, Garik Markarian and Andrew Staniforth

Deep Learning Applications of Short-Range Radars, Avik Santra, and Souvik Hazra

Deep Learning for Radar and Communications Automatic Target Recognition, Uttam K. Majumder, Erik P. Blasch, and David A. Garren

Design and Analysis of Modern Tracking Systems, Samuel Blackman and Robert Popoli

Detectability of Spread-Spectrum Signals, George Dillard

Detecting and Classifying Low Probability of Intercept Radar, Second Edition, Phillip E. Pace

FMCW Radar Design, M. Jankiraman

Fourier Transforms in Radar and Signal Processing, Second Edition, David Brandwood

Fundamentals of Short-Range FM Radar, Igor V. Komarov and Sergey M. Smolskiy

Handbook of Computer Simulation in Radio Engineering, Communications, and Radar, Sergey A. Leonov and Alexander I. Leonov

Handbook of Radar Measurement, David K. Barton

High-Resolution Radar, Second Edition, Donald R. Wehner

High Resolution Radar Cross-Section Imaging, Dean L. Mensa

Highly Integrated Low-Power Radars, Sergio Saponara, Maria Greco, Egidio Ragonese, Giuseppe Palmisano, and Bruno Neri

Introduction to LabVIEW™ FPGA for RF, Radar, and Electronic Warfare Applications, Terry Stratoudakis

An Introduction to Passive Radar, Second Edition, Hugh D. Griffiths and Christopher J. Baker

Introduction to Radar using Python and MATLAB®, Lee Andrew Harrison

Introduction to RF Equipment and System Design, Pekka Eskelinen

Linear Systems and Signals: A Primer, JC Olivier

Meter-Wave Synthetic Aperture Radar for Concealed Object Detection, Hans Hellsten

Methods and Techniques of Radar Recognition, Victor Nebabin

The Micro-Doppler Effect in Radar, Second Edition, Victor C. Chen

Microwave Radar: Imaging and Advanced Concepts, Roger J. Sullivan

Millimeter-Wave Radar Clutter, Nicholas C. Currie, Robert D. Hayes and Robert Trebits

Millimeter-Wave Radar Targets and Clutter, Gennadiy P. Kulemin

MIMO Radar: Theory and Application, Jamie Bergin and Joseph R. Guerci

Modern Radar Systems, Second Edition, Hamish Meikle

Modern Radar System Analysis, David K. Barton

Modern Radar System Analysis Software and User's Manual, Version 3.0, David K. Barton

Monopulse Principles and Techniques, Second Edition, Samuel M. Sherman and David K. Barton

MTI and Pulsed Doppler Radar with MATLAB®, Second Edition, D. Curtis Schleher

Multifunction Array Radar, Dale Billetter

Multiple-Target Tracking with Radar Applications, Samuel S. Blackman

Multitarget-Multisensor Tracking, Applications and Advances, Volume II, Yaakov Bar-Shalom and Robert F. Popoli, editors

Multitarget-Multisensor Tracking: Applications and Advances Volume III, Yaakov Bar-Shalom and William Dale Blair, editors

Non-Line-of-Sight Radar, Brian C. Watson and Joseph R. Guerci

Photonic Aspects of Modern Radar, Edward N. Toughlian and Henry Zmuda

Practical Phased Array Antenna Systems, Eli Brookner

Practical Simulation of Radar Antennas and Radomes, Herbert L. Hirsch

Precision FMCW Short-Range Radar for Industrial Applications, Boris A. Atayants, Viacheslav M. Davydochkin, Victor V. Ezerskiy, Valery S. Parshin, and Sergey M. Smolskiy

Principles of High-Resolution Radar, August W. Rihaczek

Principles of Radar and Sonar Signal Processing, François Le Chevalier

Radar Cross-Section Analysis & Control, Asoke K. Bhattacharyya and D. L. Sengupta

Radar Cross Section, Second Edition, Eugene F. Knott, et al.

Radar Detection and Tracking Systems, Shaheen A. Hovanessian

Radar Engineer's Sourcebook, William C. Morchin

Radar Equations for Modern Radar, David K. Barton

Radar Evaluation Handbook, David K. Barton, et al.

Radar for Fully Autonomous Driving, Matt Markel, Editor

Radar Meteorology, Henri Sauvageot

Radar Range Performance Analysis, Lamont V. Blake

Radar Reflectivity of Land and Sea, Third Edition, Maurice W. Long

Radar Resolution and Complex-Image Analysis, August W. Rihaczek and Stephen J. Hershkowitz

Radar RF Circuit Design, Second Edition, Nickolas Kingsley and Joseph R. Guerci

Radar Signal Processing and Adaptive Systems, Ramon Nitzberg

Radar Signals: An Introduction to Theory and Application, Charles E. Cook

Radar System Analysis, Design, and Simulation, Eyung W. Kang

Radar System Analysis and Modeling, David K. Barton

Radar System Design and Analysis, Shaheen A. Hovanessian

Radar System Performance Modeling, Second Edition, G. Richard Curry

Radar Technology, Eli Brookner

Radar Technology Encyclopedia, David K. Barton and Sergey A. Leonov, editors

Radar Vulnerability to Jamming, Stephen Lothes

Radars, Volume 1: Monopolse Radar, David K. Barton

Radars, Volume 2: The Radar Equation, David K. Barton

Radars, Volume 3: Pulse Compression, David K. Barton

Radars, Volume 4: Radar Resolution & Multipath Effects, David K. Barton

Radars, Volume 5: Radar Clutter, David K. Barton

Radars, Volume 6: Frequency Agility and Diversity, David K. Barton

Radars, Volume 7: CW and Doppler Radar, David K. Barton

Radio Wave Propagation Fundamentals, Second Edition, Artem Saakian

Range-Doppler Radar Imaging and Motion Compensation, Jae Sok Son, et al.

Robotic Navigation and Mapping with Radar, Martin Adams, John Mullane, Ebi Jose, and Ba-Ngu Vo

Russian-English and English-Russian Dictionary of Radar and Electronics, William F. Barton and Sergey A. Leonov

Secondary Surveillance Radar, Michael C. Stevens

Signal Detection and Estimation, Second Edition, Mourad Barkat

Signal Processing in Noise Waveform Radar, Krzysztof Kulpa

Signal Processing for Passive Bistatic Radar, Mateusz Malanowski

Solid-State Radar Transmitters, Edward D. Ostroff

Space-Based Radar Handbook, Leopold J. Cantafio, Richard K. Moore, and Earl E. Swartzlander Jr.

Space-Time Adaptive Processing for Radar, Second Edition, Joseph R. Guerci

Spaceborne Weather Radar, Toshiaki Kozu and Robert Meneghini

Special Design Topics in Digital Wideband Receivers, James Tsui

Statistical Signal Characterization, Herbert L. Hirsch

Surface-Based Air Defense System Analysis, Robert H. MacFadzean

Synthetic Array and Imaging Radars, Shaheen A. Hovanessian

Systems Engineering of Phased Arrays, Rick Sturdivant, Clifton Quan, and Enson Chang

Techniques of Radar Reflectivity Measurement, Nicholas C. Currie

Theory and Practice of Radar Target Identification, August W. Rihaczek and Stephen J. Hershkowitz

Time-Frequency Signal Analysis with Applications, Ljubiša Stanković, Miloš Daković, and Thayananthan Thayaparan

Time-Frequency Transforms for Radar Imaging and Signal Analysis, Victor C. Chen and Hao Ling

Transmit Receive Modules for Radar and Communication Systems, Rick Sturdivant and Mike Harris

www.ingramcontent.com/pod-product-compliance
Lightning Source LLC
LaVergne TN
LVHW080644210926
848485LV00003B/320